# Computational Genomics
# with R

# Chapman & Hall/CRC
# Computational Biology Series

About the Series:
This series aims to capture new developments in computational biology, as well as high-quality work summarizing or contributing to more established topics. Publishing a broad range of reference works, textbooks, and handbooks, the series is designed to appeal to students, researchers, and professionals in all areas of computational biology, including genomics, proteomics, and cancer computational biology, as well as interdisciplinary researchers involved in associated fields, such as bioinformatics and systems biology.

**Introduction to Bioinformatics with R: A Practical Guide for Biologists**
*Edward Curry*

**Analyzing High-Dimensional Gene Expression and DNA Methylation Data with R**
*Hongmei Zhang*

**Introduction to Computational Proteomics**
*Golan Yona*

**Glycome Informatics: Methods and Applications**
*Kiyoko F. Aoki-Kinoshita*

**Computational Biology: A Statistical Mechanics Perspective**
*Ralf Blossey*

**Computational Hydrodynamics of Capsules and Biological Cells**
*Constantine Pozrikidis*

**Computational Systems Biology Approaches in Cancer Research**
*Inna Kuperstein, Emmanuel Barillot*

**Clustering in Bioinformatics and Drug Discovery**
*John David MacCuish, Norah E. MacCuish*

**Metabolomics: Practical Guide to Design and Analysis**
*Ron Wehrens, Reza Salek*

**An Introduction to Systems Biology: Design Principles of Biological Circuits**
2nd Edition
*Uri Alon*

**Computational Biology: A Statistical Mechanics Perspective**
Second Edition
*Ralf Blossey*

**Stochastic Modelling for Systems Biology**
Third Edition
*Darren J. Wilkinson*

**Computational Genomics with R**
*Altuna Akalin*

*For more information about this series please visit:*
*https://www.routledge.com/Chapman--HallCRC-Computational-Biology-Series/book-series/CRCCBS*

# Computational Genomics
# with R

Altuna Akalin

with the assistance of
Verdan Franke
Bora Uyar
Jonathan Ronen

CRC Press
Taylor & Francis Group
Boca Raton  London  New York

CRC Press is an imprint of the
Taylor & Francis Group, an **informa** business

A CHAPMAN & HALL BOOK

First edition published 2021
by CRC Press
6000 Broken Sound Parkway NW, Suite 300, Boca Raton, FL 33487-2742

and by CRC Press
2 Park Square, Milton Park, Abingdon, Oxon, OX14 4RN

© 2021 Taylor & Francis Group, LLC

CRC Press is an imprint of Taylor & Francis Group, LLC

ISBN: **9781498781855** (hbk)
ISBN: **9780429084317** (ebk)

Typeset in Computer Modern font
by KnowledgeWorks Global Ltd.

To my family,
Anna, Julia and Gabriel

# Contents

# Preface

The aim of this book is to provide the fundamentals for data analysis for genomics. We developed this book based on the computational genomics courses we are giving every year. We have had invariably an interdisciplinary audience with backgrounds from physics, biology, medicine, math, computer science or other quantitative fields. We want this book to be a starting point for computational genomics students and a guide for further data analysis in more specific topics in genomics. This is why we tried to cover a large variety of topics from programming to basic genome biology. As the field is interdisciplinary, it requires different starting points for people with different backgrounds. A biologist might skip sections on basic genome biology and start with R programming, whereas a computer scientist might want to start with genome biology. In the same manner, a more experienced person might want to refer to this book when needing to do a certain type of analysis, but having no prior experience.

The online version of this book is licensed under the Creative Commons Attribution-NonCommercial-ShareAlike 4.0 International License[1].

## Who is this book for?

The book contains practical and theoretical aspects of computational genomics. Biology and medicine generate more data than ever before. Therefore, we need to educate more people with data analysis skills and understanding of computational genomics. Since computational genomics is interdisciplinary, this book aims to be accessible for biologists, medical scientists, computer scientists and people from other quantitative backgrounds. We wrote this book for the following audiences:

- Biologists and medical scientists who generate the data and are keen on analyzing it themselves.
- Students and researchers who are formally starting to do research on or using computational genomics do not have extensive domain-specific knowledge, but

---

[1] http://creativecommons.org/licenses/by-nc-sa/4.0/

have at least a beginner-level understanding in a quantitative field, for example, math, stats.
- Experienced researchers looking for recipes or quick how-to's to get started in specific data analysis tasks related to computational genomics.

**What will you get out of this?**

This resource describes the skills and provides how-to's that will help readers analyze their own genomics data.

After reading:

- If you are not familiar with R, you will get the basics of R and dive right in to specialized uses of R for computational genomics.
- You will understand genomic intervals and operations on them, such as overlap.
- You will be able to use R and its vast package library to do sequence analysis, such as calculating GC content for given segments of a genome or find transcription factor binding sites.
- You will be familiar with visualization techniques used in genomics, such as heatmaps, meta-gene plots, and genomic track visualization.
- You will be familiar with supervised and unsupervised learning techniques which are important in data modeling and exploratory analysis of high-dimensional data.
- You will be familiar with analysis of different high-throughput sequencing datasets (RNA-seq, ChIP-seq, BS-seq and multi-omics integration) mostly using R-based tools.

**Structure of the book**

The book is designed with the idea that practical and conceptual understanding of data analysis methods is as important, if not more important, than the theoretical understanding, such as detailed derivation of equations in statistics or machine learning. That is why we first try to give a conceptual explanation of the concepts then we try to give essential parts of the mathematical formulas for more detailed understanding. In this spirit, we always show and explain the code for a particular data analysis task. We also give additional references such as books, websites, video lectures and scientific papers for readers who desire to gain deeper theoretical understanding of data analysis-related methods or concepts.

Chapter 1: "Introduction to Genomics" introduces the basic concepts in genome biology and genomics. Understanding these concepts is important for computational genomics.

Chapter 2: "Introduction to R for Genomic Data Analysis" provides the basic R skills necessary to follow the book in addition to common data analysis paradigms we observe in genomic data analysis. Chapter 3: "Statistics for Genomics", Chapter 4: "Exploratory Data Analysis with Unsupervised Machine Learning" and Chapter 5: "Predictive Modeling with Supervised Machine Learning" introduce the necessary quantitative skills that one will need when analyzing high-dimensional genomics data.

Chapter 6: "Operations on Genomic Intervals and Genome Arithmetic" introduces the fundamental tools for dealing with genomic intervals and their relationship to each other over the genome. In addition, the chapter introduces a variety of genomic data visualization methods. The skills introduced in this chapter are key skills that are needed to work with processed genomic data which are available through public databases such as Ensembl and the UCSC browser.

The next chapters deal with specific analysis of high-throughput sequencing data and integrating different kinds of datasets. Chapter 7: "Quality Check, Processing and Alignment of High-throughput Sequencing Reads" introduces quality checks that need to be done on sequencing reads and different ways to process them further. Chapters 8, 9 and 10 deal with RNA-seq analysis, ChIP-seq analysis and BS-seq analysis. The last chapter, Chapter 11:"Multi-omics Analysis" deals with methods for integrating multiple omics datasets.

Most chapters have exercises that reinforce some of the important points introduced in the chapters. The exercises are classified into beginner, intermediate and advanced categories. If you are well versed in a certain subject you might want to skip beginner-level exercises.

To sum it up, this book is a comprehensive guide for computational genomics. Some sections are there for the sake of the wide interdisciplinary audience and completeness, and not all sections will be equally useful to all readers of this broad audience.

## Software information and conventions

Package names and inline code and file names are formatted in a typewriter font (e.g. `methylKit`). Function names are followed by parentheses (e.g. `genomation::ScoreMatrix()`). The double-colon operator `::` means accessing an object from a package.

**Assignment operator convention**

Traditionally, <- is the preferred assignment operator. However, throughout the book we use = and <- as the assignment operator interchangeably.

**Packages needed to run the book code**

This book is primarily about using R packages to analyze genomics data, therefore if you want to reproduce the analysis in this book you need to install the relevant packages in each chapter using install.packages or BiocManager::install functions. In each chapter, we load the necessary packages with the library() or require() function when we use the needed functions from respective packages. By looking at calls, you can see which packages are needed for that code chunk or chapter. If you need to install all the package dependencies for the book, you can run the following command and have a cup of tea while waiting.

```
if (!requireNamespace("BiocManager", quietly = TRUE))
    install.packages("BiocManager")
BiocManager::install(c('qvalue','plot3D','ggplot2','pheatmap','cowplot',
                'cluster', 'NbClust', 'fastICA', 'NMF','matrixStats',
                'Rtsne', 'mosaic', 'knitr', 'genomation',
                'ggbio', 'Gviz', 'DESeq2', 'RUVSeq',
                'gProfileR', 'ggfortify', 'corrplot',
                'gage', 'EDASeq', 'citr', 'formatR',
                'svglite', 'Rqc', 'ShortRead', 'QuasR',
                'methylKit','FactoMineR', 'iClusterPlus',
                'enrichR','caret','xgboost','glmnet',
                'DALEX','kernlab','pROC','nnet','RANN',
                'ranger','GenomeInfoDb', 'GenomicRanges',
                'GenomicAlignments', 'ComplexHeatmap', 'circlize',
                'rtracklayer', 'BSgenome.Hsapiens.UCSC.hg38',
                'BSgenome.Hsapiens.UCSC.hg19','tidyr',
                'AnnotationHub', 'GenomicFeatures', 'normr',
                'MotifDb', 'TFBSTools', 'rGADEM', 'JASPAR2018'
                ))
```

**Data for the book**

We rely on data from different R and Bioconductor packages throughout the book. For the datasets that do not ship with those packages, we created our own package **compGenomRData**[2]. You can install this package via `devtools::in-stall_github("compgenomr/compGenomRData")`. We use the `system.file()` function to get the path to the files. We noticed many inexperienced users are confused about this function. This function just outputs the full path to the file that is installed with the data package.

## Exercises in the book

There is a set of exercises at the end of each chapter. The exercises are separated in thematic sections that follow the major sections in the chapter. In addition, each exercise is classified based on its difficulty as "Beginner", "Intermediate" and "Advanced". Beginner-level exercises can usually be done by refactoring the code in the chapter. Advanced-level exercises usually require a combination of code from different sections or chapters. The intermediate level is somewhere in between. The solutions to the exercises are available at `https://github.com/compgenomr/exercises`.

## Reproducibility statement

This book is compiled with R 4.0.0 and the following packages. We only list the main packages and their versions but not their dependencies.

```
## qvalue_2.20.0 | plot3D_1.3 | ggplot2_3.3.1 | pheatmap_1.0.12
## cowplot_1.0.0 | cluster_2.1.0 | NbClust_3.0 | fastICA_1.2.2
## NMF_0.23.0 | matrixStats_0.56.0 | Rtsne_0.15 | mosaic_1.7.0
## knitr_1.28 | genomation_1.20.0 | ggbio_1.36.0 | Gviz_1.32.0
## DESeq2_1.28.1 | RUVSeq_1.22.0 | gProfileR_0.7.0 | ggfortify_0.4.10
## corrplot_0.84 | gage_2.37.0 | EDASeq_2.22.0 | citr_0.3.2
## formatR_1.7 | svglite_1.2.3 | Rqc_1.22.0 | ShortRead_1.46.0
## QuasR_1.28.0 | methylKit_1.14.2 | FactoMineR_2.3 | iClusterPlus_1.24.0
## enrichR_2.1 | caret_6.0.86 | xgboost_1.0.0.2 | glmnet_4.0
## DALEX_1.2.1 | kernlab_0.9.29 | pROC_1.16.2 | nnet_7.3.14
## RANN_2.6.1 | ranger_0.12.1 | GenomeInfoDb_1.24.0 | GenomicRanges_1.40.0
## GenomicAlignments_1.24.0 | ComplexHeatmap_2.4.2 | circlize_0.4.9 |
        rtracklayer_1.48.0
## tidyr_1.1.0 | AnnotationHub_2.20.0 | GenomicFeatures_1.40.0 | normr_1.14.0
## MotifDb_1.30.0 | TFBSTools_1.26.0 | rGADEM_2.36.0 | JASPAR2018_1.1.1
## BSgenome.Hsapiens.UCSC.hg38_1.4.3 | BSgenome.Hsapiens.UCSC.hg19_1.4.3
```

---

[2]`https://github.com/compgenomr/compGenomRData`

## Acknowledgements

I wish to thank the R and Bioconductor communities for developing and maintaining libraries for genomic data analysis. Without their constant work and dedication, writing such a book would not be possible.

I also wish to thank all my past and present mentors, colleagues and employers. The interaction with them provided the motivation to write such a book, and organize and teach hands-on courses on computational genomics.

I wish to thank John Kimmel, the editor from Chapman & Hall/CRC, who helped me publish this book. It was a pleasure to work with him. He generously agreed to let me keep the online version of this book, so I can continue updating it after it is printed.

This has been a long journey for me. I started writing parts of this book as early as 2013. If it wasn't for Vedran Franke, Bora Uyar and Jonathan Ronen, it would have taken even longer. They kindly agreed to contribute the missing chapters and they did a great job. I am thankful for their contributions.

The following people kindly contributed fixes for typos and code, and various suggestions: Thomas Schalch, Alex Gosdschan, Rodrigo Ogava, Fei Zhao, Jonathan Kitt, Janani Ravi, Christian Schudoma, Samuel Sledzieski, Dania Hamo and Sarvesh Nikumbh.

Altuna Akalin
Berlin, Germany

# About the Authors

*Dr. Altuna Akalin*[3] organized the book structure, wrote most of the book and edited the rest. He is a bioinformatics scientist and the head of Bioinformatics and Omics Data Science Platform at the Berlin Institute for Medical Systems Biology, Max Delbrück Center in Berlin. He has been developing computational methods for analyzing and integrating large-scale genomics data sets since 2002. He is interested in using machine learning and statistics to uncover patterns related to important biological variables such as disease state and type. He lived in the USA, Norway, Turkey, Japan and Switzerland in order to pursue research work and education related to computational genomics. The underlying aim of his current work is utilizing complex molecular signatures to provide decision support systems for disease diagnostics and biomarker discovery. In addition to the research efforts and the managing of a scientific lab, since 2015, he has been organizing and teaching computational genomics courses in Berlin with participants from across the world. This book is mostly a result of material developed for those and previous teaching efforts at Weill Cornell Medical College in New York and Friedrich Miescher Institute in Basel, Switzerland.

Dr. Akalin and the following contributing authors have decades of combined experience in data analysis for genomics. They are developers of Bioconductor packages such as **methylKit**[4], **genomation**[5], **RCAS**[6] and **netSmooth**[7]. In addition, they have played key roles in developing end-to-end genomics data analysis pipelines for RNA-seq, ChIP-seq, Bisulfite-seq, and single cell RNA-seq called PiGx[8].

## Contributing authors

*Dr. Bora Uyar*[9] contributed Chapter 8, "RNA-seq Analysis". He started his bioinformatics training in Sabanci University (Istanbul/Turkey), from which he got his undergraduate degree. Later, he obtained an MSc from Simon Fraser University (Vancouver/Canada), then a PhD from the European Molecular Biology Laboratory in Heidelberg/Germany. Since 2015, he has been working as a bioinformatics scientist at the Bioinformatics Platform and Omics Data Science Platform at the

---

[3] https://github.com/al2na
[4] https://bioconductor.org/packages/release/bioc/html/methylKit.html
[5] https://bioconductor.org/packages/release/bioc/html/genomation.html
[6] https://bioconductor.org/packages/release/bioc/html/RCAS.html
[7] https://bioconductor.org/packages/release/bioc/html/netSmooth.html
[8] http://bioinformatics.mdc-berlin.de/pigx/
[9] https://github.com/borauyar

Berlin Institute for Medical Systems Biology. He has been contributing to the bioinformatics platform through research, collaborations, services and data analysis method development. His current primary research interest is the integration of multiple types of omics datasets to discover prognostic/diagnostic biomarkers of cancers.

*Dr. Vedran Franke*[10] contributed Chapter 9, "ChIP-seq Analysis". He received his PhD from the University of Zagreb. His work focused on the biogenesis and function of small RNA molecules during early embryogenesis, and establishment of pluripotency. Prior to his PhD, he worked as a scientific researcher under Boris Lenhard at the University of Bergen, Norway, focusing on principles of gene enhancer functions. He continues his research in the Bioinformatics and Omics Data Science Platform at the Berlin Institute for Medical System Biology. He develops tools for multi-omics data integration, focusing on single-cell RNA sequencing, and epigenomics. His integrated knowledge of cellular physiology along with his proficiency in data analysis enable him to find creative solutions to difficult biological problems.

*Dr. Jonathan Ronen*[11] contributed Chapter 11, "Multi-omics Analysis". Dr. Ronen got his MSc in control engineering from the Norwegian University of Science and Technology in 2010. He then worked as a software developer in Oslo, Brussels, and Munich. During that time, he was also on the founding team of www.holderdeord.no, a website that links votes in the Norwegian parliament to pledges made in party manifestos. In 2014–2015, he worked as a data scientist in New York University's Social Media and Political Participation lab. During that time, he also launched www.lahadam.co.il, a website which tracked Israeli politicians' Facebook posts. He obtained a PhD in computational biology in 2020, where he has published tools for imputation for single cell RNA-seq using priors, and integrative analysis of multi-omics data using deep learning.

---

[10]https://github.com/frenkiboy
[11]https://github.com/jonathanronen

# 1

## Introduction to Genomics

The aim of this chapter is to provide the reader with some of the fundamentals required for understanding genome biology. By no means, is this a complete overview of the subject, but just a summary that will help the non-biologist reader understand the recurring biological concepts in computational genomics. Readers that are well-versed in genome biology and modern genome-wide quantitative assays should feel free to skip this chapter or skim it through.

### 1.1 Genes, DNA and central dogma

A central concept that will come up again and again is "the gene". Before we can explain that, we need to introduce a few other concepts that are important to understand the gene concept. The human body is made up of billions of cells. These cells specialize in different tasks. For example, in the liver there are cells that help produce enzymes to break toxins. In the heart, there are specialized muscle cells that make the heart beat. Yet, all these different kinds of cells come from a single-celled embryo. All the instructions to make different kinds of cells are contained within that single cell and with every division of that cell, those instructions are transmitted to new cells. These instructions can be coded into a string – a molecule of DNA, a polymer made of recurring units called nucleotides. The four nucleotides in DNA molecules, Adenine, Guanine, Cytosine and Thymine (coded as four letters: A, C, G, and T) in a specific sequence, store the information for life. DNA is organized in a double-helix form where two complementary polymers interlace with each other and twist into the familiar helical shape.

### 1.1.1 What is a genome?

The full DNA sequence of an organism, which contains all the hereditary information, is called a genome. The genome contains all the information to build and maintain an organism. Genomes come in different sizes and structures. Our genome is not only a naked stretch of DNA. In eukaryotic cells, DNA is wrapped around proteins (histones) forming higher-order structures like nucleosomes which make up chromatins and chromosomes (see Figure 1.1).

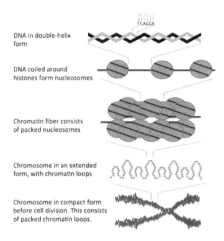

DNA in double-helix
form

DNA coiled around
histones form nucleosomes

Chromatin fiber consists
of packed nucleosomes

Chromosome in an extended
form, with chromatin loops

Chromosome in compact form
before cell division. This consists
of packed chromatin loops.

**FIGURE 1.1:** Chromosome structure in animals.

There might be several chromosomes depending on the organism. However, in some species (such as most prokaryotes) DNA is stored in a circular form. The size of the genome between species differs too. The human genome has 46 chromosomes and over 3 billion base-pairs, whereas the wheat genome has 42 chromosomes and 17 billion base-pairs; both genome size and chromosome numbers are variable between different organisms. Genome sequences of organisms are obtained using sequencing technology. With this technology, fragments of the DNA sequence from the genome, called reads, are obtained. Larger chunks of the genome sequence are later obtained by stitching the initial fragments to larger ones by using the overlapping reads. The latest sequencing technologies made genome sequencing cheaper and faster. These technologies output more reads, longer reads and more accurate reads.

The estimated cost of the first human genome was $300 million in 1999–2000; today a high-quality human genome can be obtained for $1500. Since the costs are going down, researchers and clinicians can generate more data. This drives up the costs for data storage and also drives up the demand for qualified people to analyze genomic data. This was one of the motivations behind writing this book.

### 1.1.2 What is a gene?

In the genome, there are specific regions containing the precise information that encodes for physical products of genetic information. A region in the genome with this information is traditionally called a "gene". However, the precise definition of the gene is still developing. According to the classical textbooks in molecular biology, a gene is a segment of a DNA sequence corresponding to a single protein or to a single catalytic and structural RNA molecule (Alberts et al., 2002). A modern definition is: "A region (or regions) that includes all of the sequence elements necessary to encode a functional transcript" (Eilbeck et al., 2005). No matter how variable the definitions are, all agree on the fact that genes are basic units of heredity in all living organisms.

All cells use their hereditary information in the same way most of the time; the DNA is replicated to transfer the information to new cells. If activated, the genes are transcribed into messenger RNAs (mRNAs) in the nucleus (in eukaryotes), followed by mRNAs (if the gene is protein coding) getting translated into proteins in the cytoplasm. This is essentially a process of information transfer between information-carrying polymers; DNA, RNA and proteins, known as the "central dogma" of molecular biology (see Figure 1.2 for a summary). Proteins are essential elements for life. The growth and repair, functioning and structure of all living cells depend on them. This is why the gene is a central concept in genome biology, because a gene can encode information for proteins and other functional molecules. How genes are controlled and activated dictates everything about an organism. From the identity of a cell to response to an infection, how cells develop and behave against certain stimuli is governed by the activity of the genes and the functional molecules they encode. The liver cell becomes a liver cell because certain genes are activated and their functional products are produced to help the liver cell achieve its tasks.

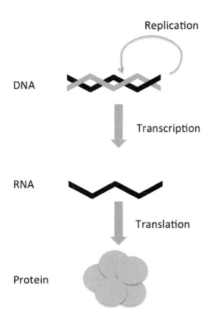

**FIGURE 1.2:** Central Dogma: replication, transcription, translation.

### 1.1.3 How are genes controlled? Transcriptional and post-transcriptional regulation

In order to answer this question, we have to dig a little deeper into the transcription concept we introduced via the central dogma. The first step in a process of information transfer - the production of an RNA copy of a part of the DNA sequence - is called transcription. This task is carried out by the RNA polymerase enzyme. RNA polymerase-dependent initiation of transcription is enabled by the existence of a specific region in the sequence of DNA - a core promoter. Core promoters are regions of DNA that promote transcription and are found upstream from the start site of transcription. In eukaryotes, several proteins, called general transcription factors, recognize and bind to core promoters and form a pre-initiation complex. RNA polymerases recognize these complexes and initiate synthesis of RNAs, the polymerase travels along the template DNA and makes an RNA copy (Hager et al., 2009). After mRNA is produced it is often spliced by spliceosome. The sections, called 'introns', are removed and sections called 'exons' left in. Then, the remaining mRNA is translated into proteins. Which exons will be part of the final mature transcript can also be regulated and creates diversity in protein structure and function (See Figure 1.3).

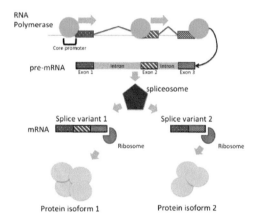

**FIGURE 1.3:** Transcription can be followed by splicing, which creates different transcript isoforms. This will in return create different protein isoforms since the information required to produce the protein is encoded in the transcripts. Differences in transcripts of the same gene can give rise to different protein isoforms.

Contrary to protein coding genes, non-coding RNA (ncRNAs) genes are processed and assume their functional structures after transcription and without going into translation, hence the name: non-coding RNAs. Certain ncRNAs can also be spliced but still not translated. ncRNAs and other RNAs in general can form complementary base-pairs within the RNA molecule which gives them additional complexity. This self-complementarity-based structure, termed the RNA secondary structure, is often necessary for functions of many ncRNA species.

In summary, the set of processes, from transcription initiation to production of the functional product, is referred to as gene expression. Gene expression quantification and regulation is a fundamental topic in genome biology.

### 1.1.4 What does a gene look like?

Before we move forward, it will be good to discuss how we can visualize genes. As someone interested in computational genomics, you will frequently encounter a gene on a computer screen, and how it is represented on the computer will be equivalent to what you imagine when you hear the word "gene". In the online databases, the genes will appear as a sequence of letters or as a series of connected boxes showing exon-intron structure, which may include the direction of transcription as well (see Figure 1.4). You will encounter more with the latter, so this is likely what will pop into your mind when you think of genes.

As we have mentioned, DNA has two strands. A gene can be located on either of them, and the direction of transcription will depend on that. In Figure 1.4, you can see arrows on introns (lines connecting boxes) indicating the direction of the gene.

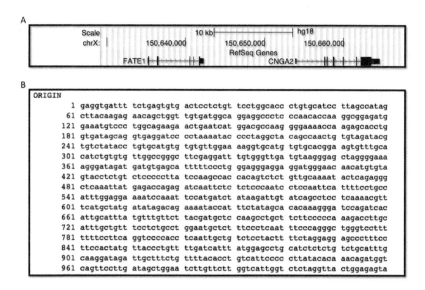

**FIGURE 1.4:** A) Representation of a gene in the UCSC browser. Boxes indicate exons, and lines indicate introns. B) Partial sequence of FATE1 gene as shown in the NCBI GenBank database.

## 1.2    Elements of gene regulation

The mechanisms regulating gene expression are essential for all living organisms as they dictate where and how much of a gene product (it may be protein or ncRNA) should be manufactured. This regulation could occur at the pre- and co-transcriptional level by controlling how many transcripts should be produced and/or which version of the transcript should be produced by regulating splicing. The same gene could encode for different versions of the same protein via splicing regulation.This process defines which parts of the gene will go into the final mRNA that will code for the protein variant. In addition, gene products can be regulated post-transcriptionally where certain molecules bind to RNA and mark them for degradation even before they can be used in protein production.

Gene regulation drives cellular differentiation; a process during which different tissues and cell types are produced. It also helps cells maintain differentiated states of cells/tissues. As a result of this process, at the final stage of differentiation, different kinds of cells maintain different expression profiles, although they contain the same genetic material. As mentioned above, there are two main types of regulation and next we will provide information on those.

### 1.2.1    Transcriptional regulation

The rate of transcription initiation is the primary regulatory element in gene expression regulation. The rate is controlled by core promoter elements as well as distant-acting regulatory elements such as enhancers. On top of that, processes like histone modifications and/or DNA methylation have a crucial regulatory impact on transcription. If a region is not accessible for the transcriptional machinery, e.g. in the case where the chromatin structure is compacted due to the presence of specific histone modifications, or if the promoter DNA is methylated, transcription may not start at all. Last but not least, gene activity is also controlled post-transcriptionally by ncRNAs such as microRNAs (miRNAs), as well as by cell signaling, resulting in protein modification or altered protein-protein interactions.

#### 1.2.1.1    Regulation by transcription factors through regulatory regions

Transcription factors are proteins that recognize a specific DNA motif to bind on a regulatory region and regulate the transcription rate of the gene associated with that regulatory region (see Figure 1.5 for an illustration). These factors bind to a variety of regulatory regions summarized in Figure 1.5, and their concerted action controls the transcription rate. Apart from their binding preference, their concentration, and the availability of synergistic or competing transcription factors will also affect the transcription rate.

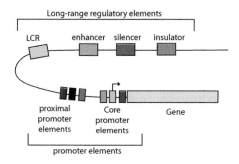

**FIGURE 1.5:** Representation of regulatory regions in animal genomes.

*1.2.1.1.1 Core and proximal promoters*

Core promoters are the immediate neighboring regions around the transcription start site (TSS) that serve as a docking site for the transcriptional machinery and pre-initiation complex (PIC) assembly. The textbook model for transcription initiation is as follows: The core promoter has a TATA motif (referred as TATA-box) 30 bp upstream of an initiator sequence (Inr), which also contains TSS. Firstly, transcription factor TFIID binds to the TATA-box. Next, general transcription factors are recruited and transcription is initiated on the initiator sequence. Apart from the TATA-box and Inr, there are a number of sequence elements on the animal core promoters that are associated with transcription initiation and PIC assembly, such as downstream promoter elements (DPEs), the BRE elements and CpG islands. DPEs are found 28-32 bp downstream of the TSS in TATA-less promoters of *Drosophila melanogaster*. They generally co-occur with the Inr element, and are thought to have a similar function to the TATA-box. The BRE element is recognized by the TFIIB protein and lie upstream of the TATA-box. CpG islands are CG dinucleotide-enriched segments of vertebrate genomes, despite the general depletion of CG dinucleotides in those genomes. 50 to 70% of promoters in the human genome are associated with CpG islands.

Proximal promoter elements are typically right upstream of the core promoters, usually contain binding sites for activator transcription factors, and provide additional control over gene expression.

*1.2.1.1.2 Enhancers*

Proximal regulation is not the only or the most important mode of gene regulation. Most of the transcription factor binding sites in the human genome are found in intergenic regions or in introns. This indicates the widespread usage of distal regulatory elements in animal genomes. On a molecular function level, enhancers are similar to proximal promoters; they contain binding sites for the same transcriptional activators and they basically enhance the gene expression. However, they are often highly modular and several of them can affect the same promoter at the same time or in different time-points or tissues. In addition, their activity is independent of their orientation and their distance to the promoter they interact with. A number of studies showed that enhancers can act upon their target genes over several kilobases away. According to a popular model, enhancers achieve this by looping the DNA and coming into contact with their target genes.

### 1.2.1.1.3   Silencers

Silencers are similar to enhancers; however their effect is opposite of enhancers on the transcription of the target gene, and results in decreasing their level of transcription. They contain binding sites for repressive transcription factors. Repressor transcription factors can either block the binding of an activator , directly compete for the same binding site, or induce a repressive chromatin state in which no activator binding is possible. Silencer effects, similar to those of enhancers, are independent of orientation and distance to target genes. In contradiction to this general view, in *Drosophila* there are two types of silencers, long-range and short-range. Short-range silencers are close to promoters and long-range silencers can silence multiple promoters or enhancers over kilobases away. Like enhancers, silencers bound by repressors may also induce changes in DNA structure by looping and creating higher-order structures. One class of such repressor proteins, which is thought to initiate higher-order structures by looping, is Polycomb group proteins (PcGs).

### 1.2.1.1.4   Insulators

Insulator regions limit the effect of other regulatory elements to certain chromosomal boundaries; in other words, they create regulatory domains untainted by the regulatory elements in regions outside that domain. Insulators can block enhancer-promoter communication and/or prevent spreading of repressive chromatin domains. In vertebrates and insects, some of the well-studied insulators are bound by CTCF (CCCTC-binding factor). Genome-wide studies from different mammalian tissues confirm that CTCF binding is largely invariant of cell type, and CTCF motif locations are conserved in vertebrates. At present, there are two models that explain the insulator function; the most prevalent model claims insulators create physically separate domains by modifying chromosome structure. This is thought to be achieved by CTCF-driven chromatin looping and recent evidence shows that CTCF can induce a higher-order chromosome structure through creating loops of chromatins. According to the second model, an insulator-bound activator cannot bind an enhancer; thus enhancer-blocking activity is achieved and insulators can also recruit an active histone domain, creating an active domain for enhancers to function.

Locus control regions (LCRs) are clusters of different regulatory elements that control an entire set of genes on a locus. LCRs help genes achieve their temporal and/or tissue-specific expression programs. LCRs may be composed of multiple cis-regulatory elements, such as insulators and enhancers, and they act upon their targets even from long distances. However, LCRs function in an orientation-dependent manner, for example the activity of beta-globin LCR is lost if inverted. The mechanism of LCR function otherwise seems similar to other long-range regulators described above. The evidence is mounting in the direction of a model where DNA-looping creates a chromosomal structure in which target genes are clustered together, which seems to be essential for maintaining an open chromatin domain.

## 1.2.1.2  Epigenetic regulation

Epigenetics in biology usually refers to constructions (chromatin structure, DNA methylation, etc.) other than DNA sequence that influence gene regulation. In essence, epigenetic regulation is the regulation of DNA packing and structure, the consequence of which is gene expression regulation. A typical example is that DNA packing inside the nucleus can directly influence gene expression by creating accessible regions for transcription factors to bind. There are two main mechanisms in epigenetic regulation: i) DNA modifications and ii) histone modifications. Below, we will introduce these two mechanisms.

DNA methylation is usually associated with gene silencing. DNA methyltransferase enzyme catalyzes the addition of a methyl group to cytosine of CpG dinucleotides (while in mammals the addition of methyl group is largely restricted to CpG dinucleotides, methylation can occur in other bases as well). This covalent modification either interferes with transcription factor binding on the region, or methyl-CpG binding proteins induce the spread of repressive chromatin domains, thus the gene is silenced if its promoter has methylated CG dinucleotides. DNA methylation usually occurs in repeat sequences to repress transposable elements. These elements, when active, can jump around and insert them to random parts of the genome, potentially disrupting the genomic functions.

DNA methylation is also related to a key core and proximal promoter element: CpG islands. CpG islands are usually unmethylated, however, for some genes, CpG island methylation accompanies their silenced expression. For example, during X-chromosome inactivation, many CpG islands are heavily methylated and the associated genes are silenced. In addition, in embryonic stem cell differentiation, pluripotency-associated genes are silenced due to DNA methylation. Apart from methylation, there are other kinds of DNA modifications present in mammalian genomes, such as hydroxy-methylation and formylcytosine. These are other modifications under current research that are either intermediate or stable modifications with distinct functional associations. There are at least a dozen distinct DNA modifications observed when we look across all studied species (Sood et al., 2019).

#### 1.2.1.2.2    *Histone modifications*

Histones are proteins that constitute a nucleosome. In eukaryotes, eight histone proteins are wrapped by DNA and make up the nucleosome. They help supercoiling of DNA and inducing high-order structure called chromatin. In chromatin, DNA is either densely packed (called heterochromatin or closed chromatin), or it is loosely packed (called euchromatin or open chromatin). Heterochromatin is thought to harbor inactive genes since DNA is densely packed and transcriptional machinery cannot access it. On the other hand, euchromatin is more accessible for transcriptional machinery and might therefore harbor active genes. Histones have long and unstructured N-terminal tails which can be covalently modified. The most studied modifications include acetylation, methylation and phosphorylation (Strahl and Allis, 2000). Using their tails, histones interact with neighboring nucleosomes and the modifications on the tail affect the nucleosomes' affinity to bind DNA and therefore influence DNA packaging around nucleosomes. Different modifications on histones are used in different combinations to program the activity of the genes during differentiation. Histone modifications have a distinct nomenclature, for example: H3K4me3 means the lysine (K) on the 4th position of histone H3 is tri-methylated.

**TABLE 1.1:** Histone modifications and their effects. If more than one histone modification has the same effect, they are separated by commas.

| Modifications | Effect |
| --- | --- |
| H3K9ac | Active promoters and enhancers |
| H3K14ac | Active transcription |
| H3K4me3/me2/me1 | Active promoters and enhancers, H3K4me1 and H3K27ac is enhancer-specific |
| H3K27ac | H3K27ac is enhancer-specific |
| H3K36me3 | Active transcribed regions |
| H3K27me3/me2/me1 | Silent promoters |
| H3K9me3/me2/me1 | Silent promoters |

Histone modifications are associated with a number of different transcription-related conditions; some of them are summarized in Table 1.1. Histone modifications can indicate where the regulatory regions are and they can also indicate activity of the genes. From a gene regulatory perspective, maybe the most important modifications are the ones associated with enhancers and promoters.

Furthermore, certain proteins can influence chromatin structure by interacting with histones. Some of these proteins, like those of the Polycomb Group (PcG) and CTCF, are discussed above in the insulators and silencer sections. In vertebrates and insects, PcGs are responsible for maintaining the silent state of developmental genes, and trithorax group proteins (trxG) for maintaining their active state (Henikoff, 2008; Schwartz and Pirrotta, 2007). PcGs and trxGs induce repressed or active states by catalyzing histone modifications or DNA methylation. Both the proteins bind PREs that can be on promoters or several kilobases away. Another protein that induces histone modifications is CTCF. CTCF is associated with boundaries between active and repressive histone marks (Phillips and Corces, 2009). This is due to the role of CTCF in regulating the 3D genome structure. Two CTCF binding sites that are far away from each other in linear distance can bind together in 3D space thus forming chromatin loops.

- **Want to know more?**
  - * Transcriptional regulatory elements in the human genome: http://www.ncbi.nlm.nih.gov/pubmed/16719718
  - * On metazoan promoters: Types and transcriptional properties: http://www.ncbi.nlm.nih.gov/pubmed/22392219
  - * General principles of regulatory sequence function: http://www.nature.com/nrg/journal/v15/n7/abs/nrg3684.html
  - * DNA methylation: Roles in mammalian development: http://www.nature.com/doifinder/10.1038/nrg3354
  - * Histone modifications and organization of the genome: http://www.nature.com/nrg/journal/v12/n1/full/nrg2905.html
  - * DNA methylation and histone modifications are linked: http://www.nature.com/nrg/journal/v10/n5/abs/nrg2540.html

### 1.2.2 Post-transcriptional regulation

#### 1.2.2.1 Regulation by non-coding RNAs

Recent years have witnessed an explosion in non-coding RNA (ncRNA)-related research. Many publications implicated ncRNAs as important regulatory elements. Plants and animals produce many different types of ncRNAs such as long noncoding RNAs (lncRNAs), small interferring RNAs (siRNAs), microRNAs (miRNAs), promoter-associated RNAs (PARs) and small nucleolar RNAs (snoRNAs) (Morris and Mattick, 2014). lncRNAs are typically >200-bp long, they are involved in epigenetic regulation by interacting with chromatin remodeling factors and they function in gene regulation. siRNAs are short double-stranded RNAs which are involved in gene regulation and transposon control; they silence their target genes by cooperating with Argonaute proteins. miRNAs are short single-stranded RNA molecules that interact with their target genes by using their complementary sequence and mark them for quicker degradation. PARs may regulate gene expression as well: they are approximately 18-to -200-bp-long ncRNAs originating from promoters of coding genes (Morris and Mattick, 2014). snoRNAs are also shown to play roles in gene regulation, although they are mostly believed to guide ribosomal RNA modifications (Morris and Mattick, 2014).

#### 1.2.2.2 Splicing regulation

Splicing is regulated by regulatory elements on the pre-mRNA and proteins binding to those elements. Regulatory elements are categorized as splicing enhancers and repressors. They can be located either in exons or introns. Depending on their activity and their locations there are four types of regulatory elements for splicing:

- exonic splicing enhancers (ESEs)
- exonic splicing silencers (ESSs)
- intronic splicing enhancers (ISEs)
- intronic splicing silencers (ISSs).

The majority of splicing repressors are heterogeneous nuclear ribonucleoproteins (hnRNPs). If splicing repressor protein bind silencer elements, they reduce the chance of a nearby site being used as a splice junction. On the contrary, splicing enhancers are sites to which splicing activator proteins bind and binding on that region increases the probability that a nearby site will be used as a splice junction (Wang and Burge, 2008). Most of the activator proteins that bind to splicing enhancers are members of the SR protein family. Such proteins can recognize specific RNA recognition motifs. By regulating splicing exons can be skipped or included, which creates protein diversity (Wang and Burge, 2008).

## 1.3 Shaping the genome: DNA mutation

Human and chimpanzee genomes are 98.8% similar. The 1.2% difference is what separates us from chimpanzees. The further you move away from human species in terms of evolutionary distance, the higher the difference gets. However, even between the members of the same species, differences in genome sequences exist. These differences are due to a process called mutation which drives differences between individuals but also provides the fuel for evolution as the source of the genetic variation. Individuals with beneficial mutations can adapt to their surroundings better than others and in time, these mutations, which are beneficial for survival, spread in the population due to a process called "natural selection". Selection acts upon individuals with beneficial features, which gives them an edge for survival in a given environment. Genetic variation created by the mutations in individuals provides the material on which selection can act. If the selection process goes for a long time in a relatively isolated environment that requires adaptation, this population can evolve into a different species given enough time. This is the basic idea behind evolution in a nutshell, and without mutations providing the genetic variation, there would be no evolution.

Mutations in the genome occur due to multiple reasons. First, DNA replication is not an error-free process. Before a cell division, the DNA is replicated with 1 mistake per $10^8$ to $10^{10}$ base-pairs. Second, mutagens such as UV light can induce mutations on the genome. The third factor that contributes to mutation is imperfect DNA repair. Every day, any human cell suffers multiple instances of DNA damage. DNA repair enzymes are there to cope with this damage but they are also not error-free, depending on which DNA repair mechanism is used (there are multiple), mistakes will be made at varying rates.

Mutations are classified by how many bases they affect, their effect on DNA structure and gene function. By their effect on DNA structure the mutations are classified as follows:

- **Base substitution**: A base is changed with another.
- **Deletion**: One or more bases is deleted.
- **Insertion**: New base or bases inserted into the genome.
- **Microsatellite mutation**: Small insertions or deletions of small tandemly repeating DNA segments.
- **Inversion**: A DNA fragment changes its orientation 180 degrees.
- **Translocation**: A DNA fragment moves to another location in the genome.

Mutations can also be classified by their size as follows:

- **Point mutations**: Mutations that involve one base. Substitutions, deletions and insertions are point mutations. They are also termed as single nucleotide polymorphisms (**SNPs**).
- **Small-scale mutations**: Mutations that involve several bases.
- **Large-scale mutations**: Mutations which involve larger chromosomal regions. Transposable element insertions (where a segment of the genome jumps to another region in the genome) and segmental duplications (a large region is copied multiple times in tandem) are typical large scale mutations.
- **Aneuploidies**: Insertions or deletions of whole chromosomes.
- **Whole-genome polyploidies**: Duplications involving whole genome.

Mutations can be classified by their effect on gene function as follows:

- **Gain-of-function mutations**: A type of mutation in which the altered gene product possesses a new molecular function or a new pattern of gene expression.
- **Loss-of-function mutations**: A mutation that results in reduced or abolished protein function. This is the more common type of mutation.

- **Want to know more?**

    * Interactive tutorial on mutations: http://www.dnaftb.org/27/

    * Tutorial on mutation and health maintained by NIH: https://ghr.nlm.nih.gov/primer#mutationsanddisorders

    * Visualizations for different types of mutations: https://www.yourgenome.org/facts/what-types-of-mutation-are-there

    * Review article on mutations regarding human disease: https://www.nature.com/articles/nrg3241

## 1.4  High-throughput experimental methods in genomics

Most of the biological phenomena described above relating to transcription, gene regulation or DNA mutation can be measured over the entire genome using high-throughput experimental techniques, which are quickly becoming the standard for studying genome biology. In addition, their applications in the clinic are also gaining momentum as there are already diagnostic tests that are based on these techniques.

Some of the things that can be measured by high-throughput assays are as follows:

- *Which genes are expressed and how much?*
- *Where does a transcription factor bind?*
- *Which bases are methylated in the genome?*
- *Which transcripts are translated?*
- *Where does RNA-binding proteins bind?*
- *Which microRNAs are expressed?*
- *Which parts of the genome are in contact with each other?*
- *Where are the mutations in the genome located?*
- *Which parts of the genome are nucleosome-free?*

There are many more questions one can answer using modern genome-wide techniques and every other day a new variant of the existing techniques comes along to answer a new question. However, one has to keep in mind that these methods are at varying degrees of maturity and they all come with technical limitations and are not noise-free. Despite this, they are extremely useful for research and clinical purposes. And, thanks to these methods, we are able to sequence and annotate genomes on a massive scale.

### 1.4.1  The general idea behind high-throughput techniques

High-throughput methods aim to quantify or locate all or most of the genome that harbors the biological feature (expressed genes, binding sites, etc.) of interest. Most of the methods rely on some sort of enrichment of the targeted biological feature. For example, if you want to measure expression of protein coding genes you need to be able to extract mRNA molecules with special post-transcriptional alterations that protein-coding genes acquire, as done in many RNA sequencing (RNA-seq) experiments. If you are looking for transcription factor binding, you need to enrich for the DNA fragments that are bound by the protein of interest, as it is done in ChIP-seq experiments. This part depends on available molecular biology and chemistry techniques, and the final product of this part is RNA or DNA fragments.

Next, you need to be able to tell where these fragments are coming from in the genome and how many of them there are. Microarrays were the standard tool for the quantification step until the spread of sequencing techniques. In microarrays, one had to design complementary bases, called "oligos" or "probes", to the genetic material enriched via the experimental protocol. If the enriched material is complementary to the oligos, a light signal will be produced and the intensity of the signal will be proportional to the amount of the genetic material pairing with that oligo. There will be more probes available for hybridization (process of complementary bases forming bonds), so the more fragments available, stronger the signal. For this to be able to work, you need to know at least part of your genome sequence, and design probes. If you want to measure gene expression, your probes should overlap with genes and should be unique enough to not to bind sequences from other genes. This technology is now being replaced with sequencing technology, where you directly sequence your genetic material. If you have the sequence of your fragments, you can align them back to the genome, see where they are coming from, and count them. This is a better technology where the quantification is based on the real identity of fragments rather than based on hybridization to designed probes.

In summary, HT techniques have the following steps, and this also summarized in Figure 1.6:

- Extraction: This is the step where you extract the genetic material of interest, RNA or DNA.
- Enrichment: In this step, you enrich for the event you are interested in. For example, protein binding sites. In some cases such as whole-genome DNA sequencing, there is no need for enrichment step. You just get fragments of genomic DNA and sequence them.
- Quantification: This is where you quantify your enriched material. Depending on the protocol you may need to quantify a control set as well, where you should see no enrichment or only background enrichment.

### 1.4.2  High-throughput sequencing

High-throughput sequencing, or massively parallel sequencing, is a collection of methods and technologies that can sequence DNA thousands/millions of fragments at a time. This is in contrast to older technologies that can produce a limited number of fragments at a time. Here, throughput refers to the number of sequenced bases per hour. The older low-throughput sequencing methods have ~100 times less throughput compared to modern high-throughput methods. The increased throughput gives the ability to measure biological features on a genome-wide scale in a shorter time frame.

Similar to other high-throughput methods, sequencing-based methods also require

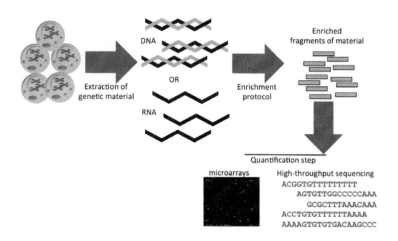

**FIGURE 1.6:** Common steps of high-throughput assays in genome biology.

an enrichment step. This step enriches for the features we are interested in. The main difference of the sequencing-based methods is the quantification step. In high-throughput sequencing, enriched fragments are put through the sequencer which outputs the sequences for the fragments. Due to limitations in current leading technologies, only a limited number of bases can be sequenced from the input fragments. However, the length is usually enough to uniquely map the reads to the genome and quantify the input fragments.

### 1.4.2.1 High-throughput sequencing data

If there is a genome available, the reads are aligned to the genome and based on the library preparation protocol, different strategies are applied for analysis. A sequencing library is composed of fragments of RNA or DNA ready to be sequenced. The library preparation primarily depends on the experiment of interest. There are a number of library preparation protocols aimed at quantifying different signals from the genome. Some of the potential analysis strategies for different library-prep protocols and processed output of read alignments are depicted in Figure 1.7. For example, we may be interested in quantifying the gene expression. The experimental protocol, called RNA sequencing, RNA-seq, enriches for fragments of RNA that are coming from protein coding genes. Upon alignment, we can calculate the coverage profile which gives us a read count per base along the genome. This information can be stored in a text file or specialized file formats to be used in subsequent analysis or visualization. We can also just count how many reads overlap with exons of each gene and record read counts per gene for further analysis. This essentially produces a table with gene names and read counts for different samples. As we will see in later chapters, this is an essential information for statistical models for RNA-seq data. Furthermore, we can stack up the reads and count how many times a base position in a read mismatches the base in the genome. Read aligners allow for mismatches, and for this reason we can see reads with mismatches. This information can be used to identify SNPs, and can be stored again in a tabular format with the information of position and mismatch type and number of reads supporting the mismatch. The original algorithms are a bit more complicated than just counting mismatches but the general idea is the same; what they are doing differently is trying to minimize false positive rates by using filters, so that not every mismatch is recorded as a SNP.

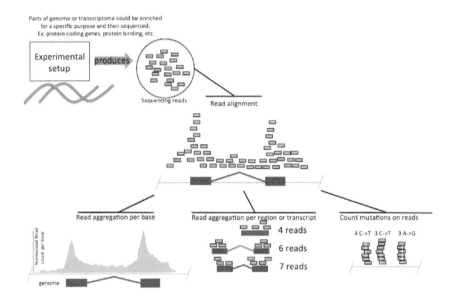

**FIGURE 1.7:** High-throughput sequencing summary.

#### 1.4.2.2   Future of high-throughput sequencing

The sequencing technology is still evolving. Obtaining longer single-molecule reads, and preferably, being able to call base modifications on the fly is the next frontier. With longer reads, the genome assembly will be easier for the regions that have high repeat content. With single-molecule sequencing, we will be able to tell how many transcripts are present in a given cell population without relying on fragment amplification methods which can introduce biases.

Another recent development is single-cell sequencing. Current technologies usually work on genetic material from thousands to millions of cells. This means that the results you receive represent the population of cells that were used in the experiment. However, there is a lot of variation between the same type of cells, but this variation is not observed at all. Newer sequencing techniques can work on single cells and give quantitative information on each cell.

## 1.5 Visualization and data repositories for genomics

There are ~100 animal genomes sequenced as of 2016. On top these, there are many research projects from either individual labs or consortia that produce petabytes of auxiliary genomics data, such as ChIP-seq, RNA-seq, etc.

There are two requirements to be able to visualize genomes and their associated data: 1) you need to be able to work with a species that has a sequenced genome and 2) you want to have annotation on that genome, meaning, at the very least, you want to know where the genes are. Most genomes after sequencing are quickly annotated with gene-predictions or known gene sequences are mapped on to them, and you can also have conservation to other species to filter functional elements. If you are working with a model organism or human, you will also have a lot of auxiliary information to help demarcate the functional regions such as regulatory regions, ncRNAs, and SNPs that are common in the population. Or you might have disease- or tissue-specific data available. The more the organism is worked on, the more auxiliary data you will have.

### 1.5.0.1 Accessing genome sequences and annotations via genome browsers

As someone who intends to work with genomics, you will need to visualize a large amount of data to make biological inferences or simply check regions of interest in the genome visually. Looking at the genome case by case with all the additional datasets is a necessary step to develop a hypothesis and understand the data.

Many genomes and their associated data are available through genome browsers. A genome browser is a website or an application that helps you visualize the genome and all the available data associated with it. Via genome browsers, you will be able to see where genes are in relation to each other and other functional elements.

You will be able to see gene structure. You will be able to see auxiliary data such as conservation, repeat content and SNPs. Here we review some of the popular browsers.

**UCSC genome browser:** This is an online browser hosted by University of California, Santa Cruz at `http://genome.ucsc.edu/`. This is an interactive website that contains genomes and annotations for many species. You can search for genes or genome coordinates for the species of your interest. It is usually very responsive and allows you to visualize large amounts of data. In addition, it has multiple other tools that can be used in connection with the browser. One of the most useful tools is the *UCSC Table Browser*, which lets you download all the data you see on the browser, including sequence data, in multiple formats. Users can upload data or provide links to the data to visualize user-specific data.

**Ensembl:** This is another online browser maintained by the European Bioinformatics Institute and the Wellcome Trust Sanger Institute in the UK, `http://www.ensembl.org`. Similar to the UCSC browser, users can visualize genes or genomic coordinates from multiple species and it also comes with auxiliary data. Ensembl is associated with the *Biomart* tool which is similar to UCSC Table browser, and can download genome data including all the auxiliary data set in multiple formats.

**IGV:** Integrated genomics viewer (IGV) is a desktop application developed by Broad institute (`https://www.broadinstitute.org/igv/`). It is developed to deal with large amounts of high-throughput sequencing data, which is harder to view in online browsers. IGV can integrate your local sequencing results with online annotation on your desktop machine. This is useful when viewing sequencing data, especially alignments. Other browsers mentioned above have similar features, however you will need to make your large sequencing data available online somewhere before it can be viewed by browsers.

### 1.5.0.2   Data repositories for high-throughput assays

Genome browsers contain lots of auxiliary high-throughput data. However, there are many more public high-throughput data sets available and they are certainly not available through genome browsers. Normally, every high-throughput dataset associated with a publication should be deposited in public archives. There are two major public archives we use to deposit data. One of them is *Gene Expression Omnibus (GEO)* hosted at `http://www.ncbi.nlm.nih.gov/geo/`, and the other one is *European Nucleotide Archive (ENA)* hosted at `http://www.ebi.ac.uk/ena`. These repositories accept high-throughput datasets and users can freely download and use these public data sets for their own research. Many data sets in these repositories are in their raw format, for example, the format the sequencer provides mostly. Some data sets will also have processed data but that is not a norm.

Apart from these repositories, there are multiple multi-national consortia dedicated to certain genome biology or disease-related problems and they maintain their own databases and provide access to processed and raw data. Some of these consortia are mentioned below.

| Consortium | What is it for? |
| --- | --- |
| ENCODE[1] | Transcription factor binding sites, gene expression and epigenomics data for cell lines |
| Epigenomics Roadmap[2] | Epigenomics data for multiple cell types |
| The Cancer Genome Atlas (TCGA)[3] | Expression, mutation and epigenomics data for multiple cancer types |
| 1000 genomes project[4] | Human genetic variation data obtained by sequencing 1000s of individuals |

---

[1]https://www.encodeproject.org/

[2]http://www.roadmapepigenomics.org/

[3]http://cancergenome.nih.gov/

[4]http://www.1000genomes.org/

# 2

## Introduction to R for Genomic Data Analysis

The aim of computational genomics is to provide biological interpretation and insights from high-dimensional genomics data. Generally speaking, it is similar to any other kind of data analysis endeavor but oftentimes doing computational genomics will require domain-specific knowledge and tools.

As new high-throughput experimental techniques are on the rise, data analysis capabilities are sought-after features for researchers. The aim of this chapter is to first familiarize readers with data analysis steps and then provide basics of R programming within the context of genomic data analysis. R is a free statistical programming language that is popular among researchers and data miners to build software and analyze data. Although basic R programming tutorials are easily accessible, we are aiming to introduce the subject with the genomic context in the background. The examples and narrative will always be from real-life situations when you try to analyze genomic data with R. We believe tailoring material to the context of genomics makes a difference when learning this programming language for the sake of analyzing genomic data.

## 2.1  Steps of (genomic) data analysis

Regardless of the analysis type, data analysis has a common pattern. We will discuss this general pattern and how it applies to genomics problems. The data analysis steps typically include data collection, quality check and cleaning, processing, modeling, visualization, and reporting. Although one expects to go through these steps in a linear fashion, it is normal to go back and repeat the steps with different parameters or tools. In practice, data analysis requires going through the same steps over and over again in order to be able to do a combination of the following: a) answer other related questions, b) deal with data quality issues that are later realized, and c) include new data sets to the analysis.

We will now go through a brief explanation of the steps within the context of genomic data analysis.

### 2.1.1   Data collection

Data collection refers to any source, experiment or survey that provides data for the data analysis question you have. In genomics, data collection is done by high-throughput assays, introduced in Chapter 1. One can also use publicly available data sets and specialized databases, also mentioned in Chapter 1. How much data and what type of data you should collect depends on the question you are trying to answer and the technical and biological variability of the system you are studying.

### 2.1.2   Data quality check and cleaning

In general, data analysis almost always deals with imperfect data. It is common to have missing values or measurements that are noisy. Data quality check and cleaning aims to identify any data quality issue and clean it from the dataset.

High-throughput genomics data is produced by technologies that could embed technical biases into the data. If we were to give an example from sequencing, the sequenced reads do not have the same quality of bases called. Towards the ends of the reads, you could have bases that might be called incorrectly. Identifying those low-quality bases and removing them will improve the read mapping step.

### 2.1.3   Data processing

This step refers to processing the data into a format that is suitable for exploratory analysis and modeling. Oftentimes, the data will not come in a ready-to-analyze format. You may need to convert it to other formats by transforming data points (such as log transforming, normalizing, etc.), or subset the data set with some arbitrary or pre-defined condition. In terms of genomics, processing includes multiple steps. Following the sequencing analysis example above, processing will include aligning reads to the genome and quantification over genes or regions of interest. This is simply counting how many reads are covering your regions of interest. This quantity can give you ideas about how much a gene is expressed if your experimental protocol was RNA sequencing. This can be followed by some normalization to aid the next step.

### 2.1.4   Exploratory data analysis and modeling

This phase usually takes in the processed or semi-processed data and applies machine learning or statistical methods to explore the data. Typically, one needs to see a relationship between variables measured, and a relationship between samples based on the variables measured. At this point, we might be looking to see if the samples are grouped as expected by the experimental design, or are there outliers or any other anomalies? After this step you might want to do additional cleanup or re-processing to deal with anomalies.

Another related step is modeling. This generally refers to modeling your variable of interest based on other variables you measured. In the context of genomics, it could be that you are trying to predict disease status of the patients from expression of genes you measured from their tissue samples. Then your variable of interest is the disease status. This kind of approach is generally called "predictive modeling", and could be solved with regression-based machine learning methods.

Statistical modeling would also be a part of this modeling step. This can cover predictive modeling as well, where we use statistical methods such as linear regression. Other analyses such as hypothesis testing, where we have an expectation and we are trying to confirm that expectation, is also related to statistical modeling. A good example of this in genomics is the differential gene expression analysis. This can be formulated as comparing two data sets, in this case expression values from condition A and condition B, with the expectation that condition A and condition B have similar expression values. You will see more on this in Chapter 3.

### 2.1.5   Visualization and reporting

Visualization is necessary for all the previous steps more or less. But in the final phase, we need final figures, tables, and text that describe the outcome of your analysis. This will be your report. In genomics, we use common data visualization methods as well as specific visualization methods developed or popularized by genomic data analysis. You will see many popular visualization methods in Chapters 3 and 6.

### 2.1.6   Why use R for genomics ?

R, with its statistical analysis heritage, plotting features, and rich user-contributed packages is one of the best languages for the task of analyzing genomic data. High-dimensional genomics datasets are usually suitable to be analyzed with core R packages and functions. On top of that, Bioconductor and CRAN have an array of specialized tools for doing genomics-specific analysis. Here is a list of computational genomics tasks that can be completed using R.

#### 2.1.6.1   Data cleanup and processing

Most of general data cleanup, such as removing incomplete columns and values, reorganizing and transforming data, can be achieved using R. In addition, with the help of packages, R can connect to databases in various formats such as mySQL, mongoDB, etc., and query and get the data into the R environment using database specific tools.

On top of these, genomic data-specific processing and quality check can be achieved via R/Bioconductor packages. For example, sequencing read quality checks and even HT-read alignments can be achieved via R packages.

#### 2.1.6.2   General data analysis and exploration

Most genomics data sets are suitable for application of general data analysis tools. In some cases, you may need to preprocess the data to get it to a state that is suitable for application of such tools. Here is a non-exhaustive list of what kind of things can be done via R. You will see popular data analysis methods in Chapters 3, 4 and 5.

- Unsupervised data analysis: clustering (k-means, hierarchical), matrix factorization (PCA, ICA, etc.)
- Supervised data analysis: generalized linear models, support vector machines, random forests

#### 2.1.6.3   Genomics-specific data analysis methods

R/Bioconductor gives you access to a multitude of other bioinformatics-specific algorithms. Here are some of the things you can do. We will touch upon many of the following methods in Chapter 6 and onwards.

- Sequence analysis: TF binding motifs, GC content and CpG counts of a given DNA sequence
- Differential expression (or arrays and sequencing-based measurements)
- Gene set/pathway analysis: What kind of genes are enriched in my gene set?
- Genomic interval operations such as overlapping CpG islands with transcription start sites, and filtering based on overlaps
- Overlapping aligned reads with exons and counting aligned reads per gene

#### 2.1.6.4   Visualization

Visualization is an important part of all data analysis techniques including computational genomics. Again, you can use core visualization techniques in R and also genomics-specific ones with the help of specific packages. Here are some of the things you can do with R.

- Basic plots: Histograms, scatter plots, bar plots, box plots, heatmaps
- Ideograms and circos plots for genomics provide visualization of different features over the whole genome.
- Meta-profiles of genomic features, such as read enrichment over all promoters
- Visualization of quantitative assays for given locus in the genome

## 2.2   Getting started with R

Download and install R (http://cran.r-project.org/) and RStudio (http://www.rstudio.com/) if you do not have them already. Rstudio is optional but it is a great tool if you are just starting to learn R. You will need specific data sets to run the code snippets in this book; we have explained how to install and use the data in the Data for the book section in the Preface. If you have not used Rstudio before, we recommend running it and familiarizing yourself with it first. To put it simply, this interface combines multiple features you will need while analyzing data. You can see your code, how it is executed, the plots you make, and your data all in one interface.

### 2.2.1 Installing packages

R packages are add-ons to base R that help you achieve additional tasks that are not directly supported by base R. It is by the action of these extra functionality that R excels as a tool for computational genomics. The Bioconductor project (http://bioconductor.org/) is a dedicated package repository for computational biology-related packages. However main package repository of R, called CRAN, also has computational biology-related packages. In addition, R-Forge (http://r-forge.r-project.org/), GitHub (https://github.com/), and Bitbucket (http://www.bitbucket.org) are some of the other locations where R packages might be hosted. The packages needed for the code snippets in this book and how to install them are explained in the Packages needed to run the book code section in the Preface of the book.

You can install CRAN packages using install.packages() (# is the comment character in R).

```
# install package named "randomForests" from CRAN
install.packages("randomForests")
```

You can install bioconductor packages with a specific installer script.

```
# get the installer package if you don't have
install.packages("BiocManager")

# install bioconductor package "rtracklayer"
BiocManager::install("rtracklayer")
```

You can install packages from GitHub using the install_github() function from devtools package.

```
library(devtools)
install_github("hadley/stringr")
```

Another way to install packages is from the source.

```
# download the source file
download.file(
"https://github.com/al2na/methylKit/releases/download/v0.99.2/methylKit_0.99.2.tar.g
            destfile="methylKit_0.99.2.tar.gz")
# install the package from the source file
install.packages("methylKit_0.99.2.tar.gz",
                repos=NULL,type="source")
# delete the source file
unlink("methylKit_0.99.2.tar.gz")
```

You can also update CRAN and Bioconductor packages.

```
# updating CRAN packages
update.packages()

# updating bioconductor packages
if (!requireNamespace("BiocManager", quietly = TRUE))
    install.packages("BiocManager")
BiocManager::install()
```

### 2.2.2  Installing packages in custom locations

If you will be using R on servers or computing clusters rather than your personal computer, it is unlikely that you will have administrator access to install packages. In that case, you can install packages in custom locations by telling R where to look for additional packages. This is done by setting up an `.Renviron` file in your home directory and add the following line:

```
R_LIBS=~/Rlibs
```

This tells R that the "Rlibs" directory at your home directory will be the first choice of locations to look for packages and install packages (the directory name and location is up to you, the above is just an example). You should go and create that directory now. After that, start a fresh R session and start installing packages. From now on, packages will be installed to your local directory where you have read-write access.

### 2.2.3  Getting help on functions and packages

You can get help on functions by using `help()` and `help.search()` functions. You can list the functions in a package with the `ls()` function

```
library(MASS)
ls("package:MASS") # functions in the package
ls() # objects in your R enviroment
# get help on hist() function
?hist
help("hist")
# search the word "hist" in help pages
help.search("hist")
??hist
```

#### 2.2.3.1 More help needed?

In addition, check package vignettes for help and practical understanding of the functions. All Bioconductor packages have vignettes that walk you through example analysis. Google search will always be helpful as well; there are many blogs and web pages that have posts about R. R-help mailing list (https://stat.ethz.ch/mailman/listinfo/r-help), Stackoverflow.com and R-bloggers.com are usually sources of good and reliable information.

## 2.3   Computations in R

R can be used as an ordinary calculator, and some say it is an over-grown calculator. Here are some examples. Remember that # is the comment character. The comments give details about the operations in case they are not clear.

```
2 + 3 * 5        # Note the order of operations.
log(10)          # Natural logarithm with base e
5^2              # 5 raised to the second power
3/2              # Division
sqrt(16)         # Square root
abs(3-7)         # Absolute value of 3-7
pi               # The number
exp(2)           # exponential function
# This is a comment line
```

## 2.4  Data structures

R has multiple data structures. If you are familiar with Excel, you can think of a single Excel sheet as a table and data structures as building blocks of that table. Most of the time you will deal with tabular data sets or you will want to transform your raw data to a tabular data set, and you will try to manipulate this tabular data set in some way. For example, you may want to take sub-sections of the table or extract all the values in a column. For these and similar purposes, it is essential to know the common data structures in R and how they can be used. R deals with named data structures, which means you can give names to data structures and manipulate or operate on them using those names. It will be clear soon what we mean by this if "named data structures" does not ring a bell.

### 2.4.1  Vectors

Vectors are one of the core R data structures. It is basically a list of elements of the same type (numeric, character or logical). Later you will see that every column of a table will be represented as a vector. R handles vectors easily and intuitively. You can create vectors with the `c()` function, however that is not the only way. The operations on vectors will propagate to all the elements of the vectors.

```
x<-c(1,3,2,10,5)      #create a vector named x with 5 components
x = c(1,3,2,10,5)
x
```

```
## [1]  1  3  2 10  5
```

```
y<-1:5                #create a vector of consecutive integers y
y+2                   #scalar addition
```

```
## [1] 3 4 5 6 7
```

```
2*y                   #scalar multiplication
```

```
## [1]  2  4  6  8 10
```

```
y^2                   #raise each component to the second power
```

```
## [1]  1  4  9 16 25
```

```
2^y                    #raise 2 to the first through fifth power
```

```
## [1]  2  4  8 16 32
```

```
y                      #y itself has not been unchanged
```

```
## [1] 1 2 3 4 5
```

```
y<-y*2
y                      #it is now changed
```

```
## [1]  2  4  6  8 10
```

```
r1<-rep(1,3)           # create a vector of 1s, length 3
length(r1)             #length of the vector
```

```
## [1] 3
```

```
class(r1)              # class of the vector
```

```
## [1] "numeric"
```

```
a<-1                   # this is actually a vector length one
```

The standard assignment operator in R is <-. This operator is preferentially used in books and documentation. However, it is also possible to use the = operator for the assignment. We have an example in the above code snippet and throughout the book we use <- and = interchangeably for assignment.

### 2.4.2 Matrices

A matrix refers to a numeric array of rows and columns. You can think of it as a stacked version of vectors where each row or column is a vector. One of the easiest ways to create a matrix is to combine vectors of equal length using cbind(), meaning 'column bind'.

```
x<-c(1,2,3,4)
y<-c(4,5,6,7)
m1<-cbind(x,y);m1
```

```
##       x y
## [1,] 1 4
## [2,] 2 5
## [3,] 3 6
## [4,] 4 7
```

```
t(m1)                   # transpose of m1
```

```
##    [,1] [,2] [,3] [,4]
## x    1    2    3    4
## y    4    5    6    7
```

```
dim(m1)                 # 2 by 5 matrix
```

```
## [1] 4 2
```

You can also directly list the elements and specify the matrix:

```
m2<-matrix(c(1,3,2,5,-1,2,2,3,9),nrow=3)
m2
```

```
##      [,1] [,2] [,3]
## [1,]    1    5    2
## [2,]    3   -1    3
## [3,]    2    2    9
```

Matrices and the next data structure, **data frames**, are tabular data structures. You can subset them using [] and providing desired rows and columns to subset. Figure 2.1 shows how that works conceptually.

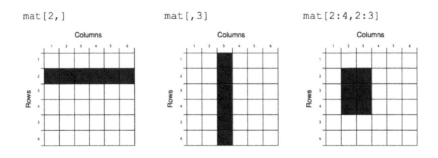

**FIGURE 2.1:** Slicing/subsetting of a matrix and a data frame.

### 2.4.3   Data frames

A data frame is more general than a matrix, in that different columns can have
different modes (numeric, character, factor, etc.). A data frame can be constructed
by the data.frame() function. For example, we illustrate how to construct a data
frame from genomic intervals or coordinates.

```
chr <- c("chr1", "chr1", "chr2", "chr2")
strand <- c("-","-","+","+")
start<- c(200,4000,100,400)
end<-c(250,410,200,450)
mydata <- data.frame(chr,start,end,strand)
#change column names
names(mydata) <- c("chr","start","end","strand")
mydata # OR this will work too
```

```
##      chr start end strand
## 1 chr1    200 250      -
## 2 chr1   4000 410      -
## 3 chr2    100 200      +
## 4 chr2    400 450      +
```

```
mydata <- data.frame(chr=chr,start=start,end=end,strand=strand)
mydata
```

```
##      chr start end strand
## 1 chr1    200 250      -
## 2 chr1   4000 410      -
## 3 chr2    100 200      +
## 4 chr2    400 450      +
```

There are a variety of ways to extract the elements of a data frame. You can extract
certain columns using column numbers or names, or you can extract certain rows
by using row numbers. You can also extract data using logical arguments, such as
extracting all rows that have a value in a column larger than your threshold.

```
mydata[,2:4] # columns 2,3,4 of data frame
```

```
##    start end strand
## 1    200 250      -
## 2   4000 410      -
## 3    100 200      +
## 4    400 450      +
```

```
mydata[,c("chr","start")] # columns chr and start from data frame
```

```
##     chr start
## 1 chr1   200
## 2 chr1  4000
## 3 chr2   100
## 4 chr2   400
```

```
mydata$start # variable start in the data frame
```

```
## [1]  200 4000  100  400
```

```
mydata[c(1,3),] # get 1st and 3rd rows
```

```
##     chr start end strand
## 1 chr1   200 250      -
## 3 chr2   100 200      +
```

```
mydata[mydata$start>400,] # get all rows where start>400
```

```
##     chr start end strand
## 2 chr1  4000 410      -
```

### 2.4.4  Lists

A list in R is an ordered collection of objects (components). A list allows you to gather a variety of (possibly unrelated) objects under one name. You can create a list with the `list()` function. Each object or element in the list has a numbered position and can have names. Below we show a few examples of how to create lists.

```
# example of a list with 4 components
# a string, a numeric vector, a matrix, and a scalar
w <- list(name="Fred",
      mynumbers=c(1,2,3),
      mymatrix=matrix(1:4,ncol=2),
      age=5.3)
w
```

```
## $name
## [1] "Fred"
##
```

```
## $mynumbers
## [1] 1 2 3
##
## $mymatrix
##      [,1] [,2]
## [1,]   1    3
## [2,]   2    4
##
## $age
## [1] 5.3
```

You can extract elements of a list using the `[[]]`, the double square-bracket, convention using either its position in the list or its name.

```
w[[3]] # 3rd component of the list
```

```
##      [,1] [,2]
## [1,]   1    3
## [2,]   2    4
```

```
w[["mynumbers"]] # component named mynumbers in list
```

```
## [1] 1 2 3
```

```
w$age
```

```
## [1] 5.3
```

### 2.4.5 Factors

Factors are used to store categorical data. They are important for statistical modeling since categorical variables are treated differently in statistical models than continuous variables. This ensures categorical data treated accordingly in statistical models.

```
features=c("promoter","exon","intron")
f.feat=factor(features)
```

An important thing to note is that when you are reading a data frame with `read.table()` or creating a data frame with `data.frame()` function, the character columns are stored as factors by default, to change this behavior you need to set `stringsAsFactors=FALSE` in `read.table()` and/or `data.frame()` function arguments.

## 2.5   Data types

There are four common data types in R, they are `numeric`, `logical`, `character` and `integer`. All these data types can be used to create vectors natively.

```
#create a numeric vector x with 5 components
x<-c(1,3,2,10,5)
x
```

```
## [1]   1  3  2 10  5
```

```
#create a logical vector x
x<-c(TRUE,FALSE,TRUE)
x
```

```
## [1]   TRUE FALSE   TRUE
```

```
# create a character vector
x<-c("sds","sd","as")
x
```

```
## [1] "sds" "sd"  "as"
```

```
class(x)
```

```
## [1] "character"
```

```
# create an integer vector
x<-c(1L,2L,3L)
x
```

```
## [1] 1 2 3
```

```
class(x)
```

```
## [1] "integer"
```

## 2.6 Reading and writing data

Most of the genomics data sets are in the form of genomic intervals associated with a score. That means mostly the data will be in table format with columns denoting chromosome, start positions, end positions, strand and score. One of the popular formats is the BED format, which is used primarily by the UCSC genome browser but most other genome browsers and tools will support the BED file format. We have all the annotation data in BED format. You will read more about data formats in Chapter 6. In R, you can easily read tabular format data with the `read.table()` function.

```r
enhancerFilePath=system.file("extdata",
                "subset.enhancers.hg18.bed",
                package="compGenomRData")
cpgiFilePath=system.file("extdata",
                "subset.cpgi.hg18.bed",
                package="compGenomRData")
# read enhancer marker BED file
enh.df <- read.table(enhancerFilePath, header = FALSE)

# read CpG island BED file
cpgi.df <- read.table(cpgiFilePath, header = FALSE)

# check first lines to see how the data looks like
head(enh.df)
```

```
##       V1     V2     V3 V4   V5 V6    V7      V8 V9
## 1 chr20 266275 267925  . 1000  .  9.11 13.1693 -1
## 2 chr20 287400 294500  . 1000  . 10.53 13.0231 -1
## 3 chr20 300500 302500  . 1000  .  9.10 13.3935 -1
## 4 chr20 330400 331800  . 1000  .  6.39 13.5105 -1
## 5 chr20 341425 343400  . 1000  .  6.20 12.9852 -1
## 6 chr20 437975 439900  . 1000  .  6.31 13.5184 -1
```

```r
head(cpgi.df)
```

```
##       V1     V2     V3      V4
## 1 chr20 195575 195851  CpG:_28
## 2 chr20 207789 208148  CpG:_32
## 3 chr20 219055 219437  CpG:_33
```

```
## 4 chr20 225831 227155 CpG:_135
## 5 chr20 252826 256323 CpG:_286
## 6 chr20 275376 276977 CpG:_116
```

You can save your data by writing it to disk as a text file. A data frame or matrix can be written out by using the write.table() function. Now let us write out cpgi.df. We will write it out as a tab-separated file; pay attention to the arguments.

```
write.table(cpgi.df,file="cpgi.txt",quote=FALSE,
            row.names=FALSE,col.names=FALSE,sep="\t")
```

You can save your R objects directly into a file using save() and saveRDS() and load them back in with load() and readRDS(). By using these functions you can save any R object whether or not it is in data frame or matrix classes.

```
save(cpgi.df,enh.df,file="mydata.RData")
load("mydata.RData")
# saveRDS() can save one object at a type
saveRDS(cpgi.df,file="cpgi.rds")
x=readRDS("cpgi.rds")
head(x)
```

One important thing is that with save() you can save many objects at a time, and when they are loaded into memory with load() they retain their variable names. For example, in the above code when you use load("mydata.RData") in a fresh R session, an object named cpg.df will be created. That means you have to figure out what name you gave to the objects before saving them. Conversely, when you save an object by saveRDS() and read by readRDS(), the name of the object is not retained, and you need to assign the output of readRDS() to a new variable (x in the above code chunk).

### 2.6.1   Reading large files

Reading large files that contain tables with base R function read.table() might take a very long time. Therefore, there are additional packages that provide faster functions to read the files. The data.table and readr packages provide this functionality. Below, we show how to use them. These functions with provided parameters will return equivalent output to the read.table() function.

```
library(data.table)
df.f=d(enhancerFilePath, header = FALSE,data.table=FALSE)
```

```
library(readr)
df.f2=read_table(enhancerFilePath, col_names = FALSE)
```

## 2.7 Plotting in R with base graphics

R has great support for plotting and customizing plots by default. This basic capability for plotting in R is referred to as "base graphics" or "R base graphics". We will show only a few below. Let us sample 50 values from the normal distribution and plot them as a histogram. A histogram is an approximate representation of a distribution. Bars show how frequently we observe certain values in our sample. The resulting histogram from the code chunk below is shown in Figure 2.2.

```
# sample 50 values from normal distribution
# and store them in vector x
x<-rnorm(50)
hist(x) # plot the histogram of those values
```

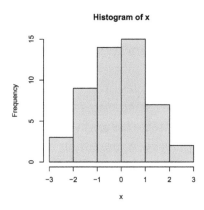

**FIGURE 2.2:** Histogram of values sampled from normal distribution.

We can modify all the plots by providing certain arguments to the plotting function. Now let's give a title to the plot using the `main` argument. We can also change the color of the bars using the `col` argument. You can simply provide the name of the color. Below, we are using `'red'` for the color. See Figure 2.3 for the result of this code chunk.

```
hist(x,main="Hello histogram!!!",col="red")
```

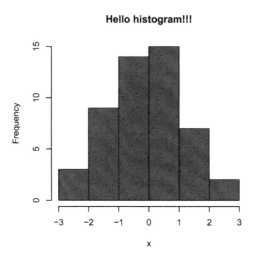

**FIGURE 2.3:** Histogram in red color.

Next, we will make a scatter plot. Scatter plots are one the most common plots you will encounter in data analysis. We will sample another set of 50 values and plot those against the ones we sampled earlier. The scatter plot shows values of two variables for a set of data points. It is useful to visualize relationships between two variables. It is frequently used in connection with correlation and linear regression. There are other variants of scatter plots which show density of the points with different colors. We will show examples of those scatter plots in later chapters. The scatter plot from our sampling experiment is shown in Figure 2.4. Notice that, in addition to main argument we used xlab and ylab arguments to give labels to the plot. You can customize the plots even more than this. See ?plot and ?par for more arguments that can help you customize the plots.

```
# randomly sample 50 points from normal distribution
y<-rnorm(50)
#plot a scatter plot
# control x-axis and y-axis labels
plot(x,y,main="scatterplot of random samples",
        ylab="y values",xlab="x values")
```

We can also plot boxplots for vectors x and y. Boxplots depict groups of numerical data through their quartiles. The edges of the box denote the 1st and 3rd quartiles, and the line that crosses the box is the median. The distance between the 1st and the

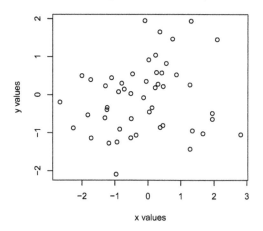

FIGURE 2.4: Scatter plot example.

3rd quartiles is called interquartile tange. The whiskers (lines extending from the boxes) are usually defined using the interquartile range for symmetric distributions as follows: `lowerWhisker=Q1-1.5[IQR]` and `upperWhisker=Q3+1.5[IQR]`.

In addition, outliers can be depicted as dots. In this case, outliers are the values that remain outside the whiskers. The resulting plot from the code snippet below is shown in Figure 2.5.

```
boxplot(x,y,main="boxplots of random samples")
```

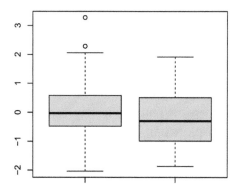

FIGURE 2.5: Boxplot example.

Next up is the bar plot, which you can plot using the `barplot()` function. We are going to plot four imaginary percentage values and color them with two colors, and this time we will also show how to draw a legend on the plot using the `legend()` function. The resulting plot is in Figure 2.6.

```r
perc=c(50,70,35,25)
barplot(height=perc,
        names.arg=c("CpGi","exon","CpGi","exon"),
        ylab="percentages",main="imagine %s",
        col=c("red","red","blue","blue"))
legend("topright",legend=c("test","control"),
       fill=c("red","blue"))
```

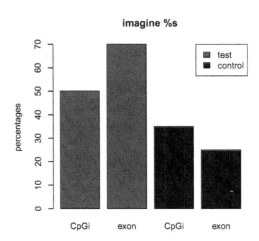

**FIGURE 2.6:** Bar plot example.

### 2.7.1  Combining multiple plots

In R, we can combine multiple plots in the same graphic. For this purpose, we use the `par()` function for simple combinations. More complicated arrangements with different sizes of sub-plots can be created with the `layout()` function. Below we will show how to combine two plots side-by-side using `par(mfrow=c(1,2))`. The `mfrow=c(nrows, ncols)` construct will create a matrix of `nrows` x `ncols` plots that are filled in by row. The following code will produce a histogram and a scatter plot stacked side by side. The result is shown in Figure 2.7. If you want to see the plots on top of each other, simply change `mfrow=c(1,2)` to `mfrow=c(2,1)`.

```
par(mfrow=c(1,2)) #

# make the plots
hist(x,main="Hello histogram!!!",col="red")
plot(x,y,main="scatterplot",
        ylab="y values",xlab="x values")
```

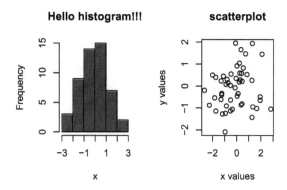

**FIGURE 2.7:** Combining two plots, a histogram and a scatter plot, with 'par()' function.

### 2.7.2  Saving plots

If you want to save your plots to an image file there are couple of ways of doing that. Normally, you will have to do the following:

1. Open a graphics device.
2. Create the plot.
3. Close the graphics device.

```
pdf("mygraphs/myplot.pdf",width=5,height=5)
plot(x,y)
dev.off()
```

Alternatively, you can first create the plot then copy the plot to a graphics device.

```
plot(x,y)
dev.copy(pdf,"mygraphs/myplot.pdf",width=7,height=5)
dev.off()
```

## 2.8 Plotting in R with ggplot2

In R, there are other plotting systems besides "base graphics", which is what we have shown until now. There is another popular plotting system called ggplot2 which implements a different logic when constructing the plots. This system or logic is known as the "grammar of graphics". This system defines a plot or graphics as a combination of different components. For example, in the scatter plot in 2.4, we have the points which are geometric shapes, we have the coordinate system and scales of data. In addition, data transformations are also part of a plot. In Figure 2.3, the histogram has a binning operation and it puts the data into bins before displaying it as geometric shapes, the bars. The ggplot2 system and its implementation of "grammar of graphics"[1] allows us to build the plot layer by layer using the predefined components.

Next we will see how this works in practice. Let's start with a simple scatter plot using ggplot2. In order to make basic plots in ggplot2, one needs to combine different components. First, we need the data and its transformation to a geometric object; for a scatter plot this would be mapping data to points, for histograms it would be binning the data and making bars. Second, we need the scales and coordinate system, which generates axes and legends so that we can see the values on the plot. And the last component is the plot annotation such as plot title and the background.

The main ggplot2 function, called ggplot(), requires a data frame to work with, and this data frame is its first argument as shown in the code snippet below. The second thing you will notice is the aes() function in the ggplot() function. This function defines which columns in the data frame map to x and y coordinates and if they should be colored or have different shapes based on the values in a different column. These elements are the "aesthetic" elements, this is what we observe in the plot. The last line in the code represents the geometric object to be plotted. These geometric objects define the type of the plot. In this case, the object is a point, indicated by the geom_point() function. Another, peculiar thing in the code is the + operation. In ggplot2, this operation is used to add layers and modify the plot. The resulting scatter plot from the code snippet below can be seen in Figure 2.8.

```
library(ggplot2)

myData=data.frame(col1=x,col2=y)

# the data is myData and I'm using col1 and col2
```

---

[1]This is a concept developed by Leland Wilkinson and popularized in R community by Hadley Wickham: https://doi.org/10.1198/jcgs.2009.07098

```
# columns on x and y axes
ggplot(myData, aes(x=col1, y=col2)) +
  geom_point() # map x and y as points
```

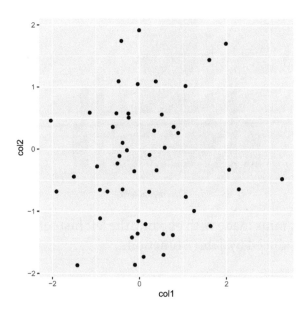

**FIGURE 2.8:** Scatter plot with ggplot2.

Now, let's re-create the histogram we created before. For this, we will start again with the ggplot() function. We are interested only in the x-axis in the histogram, so we will only use one column of the data frame. Then, we will add the histogram layer with the geom_histogram() function. In addition, we will be showing how to modify your plot further by adding an additional layer with the labs() function, which controls the axis labels and titles. The resulting plot from the code chunk below is shown in Figure 2.9.

```
ggplot(myData, aes(x=col1)) +
  geom_histogram() + # map x and y as points
  labs(title="Histogram for a random variable", x="my variable", y="Count")
```

We can also plot boxplots using ggplot2. Let's re-create the boxplot we did in Figure 2.5. This time we will have to put all our data into a single data frame with extra columns denoting the group of our values. In the base graphics case, we could just input variables containing different vectors. However, ggplot2 does not work like that and we need to create a data frame with the right format to use the ggplot()

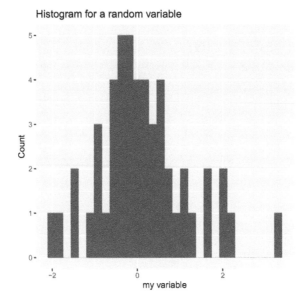

**FIGURE 2.9:** Histograms made with ggplot2, the left histogram contains additional modifications introduced by 'labs()' function.

function. Below, we first concatenate the x and y vectors and create a second column denoting the group for the vectors. In this case, the x-axis will be the "group" variable which is just a character denoting the group, and the y-axis will be the numeric "values" for the x and y vectors. You can see how this is passed to the `aes()` function below. The resulting plot is shown in Figure 2.10.

```
# data frame with group column showing which
# groups the vector x and y belong
myData2=rbind(data.frame(values=x,group="x"),
              data.frame(values=y,group="y"))

# x-axis will be group and y-axis will be values
ggplot(myData2, aes(x=group,y=values)) +
       geom_boxplot()
```

### 2.8.1 Combining multiple plots

There are different options for combining multiple plots. If we are trying to make similar plots for the subsets of the same data set, we can use faceting. This is a built-in and very useful feature of ggplot2. This feature is frequently used when investigating whether patterns are the same or different in different conditions or subsets of the data. It can be used via the `facet_grid()` function. Below, we will make two histograms faceted by the `group` variable in the input data frame. We will

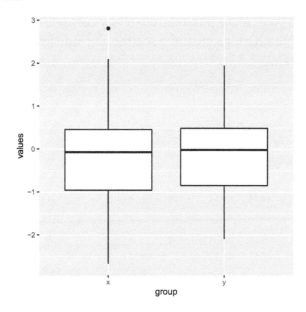

**FIGURE 2.10:** Boxplots using ggplot2.

be using the same data frame we created for the boxplot in the previous section. The resulting plot is in Figure 2.11.

```
ggplot(myData2, aes(x=values)) +
        geom_histogram() +facet_grid(.~group)
```

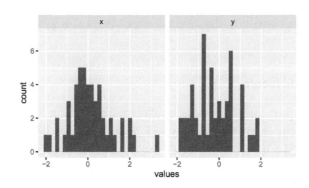

**FIGURE 2.11:** Combining two plots using ggplot2::facet_grid() function.

Faceting only works when you are using the subsets of the same data set. However, you may want to combine different types of plots from different data sets. The base R functions such as par() and layout() will not work with ggplot2 because it uses a different graphics system and this system does not recognize base R functionality for plotting. However, there are multiple ways you can combine plots from ggplot2. One way is using the cowplot package. This package aligns the individual plots in a grid and will help you create publication-ready compound plots. Below, we will show how to combine a histogram and a scatter plot side by side. The resulting plot is shown in Figure 2.12.

```
library(cowplot)
# histogram
p1 <- ggplot(myData2, aes(x=values,fill=group)) +
        geom_histogram()
# scatterplot
p2 <- ggplot(myData, aes(x=col1, y=col2)) +
  geom_point()

# plot two plots in a grid and label them as A and B
plot_grid(p1, p2, labels = c('A', 'B'), label_size = 12)
```

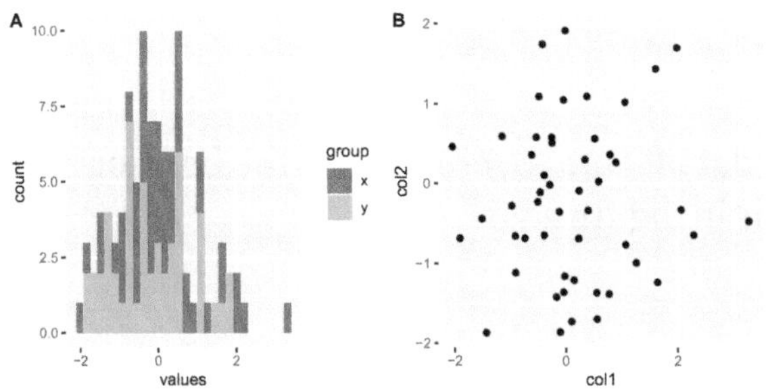

**FIGURE 2.12:** Combining a histogram and scatter plot using `cowplot` package. The plots are labeled as A and B using the arguments in `plot_grid()` function.

### 2.8.2   ggplot2 and tidyverse

ggplot2 is actually part of a larger ecosystem. You will need packages from this ecosystem when you want to use ggplot2 in a more sophisticated manner or if you need additional functionality that is not readily available in base R or other packages. For example, when you want to make more complicated plots using ggplot2, you will need to modify your data frames to the formats required by the ggplot() function, and you will need to learn about the dplyr and tidyr packages for data formatting purposes. If you are working with character strings, stringr package might have functionality that is not available in base R. There are many more packages that users find useful in tidyverse and it could be important to know about this ecosystem of R packages.

- **Want to know more ?**

  * `ggplot2` has a free online book written by Hadley Wickham: https://ggplot2-book.org/

  * The `tidyverse` packages and the ecosystem is described in their website: https://www.tidyverse.org/. There you will find extensive documentation and resources on `tidyverse` packages.

## 2.9 Functions and control structures (for, if/else, etc.)

### 2.9.1 User-defined functions

Functions are useful for transforming larger chunks of code to re-usable pieces of code. Generally, if you need to execute certain tasks with variable parameters, then it is time you write a function. A function in R takes different arguments and returns a definite output, much like mathematical functions. Here is a simple function that takes two arguments, $x$ and $y$, and returns the sum of their squares.

```
sqSum<-function(x,y){
result=x^2+y^2
return(result)
}
# now try the function out
sqSum(2,3)
```

```
## [1] 13
```

Functions can also output plots and/or messages to the terminal. Here is a function that prints a message to the terminal:

```
sqSumPrint<-function(x,y){
result=x^2+y^2
cat("here is the result:",result,"\n")
}
# now try the function out
sqSumPrint(2,3)
```

```
## here is the result: 13
```

Sometimes we would want to execute a certain part of the code only if a certain condition is satisfied. This condition can be anything from the type of an object (Ex: if the object is a matrix, execute certain code), or it can be more complicated, such as if the object value is between certain thresholds. Let us see how these if statements can be used. They can be used anywhere in your code; now we will use it in a function to decide if the CpG island is large, normal length or short.

```r
cpgi.df <- read.table("intro2R_data/data/subset.cpgi.hg18.bed", header = FALSE)
# function takes input one row
# of CpGi data frame
largeCpGi<-function(bedRow){
 cpglen=bedRow[3]-bedRow[2]+1
 if(cpglen>1500){
    cat("this is large\n")
 }
 else if(cpglen<=1500 & cpglen>700){
    cat("this is normal\n")
 }
 else{
    cat("this is short\n")
 }
}
largeCpGi(cpgi.df[10,])
largeCpGi(cpgi.df[100,])
largeCpGi(cpgi.df[1000,])
```

### 2.9.2 Loops and looping structures in R

When you need to repeat a certain task or execute a function multiple times, you can do that with the help of loops. A loop will execute the task until a certain condition is reached. The loop below is called a "for-loop" and it executes the task sequentially 10 times.

```r
for(i in 1:10){ # number of repetitions
cat("This is iteration") # the task to be repeated
print(i)
}

## This is iteration[1] 1
## This is iteration[1] 2
```

```
## This is iteration[1] 3
## This is iteration[1] 4
## This is iteration[1] 5
## This is iteration[1] 6
## This is iteration[1] 7
## This is iteration[1] 8
## This is iteration[1] 9
## This is iteration[1] 10
```

The task above is a bit pointless. Normally in a loop, you would want to do something meaningful. Let us calculate the length of the CpG islands we read in earlier. Although this is not the most efficient way of doing that particular task, it serves as a good example for looping. The code below will execute a hundred times, and it will calculate the length of the CpG islands for the first 100 islands in the data frame (by subtracting the end coordinate from the start coordinate).

**Note:**If you are going to run a loop that has a lot of repetitions, it is smart to try the loop with few repetitions first and check the results. This will help you make sure the code in the loop works before executing it thousands of times.

```
# this is where we will keep the lenghts
# for now it is an empty vector
result=c()
# start the loop
for(i in 1:100){
    #calculate the length
    len=cpgi.df[i,3]-cpgi.df[i,2]+1
    #append the length to the result
    result=c(result,len)
}
# check the results
head(result)
```

```
## [1]  277  360  383 1325 3498 1602
```

### 2.9.2.1 Apply family functions instead of loops

R has other ways of repeating tasks, which tend to be more efficient than using loops. They are known collectively as the "apply" family of functions, which include apply, lapply,mapply and tapply (and some other variants). All of these functions apply a given function to a set of instances and return the results of those functions for each instance. The difference between them is that they take different types of inputs. For example, apply works on data frames or matrices and applies the function on each row or column of the data structure. lapply works on lists or vectors and applies a function which takes the list element as an argument. Next we will demonstrate how to use `apply()` on a matrix. The example applies the sum function on the rows of a matrix; it basically sums up the values on each row of the matrix, which is conceptualized in Figure 2.13.

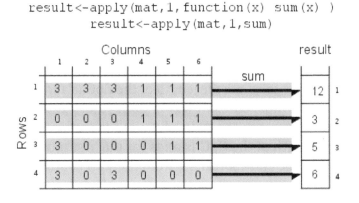

**FIGURE 2.13:** apply() concept in R.

```
mat=cbind(c(3,0,3,3),c(3,0,0,0),c(3,0,0,3),c(1,1,0,0),c(1,1,1,0),c(1,1,1,0))
result<-apply(mat,1,sum)
result
```

```
## [1] 12  3  5  6
```

```
# OR you can define the function as an argument to apply()
result<-apply(mat,1,function(x) sum(x))
result
```

```
## [1] 12  3  5  6
```

Notice that we used a second argument which equals 1, that indicates that rows of the matrix/ data frame will be the input for the function. If we change the second argument to 2, this will indicate that columns should be the input for the function that will be applied. See Figure 2.14 for the visualization of apply() on columns.

```
result<-apply(mat,2,sum)
result
```

```
## [1] 9 3 6 2 3 3
```

Next, we will use `lapply()`, which applies a function on a list or a vector. The function that will be applied is a simple function that takes the square of a given number.

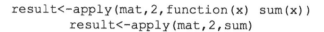

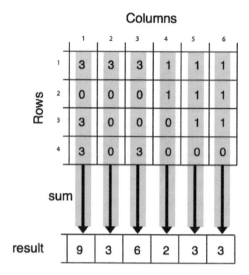

**FIGURE 2.14:** apply() function on columns.

```
input=c(1,2,3)
lapply(input,function(x) x^2)
```

```
## [[1]]
## [1] 1
##
## [[2]]
## [1] 4
##
## [[3]]
## [1] 9
```

mapply() is another member of the apply family, it can apply a function on an unlimited set of vectors/lists, it is like a version of lapply that can handle multiple vectors as arguments. In this case, the argument to the mapply() is the function to be applied and the sets of parameters to be supplied as arguments of the function. As shown in the conceptualized Figure 2.15, the function to be applied is a function that takes two arguments and sums them up. The arguments to be summed up are in the format of vectors Xs and Ys. mapply() applies the summation function to each pair in the Xs and Ys vector. Notice that the order of the input function and extra arguments are different for `mapply`.

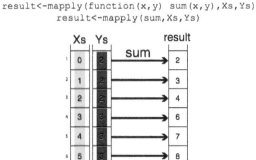

**FIGURE 2.15:** mapply() concept.

```
Xs=0:5
Ys=c(2,2,2,3,3,3)
result<-mapply(function(x,y) sum(x,y),Xs,Ys)
result
```

```
## [1] 2 3 4 6 7 8
```

#### 2.9.2.2 Apply family functions on multiple cores

If you have large data sets, apply family functions can be slow (although probably still better than for loops). If that is the case, you can easily use the parallel versions of those functions from the parallel package. These functions essentially divide your tasks to smaller chunks, run them on separate CPUs, and merge the results from those parallel operations. This concept is visualized in the figure below 2.16, mcapply runs the summation function on three different processors. Each processor executes the summation function on a part of the data set, and the results are merged and returned as a single vector that has the same order as the input parameters Xs and Ys.

#### 2.9.2.3 Vectorized functions in R

The above examples have been put forward to illustrate functions and loops in R because functions using sum() are not complicated and are easy to understand. You will probably need to use loops and looping structures with more complicated functions. In reality, most of the operations we used do not need the use of loops or looping structures because there are already vectorized functions that can achieve the same outcomes, meaning if the input arguments are R vectors, the output will be a vector as well, so no need for loops or vectorization.

```
result<-mcmapply(function(x,y) sum(x,y),
                Xs,Ys,mc.cores=3)
result<-mcmapply(sum,Xs,Ys,mc.cores=3)
```

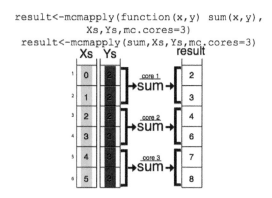

**FIGURE 2.16:** mcapply() concept.

For example, instead of using mapply() and sum() functions, we can just use the +
operator and sum up Xs and Ys.

```
result=Xs+Ys
result
```

```
## [1] 2 3 4 6 7 8
```

In order to get the column or row sums, we can use the vectorized functions
colSums() and rowSums().

```
colSums(mat)
```

```
## [1] 9 3 6 2 3 3
```

```
rowSums(mat)
```

```
## [1] 12  3  5  6
```

However, remember that not every function is vectorized in R, so use the ones that
are. But sooner or later, apply family functions will come in handy.

## 2.10   Exercises

### 2.10.1   Computations in R

1.  Sum 2 and 3 using the + operator. [Difficulty: **Beginner**]

2.  Take the square root of 36, use `sqrt()`. [Difficulty: **Beginner**]

3.  Take the log10 of 1000, use function `log10()`. [Difficulty: **Beginner**]

4.  Take the log2 of 32, use function `log2()`. [Difficulty: **Beginner**]

5.  Assign the sum of 2,3 and 4 to variable x. [Difficulty: **Beginner**]

6.  Find the absolute value of the expression `5 - 145` using the `abs()` function. [Difficulty: **Beginner**]

7.  Calculate the square root of 625, divide it by 5, and assign it to variable x.Ex: `y= log10(1000)/5`, the previous statement takes log10 of 1000, divides it by 5, and assigns the value to variable y. [Difficulty: **Beginner**]

8.  Multiply the value you get from previous exercise by 10000, assign it to variable x Ex: `y=y*5`, multiplies y by 5 and assigns the value to y. **KEY CONCEPT:** results of computations or arbitrary values can be stored in variables we can re-use those variables later on and over-write them with new values. [Difficulty: **Beginner**]

### 2.10.2   Data structures in R

10.  Make a vector of 1,2,3,5 and 10 using `c()`, and assign it to the `vec` variable. Ex: `vec1=c(1,3,4)` makes a vector out of 1,3,4. [Difficulty: **Beginner**]

11.  Check the length of your vector with length(). Ex: `length(vec1)` should return 3. [Difficulty: **Beginner**]

12.  Make a vector of all numbers between 2 and 15. Ex: `vec=1:6` makes a vector of numbers between 1 and 6, and assigns it to the `vec` variable. [Difficulty: **Beginner**]

13.  Make a vector of 4s repeated 10 times using the `rep()` function. Ex: `rep(x=2,times=5)` makes a vector of 2s repeated 5 times. [Difficulty: **Beginner**]

14.  Make a logical vector with TRUE, FALSE values of length 4, use `c()`. Ex: `c(TRUE,FALSE)`. [Difficulty: **Beginner**]

15. Make a character vector of the gene names PAX6,ZIC2,OCT4 and SOX2. Ex: `avec=c("a","b","c")` makes a character vector of a,b and c. [Difficulty: **Beginner**]

16. Subset the vector using `[]` notation, and get the 5th and 6th elements. Ex: `vec1[1]` gets the first element. `vec1[c(1,3)]` gets the 1st and 3rd elements. [Difficulty: **Beginner**]

17. You can also subset any vector using a logical vector in `[]`. Run the following:

```
myvec=1:5
# the length of the logical vector
# should be equal to length(myvec)
myvec[c(TRUE,TRUE,FALSE,FALSE,FALSE)]
myvec[c(TRUE,FALSE,FALSE,FALSE,TRUE)]
```

[Difficulty: **Beginner**]

18. `==,>,<, >=, <=` operators create logical vectors. See the results of the following operations:

```
myvec > 3
myvec == 4
myvec <= 2
myvec != 4
```

[Difficulty: **Beginner**]

19. Use the `>` operator in `myvec[ ]` to get elements larger than 2 in `myvec` which is described above. [Difficulty: **Beginner**]

20. Make a 5x3 matrix (5 rows, 3 columns) using `matrix()`. Ex: `matrix(1:6,nrow=3,ncol=2)` makes a 3x2 matrix using numbers between 1 and 6. [Difficulty: **Beginner**]

21. What happens when you use `byrow = TRUE` in your matrix() as an additional argument? Ex: `mat=matrix(1:6,nrow=3,ncol=2,byrow = TRUE)`. [Difficulty: **Beginner**]

22. Extract the first 3 columns and first 3 rows of your matrix using `[]` notation. [Difficulty: **Beginner**]

23. Extract the last two rows of the matrix you created earlier. Ex: `mat[2:3,]` or `mat[c(2,3),]` extracts the 2nd and 3rd rows. [Difficulty: **Beginner**]

24. Extract the first two columns and run `class()` on the result. [Difficulty: **Beginner**]

25. Extract the first column and run `class()` on the result, compare with the above exercise. [Difficulty: **Beginner**]

26. Make a data frame with 3 columns and 5 rows. Make sure first column is a sequence of numbers 1:5, and second column is a character vector. Ex: `df=data.frame(col1=1:3,col2=c("a","b","c"),col3=3:1) # 3x3 data frame`. Remember you need to make a 3x5 data frame. [Difficulty: **Beginner**]

27. Extract the first two columns and first two rows. **HINT:** Use the same notation as matrices. [Difficulty: **Beginner**]

28. Extract the last two rows of the data frame you made. **HINT:** Same notation as matrices. [Difficulty: **Beginner**]

29. Extract the last two columns using the column names of the data frame you made. [Difficulty: **Beginner**]

30. Extract the second column using the column names. You can use `[]` or `$` as in lists; use both in two different answers. [Difficulty: **Beginner**]

31. Extract rows where the 1st column is larger than 3. **HINT:** You can get a logical vector using the > operator , and logical vectors can be used in `[]` when subsetting. [Difficulty: **Beginner**]

32. Extract rows where the 1st column is larger than or equal to 3. [Difficulty: **Beginner**]

33. Convert a data frame to the matrix. **HINT:** Use `as.matrix()`. Observe what happens to numeric values in the data frame. [Difficulty: **Beginner**]

34. Make a list using the `list()` function. Your list should have 4 elements; the one below has 2. Ex: `mylist= list(a=c(1,2,3),b=c("apple,"orange"))` [Difficulty: **Beginner**]

35. Select the 1st element of the list you made using $ notation. Ex: `mylist$a` selects first element named "a". [Difficulty: **Beginner**]

36. Select the 4th element of the list you made earlier using $ notation. [Difficulty: **Beginner**]

37. Select the 1st element of your list using `[ ]` notation. Ex: `mylist[1]` selects the first element named "a", and you get a list with one element. `mylist["a"]` selects the first element named "a", and you get a list with one element. [Difficulty: **Beginner**]

38. Select the 4th element of your list using [ ] notation. [Difficulty: **Beginner**]

39. Make a factor using factor(), with 5 elements. Ex: `fa=factor(c("a","a","b"))`. [Difficulty: **Beginner**]

40. Convert a character vector to a factor using `as.factor()`. First, make a character vector using `c()` then use `as.factor()`. [Difficulty: **Intermediate**]

41. Convert the factor you made above to a character using `as.character()`. [Difficulty: **Beginner**]

### 2.10.3   Reading in and writing data out in R

1. Read CpG island (CpGi) data from the compGenomRData package `CpGi.table.hg18.txt`. This is a tab-separated file. Store it in a variable called `cpgi`. Use

```
cpgFilePath=system.file("extdata",
                "CpGi.table.hg18.txt",
                package="compGenomRData")
```

to get the file path within the installed compGenomRData package. [Difficulty: **Beginner**]

2. Use `head()` on CpGi to see the first few rows. [Difficulty: **Beginner**]

3. Why doesn't the following work? See sep argument at `help(read.table)`. [Difficulty: **Beginner**]

```
cpgtFilePath=system.file("extdata",
                "CpGi.table.hg18.txt",
                package="compGenomRData")
cpgtFilePath
cpgiSepComma=read.table(cpgtFilePath,header=TRUE,sep=",")
head(cpgiSepComma)
```

4. What happens when you set `stringsAsFactors=FALSE` in `read.table()`? [Difficulty: **Beginner**]

```
cpgiHF=read.table("intro2R_data/data/CpGi.table.hg18.txt",
                header=FALSE,sep="\t",
                stringsAsFactors=FALSE)
```

5.  Read only the first 10 rows of the CpGi table. [Difficulty: **Beginner/Inter-mediate**]

6.  Use `cpgFilePath=system.file("extdata","CpGi.table.hg18.txt", package="compGenomRData")` to get the file path, then use `read.table()` with argument `header=FALSE`. Use `head()` to see the results. [Difficulty: **Beginner**]

7.  Write CpG islands to a text file called "my.cpgi.file.txt". Write the file to your home folder; you can use `file="~/my.cpgi.file.txt"` in linux. `~/` denotes home folder.[Difficulty: **Beginner**]

8.  Same as above but this time make sure to use the `quote=FALSE,sep="\t"` and `row.names=FALSE` arguments. Save the file to "my.cpgi.file2.txt" and compare it with "my.cpgi.file.txt". [Difficulty: **Beginner**]

9.  Write out the first 10 rows of the `cpgi` data frame. **HINT:** Use subsetting for data frames we learned before. [Difficulty: **Beginner**]

10. Write the first 3 columns of the `cpgi` data frame. [Difficulty: **Beginner**]

11. Write CpG islands only on chr1. **HINT:** Use subsetting with `[]`, feed a logical vector using `==` operator.[Difficulty: **Beginner/Intermediate**]

12. Read two other data sets "rn4.refseq.bed" and "rn4.refseq2name.txt" with `header=FALSE`, and assign them to df1 and df2 respectively. They are again included in the compGenomRData package, and you can use the `system.file()` function to get the file paths. [Difficulty: **Beginner**]

13. Use `head()` to see what is inside the data frames above. [Difficulty: **Beginner**]

14. Merge data sets using `merge()` and assign the results to a variable named 'new.df', and use `head()` to see the results. [Difficulty: **Intermediate**]

### 2.10.4  Plotting in R

Please run the following code snippet for the rest of the exercises.

```
set.seed(1001)
x1=1:100+rnorm(100,mean=0,sd=15)
y1=1:100
```

1. Make a scatter plot using the `x1` and `y1` vectors generated above. [Difficulty: **Beginner**]

2. Use the `main` argument to give a title to `plot()` as in `plot(x,y,main="title")`. [Difficulty: **Beginner**]

3. Use the `xlab` argument to set a label for the x-axis. Use `ylab` argument to set a label for the y-axis. [Difficulty: **Beginner**]

4. Once you have the plot, run the following expression in R console. `mtext(side=3,text="hi there")` does. **HINT:** `mtext` stands for margin text. [Difficulty: **Beginner**]

5. See what `mtext(side=2,text="hi there")` does. Check your plot after execution. [Difficulty: **Beginner**]

6. Use *mtext()* and *paste()* to put a margin text on the plot. You can use `paste()` as 'text' argument in `mtext()`. **HINT:** `mtext(side=3,text=paste(...))`. See how `paste()` is used for below. [Difficulty: **Beginner/Intermediate**]

```
paste("Text","here")
```

```
## [1] "Text here"
```

```
myText=paste("Text","here")
myText
```

```
## [1] "Text here"
```

7. `cor()` calculates the correlation between two vectors. Pearson correlation is a measure of the linear correlation (dependence) between two variables X and Y. Try using the `cor()` function on the `x1` and `y1` variables. [Difficulty: **Intermediate**]

8. Try to use `mtext()`,`cor()` and `paste()` to display the correlation coefficient on your scatter plot. [Difficulty: **Intermediate**]

9. Change the colors of your plot using the `col` argument. Ex: `plot(x,y,col="red")`. [Difficulty: **Beginner**]

10. Use `pch=19` as an argument in your `plot()` command. [Difficulty: **Beginner**]

11. Use `pch=18` as an argument to your `plot()` command. [Difficulty: **Beginner**]

12. Make a histogram of `x1` with the `hist()` function. A histogram is a graphical representation of the data distribution. [Difficulty: **Beginner**]

13. You can change colors with 'col', add labels with 'xlab', 'ylab', and add a 'title' with 'main' arguments. Try all these in a histogram. [Difficulty: **Beginner**]

14. Make a boxplot of y1 with `boxplot()`.[Difficulty: **Beginner**]

15. Make boxplots of `x1` and `y1` vectors in the same plot.[Difficulty: **Beginner**]

16. In boxplot, use the `horizontal` = `TRUE` argument. [Difficulty: **Beginner**]

17. Make multiple plots with `par(mfrow=c(2,1))`

   •run `par(mfrow=c(2,1))`
   •make a boxplot
   •make a histogram [Difficulty: **Beginner/Intermediate**]

18. Do the same as above but this time with `par(mfrow=c(1,2))`. [Difficulty: **Beginner/Intermediate**]

19. Save your plot using the "Export" button in Rstudio. [Difficulty: **Beginner**]

20. You can make a scatter plot showing the density of points rather than points themselves. If you use points it looks like this:

```
x2=1:1000+rnorm(1000,mean=0,sd=200)
y2=1:1000
plot(x2,y2,pch=19,col="blue")
```

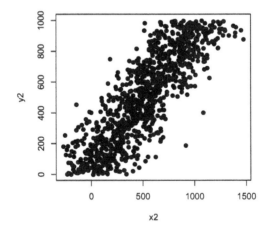

If you use the `smoothScatter()` function, you get the densities.

```
smoothScatter(x2,y2,
              colramp=colorRampPalette(c("white","blue",
                                         "green","yellow","red")))
```

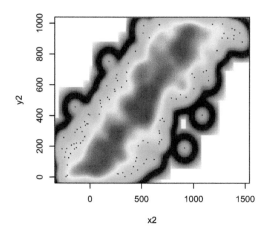

Now, plot with the `colramp=heat.colors` argument and then use a custom color scale using the following argument.

```
colramp = colorRampPalette(c("white","blue", "green","yellow","red")))
```

[Difficulty: **Beginner/Intermediate**]

### 2.10.5   Functions and control structures (for, if/else, etc.)

Read CpG island data as shown below for the rest of the exercises.

```
cpgtFilePath=system.file("extdata",
                "CpGi.table.hg18.txt",
                package="compGenomRData")
cpgi=read.table(cpgtFilePath,header=TRUE,sep="\t")
head(cpgi)
```

```
##    chrom chromStart chromEnd    name length cpgNum gcNum perCpg perGc obsExp
## 1   chr1      18598    19673 CpG: 116   1075    116   787   21.6  73.2   0.83
## 2   chr1     124987   125426  CpG: 30    439     30   295   13.7  67.2   0.64
## 3   chr1     317653   318092  CpG: 29    439     29   295   13.2  67.2   0.62
## 4   chr1     427014   428027  CpG: 84   1013     84   734   16.6  72.5   0.64
## 5   chr1     439136   440407  CpG: 99   1271     99   777   15.6  61.1   0.84
```

```
## 6  chr1      523082   523977 CpG: 94      895      94    570   21.0  63.7    1.04
```

1. Check values in the perGc column using a histogram. The 'perGc' column in the data stands for GC percent => percentage of C+G nucleotides. [Difficulty: **Beginner**]

2. Make a boxplot for the 'perGc' column. [Difficulty: **Beginner**]

3. Use if/else structure to decide if the given GC percent is high, low or medium. If it is low, high, or medium: low < 60, high>75, medium is between 60 and 75; use greater or less than operators, < or >. Fill in the values in the code below, where it is written 'YOU_FILL_IN'. [Difficulty: **Intermediate**]

```
GCper=65

  # check if GC value is lower than 60,
  # assign "low" to result
  if('YOU_FILL_IN'){
    result="low"
    cat("low")
  }
  else if('YOU_FILL_IN'){  # check if GC value is higher than 75,
                           #assign "high" to result
    result="high"
    cat("high")
  }else{ # if those two conditions fail then it must be "medium"
    result="medium"
  }

result
```

4. Write a function that takes a value of GC percent and decides if it is low, high, or medium: low < 60, high>75, medium is between 60 and 75. Fill in the values in the code below, where it is written 'YOU_FILL_IN'. [Difficulty: **Intermediate/Advanced**]

```
GCclass<-function(my.gc){

  YOU_FILL_IN


  return(result)
}
```

```
GCclass(10) # should return "low"
GCclass(90) # should return "high"
GCclass(65) # should return "medium"
```

5. Use a for loop to get GC percentage classes for gcValues below. Use the function you wrote above.[Difficulty: **Intermediate/Advanced**]

```
gcValues=c(10,50,70,65,90)
for( i in YOU_FILL_IN){
  YOU_FILL_IN
}
```

6. Use lapply to get GC percentage classes for gcValues. [Difficulty: **Intermediate/Advanced**]

```
vec=c(1,2,4,5)
power2=function(x){ return(x^2)  }
    lapply(vec,power2)
```

7. Use sapply to get values to get GC percentage classes for gcValues. [Difficulty: **Intermediate**]

8. Is there a way to decide on the GC percentage class of a given vector of GCpercentages without using if/else structure and loops ? if so, how can you do it? **HINT:** Subsetting using < and > operators. [Difficulty: **Intermediate**]

# 3

## Statistics for Genomics

This chapter will summarize statistics methods frequently used in computational genomics. As these fields are continuously evolving, the techniques introduced here do not form an exhaustive list but mostly cornerstone methods that are often and still being used. In addition, we focused on giving intuitive and practical understanding of the methods with relevant examples from the field. If you want to dig deeper into statistics and math, beyond what is described here, we included appropriate references with annotation after each major section.

### 3.1 How to summarize collection of data points: The idea behind statistical distributions

In biology and many other fields, data is collected via experimentation. The nature of the experiments and natural variation in biology makes it impossible to get the same exact measurements every time you measure something. For example, if you are measuring gene expression values for a certain gene, say PAX6, and let's assume you are measuring expression per sample and cell with any method (microarrays, rt-qPCR, etc.). You will not get the same expression value even if your samples are homogeneous, due to technical bias in experiments or natural variation in the samples. Instead, we would like to describe this collection of data some other way that represents the general properties of the data. Figure 3.1 shows a sample of 20 expression values from the PAX6 gene.

#### 3.1.1 Describing the central tendency: Mean and median

As seen in Figure 3.1, the points from this sample are distributed around a central value and the histogram below the dot plot shows the number of points in each bin. Another observation is that there are some bins that have more points than others. If we want to summarize what we observe, we can try to represent the collection of data points with an expression value that is typical to get, something that represents the general tendency we observe on the dot plot and the histogram. This value is sometimes called the central value or central tendency, and there are different ways to calculate such a value. In Figure 3.1, we see that all the values are spread around 6.13 (red line), and that is indeed what we call the mean value of

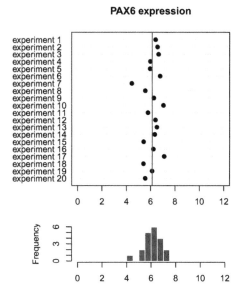

**FIGURE 3.1:** Expression of the PAX6 gene in 20 replicate experiments.

this sample of expression values. It can be calculated with the following formula $\overline{X} = \sum_{i=1}^{n} x_i/n$, where $x_i$ is the expression value of an experiment and $n$ is the number of expression values obtained from the experiments. In R, the mean() function will calculate the mean of a provided vector of numbers. This is called a "sample mean". In reality, there are many more than 20 possible PAX6 expression values (provided each cell is of the identical cell type and is in identical conditions). If we had the time and the funding to sample all cells and measure PAX6 expression we would get a collection of values that would be called, in statistics, a "population". In our case, the population will look like the left hand side of the Figure 3.2. What we have done with our 20 data points is that we took a sample of PAX6 expression values from this population, and calculated the sample mean.

The mean of the population is calculated the same way but traditionally the Greek letter $\mu$ is used to denote the population mean. Normally, we would not have access to the population and we will use the sample mean and other quantities derived from the sample to estimate the population properties. This is the basic idea behind statistical inference, which we will see in action in later sections as well. We estimate the population parameters from the sample parameters and there is some uncertainty associated with those estimates. We will be trying to assess those uncertainties and make decisions in the presence of those uncertainties.

We are not yet done with measuring central tendency. There are other ways to describe it, such as the median value. The mean can be affected by outliers easily. If certain values are very high or low compared to the bulk of the sample, this will shift mean toward those outliers. However, the median is not affected by outliers. It is

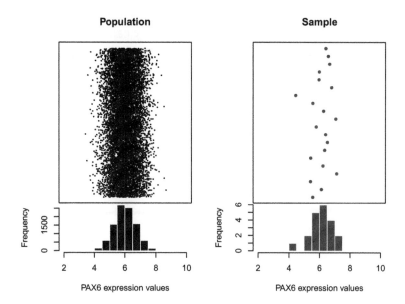

**FIGURE 3.2:** Expression of all possible PAX6 gene expression measures on all available biological samples (left). Expression of the PAX6 gene from the statistical sample, a random subset from the population of biological samples (right).

simply the value in a distribution where half of the values are above and the other half are below. In R, the `median()` function will calculate the mean of a provided vector of numbers. Let's create a set of random numbers and calculate their mean and median using R.

```
#create 10 random numbers from uniform distribution
x=runif(10)
# calculate mean
mean(x)
```

```
## [1] 0.3738963
```

```
# calculate median
median(x)
```

```
## [1] 0.3277896
```

### 3.1.2 Describing the spread: Measurements of variation

Another useful way to summarize a collection of data points is to measure how variable the values are. You can simply describe the range of the values, such as the

minimum and maximum values. You can easily do that in R with the `range()` function. A more common way to calculate variation is by calculating something called
"standard deviation" or the related quantity called "variance". This is a quantity that
shows how variable the values are. A value around zero indicates there is not much
variation in the values of the data points, and a high value indicates high variation
in the values. The variance is the squared distance of data points from the mean.
Population variance is again a quantity we usually do not have access to and is
simply calculated as follows $\sigma^2 = \sum_{i=1}^{n} \frac{(x_i-\mu)^2}{n}$, where $\mu$ is the population mean,
$x_i$ is the $i$th data point in the population and $n$ is the population size. However,
when we only have access to a sample, this formulation is biased. That means that
it underestimates the population variance, so we make a small adjustment when
we calculate the sample variance, denoted as $s^2$:

$$s^2 = \sum_{i=1}^{n} \frac{(x_i - \overline{X})^2}{n-1} \qquad \text{where } x_i \text{ is the ith data point and } \overline{X} \text{ is the sample mean.}$$

The sample standard deviation is simply the square root of the sample variance,
$s = \sqrt{\sum_{i=1}^{n} \frac{(x_i-\overline{X})^2}{n-1}}$. The good thing about standard deviation is that it has the
same unit as the mean so it is more intuitive.

We can calculate the sample standard deviation and variation with the `sd()` and
`var()` functions in R. These functions take a vector of numeric values as input
and calculate the desired quantities. Below we use those functions on a randomly
generated vector of numbers.

```
x=rnorm(20,mean=6,sd=0.7)
var(x)
```

```
## [1] 0.2531495
```

```
sd(x)
```

```
## [1] 0.5031397
```

One potential problem with the variance is that it could be affected by outliers. The
points that are too far away from the mean will have a large effect on the variance
even though there might be few of them. A way to measure variance that could be
less affected by outliers is looking at where the bulk of the distribution is. How do
we define where the bulk is? One common way is to look at the difference between
75th percentile and 25th percentile, this effectively removes a lot of potential outliers
which will be towards the edges of the range of values. This is called the interquartile

range, and can be easily calculated using R via the IQR() function and the quantiles of a vector are calculated with the quantile() function.

Let us plot the boxplot for a random vector and also calculate IQR using R. In the boxplot (Figure 3.3), 25th and 75th percentiles are the edges of the box, and the median is marked with a thick line cutting through the box.

```
x=rnorm(20,mean=6,sd=0.7)
IQR(x)


## [1] 0.5010954


quantile(x)


##         0%       25%       50%       75%      100%
## 5.437119 5.742895 5.860302 6.243991 6.558112


boxplot(x,horizontal = T)
```

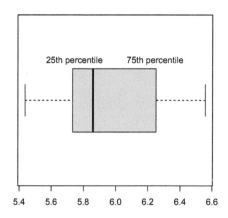

**FIGURE 3.3:** Boxplot showing the 25th percentile and 75th percentile and median for a set of points sampled from a normal distribution with mean=6 and standard deviation=0.7.

#### 3.1.2.1 Frequently used statistical distributions

The distributions have parameters (such as mean and variance) that summarize them, but also they are functions that assign each outcome of a statistical experiment to its probability of occurrence. One distribution that you will frequently encounter is the normal distribution or Gaussian distribution. The normal distribution has a typical "bell-curve" shape and is characterized by mean and standard deviation. A set of data points that follow normal distribution will mostly be close to the mean but spread around it, controlled by the standard deviation parameter. That means that if we sample data points from a normal distribution, we are more likely to sample data points near the mean and sometimes away from the mean. The probability of an event occurring is higher if it is nearby the mean. The effect of the parameters for the normal distribution can be observed in the following plot.

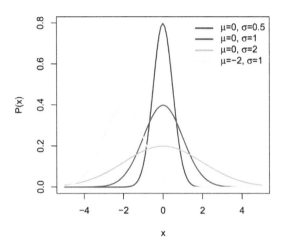

**FIGURE 3.4:** Different parameters for normal distribution and effect of those on the shape of the distribution.

The normal distribution is often denoted by $\mathcal{N}(\mu, \sigma^2)$. When a random variable $X$ is distributed normally with mean $\mu$ and variance $\sigma^2$, we write:

$$X \sim \mathcal{N}(\mu, \sigma^2)$$

The probability density function of the normal distribution with mean $\mu$ and standard deviation $\sigma$ is as follows:

$$P(x) = \frac{1}{\sigma\sqrt{2\pi}}\, e^{-\frac{(x-\mu)^2}{2\sigma^2}}$$

The probability density function gives the probability of observing a value on a normal distribution defined by the $\mu$ and $\sigma$ parameters.

Oftentimes, we do not need the exact probability of a value, but we need the probability of observing a value larger or smaller than a critical value or reference point. For example, we might want to know the probability of $X$ being smaller than or equal to -2 for a normal distribution with mean 0 and standard deviation 2: $P(X <= -2 \mid \mu = 0, \sigma = 2)$. In this case, what we want is the area under the curve shaded in dark blue. To be able to do that, we need to integrate the probability density function but we will usually let software do that. Traditionally, one calculates a Z-score which is simply $(X - \mu)/\sigma = (-2 - 0)/2 = -1$, and corresponds to how many standard deviations you are away from the mean. This is also called "standardization", the corresponding value is distributed in "standard normal distribution" where $\mathcal{N}(0, 1)$. After calculating the Z-score, we can look up the area under the curve for the left and right sides of the Z-score in a table, but again, we use software for that. The tables are outdated when you can use a computer.

Below in Figure 3.5, we show the Z-score and the associated probabilities derived from the calculation above for $P(X <= -2 \mid \mu = 0, \sigma = 2)$.

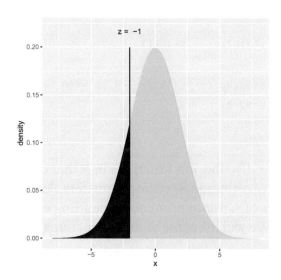

**FIGURE 3.5:** Z-score and associated probabilities for $Z = -1$.

In R, the family of *norm functions (rnorm,dnorm,qnorm and pnorm) can be used to operate with the normal distribution, such as calculating probabilities and generating random numbers drawn from a normal distribution. We show some of those capabilities below.

```
# get the value of probability density function when X= -2,
# where mean=0 and sd=2
dnorm(-2, mean=0, sd=2)
```

```
## [1] 0.1209854
```

```
# get the probability of P(X =< -2) where mean=0 and sd=2
pnorm(-2, mean=0, sd=2)
```

```
## [1] 0.1586553
```

```
# get the probability of P(X > -2) where mean=0 and sd=2
pnorm(-2, mean=0, sd=2,lower.tail = FALSE)
```

```
## [1] 0.8413447
```

```
# get 5 random numbers from normal dist with  mean=0 and sd=2
rnorm(5, mean=0 , sd=2)
```

```
## [1] -1.8109030 -1.9220710 -0.5146717  0.8216728 -0.7900804
```

```
# get y value corresponding to P(X > y) = 0.15 with  mean=0 and sd=2
qnorm( 0.15, mean=0 , sd=2)
```

```
## [1] -2.072867
```

There are many other distribution functions in R that can be used the same way. You have to enter the distribution-specific parameters along with your critical value, quantiles, or number of random numbers depending on which function you are using in the family. We will list some of those functions below.

- dbinom is for the binomial distribution. This distribution is usually used to model fractional data and binary data. Examples from genomics include methylation data.

- dpois is used for the Poisson distribution and dnbinom is used for the negative binomial distribution. These distributions are used to model count data such as sequencing read counts.

- df (F distribution) and dchisq (Chi-Squared distribution) are used in relation to the distribution of variation. The F distribution is used to model ratios of variation and Chi-Squared distribution is used to model distribution of variations. You will frequently encounter these in linear models and generalized linear models.

### 3.1.3   Precision of estimates: Confidence intervals

When we take a random sample from a population and compute a statistic, such as the mean, we are trying to approximate the mean of the population. How well this sample statistic estimates the population value will always be a concern. A confidence interval addresses this concern because it provides a range of values which will plausibly contain the population parameter of interest. Normally, we would not have access to a population. If we did, we would not have to estimate the population parameters and their precision.

When we do not have access to the population, one way to estimate intervals is to repeatedly take samples from the original sample with replacement, that is, we take a data point from the sample we replace, and we take another data point until we have sample size of the original sample. Then, we calculate the parameter of interest, in this case the mean, and repeat this process a large number of times, such as 1000. At this point, we would have a distribution of re-sampled means. We can then calculate the 2.5th and 97.5th percentiles and these will be our so-called 95% confidence interval. This procedure, resampling with replacement to estimate the precision of population parameter estimates, is known as the **bootstrap resampling** or **bootstraping**.

Let's see how we can do this in practice. We simulate a sample coming from a normal distribution (but we pretend we don't know the population parameters). We will estimate the precision of the mean of the sample using bootstrapping to build confidence intervals, the resulting plot after this procedure is shown in Figure 3.6.

```
library(mosaic)
set.seed(21)
sample1= rnorm(50,20,5) # simulate a sample


# do bootstrap resampling, sampling with replacement
boot.means=do(1000) * mean(resample(sample1))


# get percentiles from the bootstrap means
q=quantile(boot.means[,1],p=c(0.025,0.975))


# plot the histogram
hist(boot.means[,1],col="cornflowerblue",border="white",
                    xlab="sample means")
abline(v=c(q[1], q[2] ),col="red")
text(x=q[1],y=200,round(q[1],3),adj=c(1,0))
text(x=q[2],y=200,round(q[2],3),adj=c(0,0))
```

If we had a convenient mathematical method to calculate the confidence interval, we could also do without resampling methods. It turns out that if we take repeated samples from a population with sample size $n$, the distribution of means ($\overline{X}$) of those samples will be approximately normal with mean $\mu$ and standard deviation

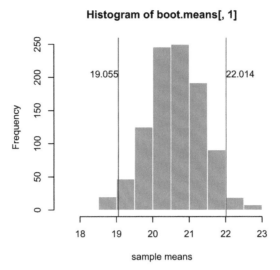

**FIGURE 3.6:** Precision estimate of the sample mean using 1000 bootstrap samples. Confidence intervals derived from the bootstrap samples are shown with red lines.

$\sigma/\sqrt{n}$. This is also known as the **Central Limit Theorem(CLT)** and is one of the most important theorems in statistics. This also means that $\frac{\overline{X}-\mu}{\sigma\sqrt{n}}$ has a standard normal distribution and we can calculate the Z-score, and then we can get the percentiles associated with the Z-score. Below, we are showing the Z-score calculation for the distribution of $\overline{X}$, and then we are deriving the confidence intervals starting with the fact that the probability of Z being between $-1.96$ and $1.96$ is $0.95$. We then use algebra to show that the probability that unknown $\mu$ is captured between $\overline{X} - 1.96\sigma/\sqrt{n}$ and $\overline{X} + 1.96\sigma/\sqrt{n}$ is $0.95$, which is commonly known as the 95% confidence interval.

$$Z = \frac{\overline{X}-\mu}{\sigma/\sqrt{n}}$$
$$P(-1.96 < Z < 1.96) = 0.95$$
$$P(-1.96 < \frac{\overline{X}-\mu}{\sigma/\sqrt{n}} < 1.96) = 0.95$$
$$P(\mu - 1.96\sigma/\sqrt{n} < \overline{X} < \mu + 1.96\sigma/\sqrt{n}) = 0.95$$
$$P(\overline{X} - 1.96\sigma/\sqrt{n} < \mu < \overline{X} + 1.96\sigma/\sqrt{n}) = 0.95$$
$$confint = [\overline{X} - 1.96\sigma/\sqrt{n}, \overline{X} + 1.96\sigma/\sqrt{n}]$$

A 95% confidence interval for the population mean is the most common interval to use, and would mean that we would expect 95% of the interval estimates to include the population parameter, in this case, the mean. However, we can pick any value such as 99% or 90%. We can generalize the confidence interval for $100(1 - \alpha)$ as follows:

$$\overline{X} \pm Z_{\alpha/2}\sigma/\sqrt{n}$$

In R, we can do this using the qnorm() function to get Z-scores associated with $\alpha/2$ and $1 - \alpha/2$. As you can see, the confidence intervals we calculated using CLT are very similar to the ones we got from the bootstrap for the same sample. For bootstrap we got $[19.21, 21.989]$ and for the CLT-based estimate we got $[19.23638, 22.00819]$.

```
alpha=0.05
sd=5
n=50
mean(sample1)+qnorm(c(alpha/2,1-alpha/2))*sd/sqrt(n)
```

```
## [1] 19.23638 22.00819
```

The good thing about CLT is, as long as the sample size is large, regardless of the population distribution, the distribution of sample means drawn from that population will always be normal. In Figure 3.7, we repeatedly draw samples 1000 times with sample size $n = 10, 30$, and $100$ from a bimodal, exponential and a uniform distribution and we are getting sample mean distributions following normal distribution.

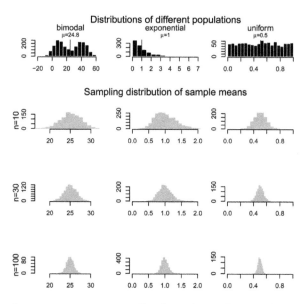

**FIGURE 3.7:** Sample means are normally distributed regardless of the population distribution they are drawn from.

However, we should note that how we constructed the confidence interval using standard normal distribution, $N(0, 1)$, only works when we know the population standard deviation. In reality, we usually have only access to a sample and have no idea about the population standard deviation. If this is the case, we should estimate the standard deviation using the sample standard deviation and use something called the *t distribution* instead of the standard normal distribution in our interval calculation. Our confidence interval becomes $\overline{X} \pm t_{\alpha/2}s/\sqrt{n}$, with t distribution parameter $d.f = n - 1$, since now the following quantity is t distributed $\frac{\overline{X}-\mu}{s/\sqrt{n}}$ instead of standard normal distribution.

The t distribution is similar to the standard normal distribution and has mean 0 but its spread is larger than the normal distribution especially when the sample size is small, and has one parameter $v$ for the degrees of freedom, which is $n - 1$ in this case. Degrees of freedom is simply the number of data points minus the number of parameters estimated.Here we are estimating the mean from the data, therefore the degrees of freedom is $n - 1$. The resulting distributions are shown in Figure 3.8.

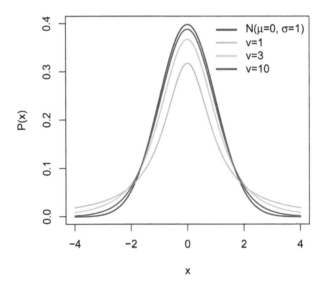

**FIGURE 3.8:** Normal distribution and t distribution with different degrees of freedom. With increasing degrees of freedom, the t distribution approximates the normal distribution better.

## 3.2 How to test for differences between samples

Oftentimes we would want to compare sets of samples. Such comparisons include if wild-type samples have different expression compared to mutants or if healthy samples are different from disease samples in some measurable feature (blood count, gene expression, methylation of certain loci). Since there is variability in our measurements, we need to take that into account when comparing the sets of samples. We can simply subtract the means of two samples, but given the variability of sampling, at the very least we need to decide a cutoff value for differences of means; small differences of means can be explained by random chance due to sampling. That means we need to compare the difference we get to a value that is typical to get if the difference between two group means were only due to sampling. If you followed the logic above, here we actually introduced two core ideas of something called "hypothesis testing", which is simply using statistics to determine the probability that a given hypothesis (Ex: if two sample sets are from the same population or not) is true. Formally, expanded version of those two core ideas are as follows:

1. Decide on a hypothesis to test, often called the "null hypothesis" ($H_0$). In our case, the hypothesis is that there is no difference between sets of samples. An "alternative hypothesis" ($H_1$) is that there is a difference between the samples.
2. Decide on a statistic to test the truth of the null hypothesis.
3. Calculate the statistic.
4. Compare it to a reference value to establish significance, the P-value. Based on that, either reject or not reject the null hypothesis, $H_0$.

### 3.2.1 Randomization-based testing for difference of the means

There is one intuitive way to go about this. If we believe there are no differences between samples, that means the sample labels (test vs. control or healthy vs. disease) have no meaning. So, if we randomly assign labels to the samples and calculate the difference of the means, this creates a null distribution for $H_0$ where we can compare the real difference and measure how unlikely it is to get such a value under the expectation of the null hypothesis. We can calculate all possible permutations to calculate the null distribution. However, sometimes that is not very feasible and the equivalent approach would be generating the null distribution by taking a smaller number of random samples with shuffled group membership.

Below, we are doing this process in R. We are first simulating two samples from two different distributions. These would be equivalent to gene expression measurements obtained under different conditions. Then, we calculate the differences in the means and do the randomization procedure to get a null distribution when we assume there is no difference between samples, $H_0$. We then calculate how often we would get the original difference we calculated under the assumption that $H_0$ is true. The resulting null distribution and the original value is shown in Figure 3.9.

```r
set.seed(100)
gene1=rnorm(30,mean=4,sd=2)
gene2=rnorm(30,mean=2,sd=2)
org.diff=mean(gene1)-mean(gene2)
gene.df=data.frame(exp=c(gene1,gene2),
               group=c( rep("test",30),rep("control",30) ) )



exp.null <- do(1000) * diff(mosaic::mean(exp ~ shuffle(group), data=gene.df))
hist(exp.null[,1],xlab="null distribution | no difference in samples",
     main=expression(paste(H[0]," :no difference in means") ),
     xlim=c(-2,2),col="cornflowerblue",border="white")
abline(v=quantile(exp.null[,1],0.95),col="red" )
abline(v=org.diff,col="blue" )
text(x=quantile(exp.null[,1],0.95),y=200,"0.05",adj=c(1,0),col="red")
text(x=org.diff,y=200,"org. diff.",adj=c(1,0),col="blue")



p.val=sum(exp.null[,1]>org.diff)/length(exp.null[,1])
p.val
```

```
## [1] 0.001
```

After doing random permutations and getting a null distribution, it is possible to get a confidence interval for the distribution of difference in means. This is simply the $2.5th$ and $97.5th$ percentiles of the null distribution, and directly related to the P-value calculation above.

### 3.2.2   Using t-test for difference of the means between two samples

We can also calculate the difference between means using a t-test. Sometimes we will have too few data points in a sample to do a meaningful randomization test, also randomization takes more time than doing a t-test. This is a test that depends on the t distribution. The line of thought follows from the CLT and we can show differences in means are t distributed. There are a couple of variants of the t-test

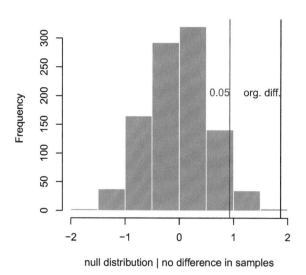

**FIGURE 3.9:** The null distribution for differences of means obtained via randomization. The original difference is marked via the blue line. The red line marks the value that corresponds to P-value of 0.05.

for this purpose. If we assume the population variances are equal we can use the following version

$$t = \frac{\bar{X}_1 - \bar{X}_2}{s_{X_1 X_2} \cdot \sqrt{\frac{1}{n_1} + \frac{1}{n_2}}}$$

where

$$s_{X_1 X_2} = \sqrt{\frac{(n_1 - 1)s_{X_1}^2 + (n_2 - 1)s_{X_2}^2}{n_1 + n_2 - 2}}$$

In the first equation above, the quantity is t distributed with $n_1 + n_2 - 2$ degrees of freedom. We can calculate the quantity and then use software to look for the percentile of that value in that t distribution, which is our P-value. When we cannot assume equal variances, we use "Welch's t-test" which is the default t-test in R and also works well when variances and the sample sizes are the same. For this test we calculate the following quantity:

$$t = \frac{\bar{X}_1 - \bar{X}_2}{s_{\bar{X}_1 - \bar{X}_2}}$$

where

$$s_{\overline{X}_1 - \overline{X}_2} = \sqrt{\frac{s_1^2}{n_1} + \frac{s_2^2}{n_2}}$$

and the degrees of freedom equals to

$$\text{d.f.} = \frac{(s_1^2/n_1 + s_2^2/n_2)^2}{(s_1^2/n_1)^2/(n_1 - 1) + (s_2^2/n_2)^2/(n_2 - 1)}$$

Luckily, R does all those calculations for us. Below we will show the use of `t.test()` function in R. We will use it on the samples we simulated above.

```
# Welch's t-test
stats::t.test(gene1,gene2)
```

```
##
##   Welch Two Sample t-test
##
## data:  gene1 and gene2
## t = 3.7653, df = 47.552, p-value = 0.0004575
## alternative hypothesis: true difference in means is not equal to 0
## 95 percent confidence interval:
##   0.872397 2.872761
## sample estimates:
## mean of x mean of y
##   4.057728  2.185149
```

```
# t-test with equal variance assumption
stats::t.test(gene1,gene2,var.equal=TRUE)
```

```
##
##   Two Sample t-test
##
## data:  gene1 and gene2
## t = 3.7653, df = 58, p-value = 0.0003905
## alternative hypothesis: true difference in means is not equal to 0
## 95 percent confidence interval:
##   0.8770753 2.8680832
## sample estimates:
## mean of x mean of y
##   4.057728  2.185149
```

A final word on t-tests: they generally assume a population where samples coming from them have a normal distribution, however it is been shown t-test can tolerate deviations from normality, especially, when two distributions are moderately skewed in the same direction. This is due to the central limit theorem, which says that the means of samples will be distributed normally no matter the population distribution if sample sizes are large.

### 3.2.3 Multiple testing correction

We should think of hypothesis testing as a non-error-free method of making decisions. There will be times when we declare something significant and accept $H_1$ but we will be wrong. These decisions are also called "false positives" or "false discoveries", and are also known as "type I errors". Similarly, we can fail to reject a hypothesis when we actually should. These cases are known as "false negatives", also known as "type II errors".

The ratio of true negatives to the sum of true negatives and false positives ($\frac{TN}{FP+TN}$) is known as specificity. And we usually want to decrease the FP and get higher specificity. The ratio of true positives to the sum of true positives and false negatives ($\frac{TP}{TP+FN}$) is known as sensitivity. And, again, we usually want to decrease the FN and get higher sensitivity. Sensitivity is also known as the "power of a test" in the context of hypothesis testing. More powerful tests will be highly sensitive and will have fewer type II errors. For the t-test, the power is positively associated with sample size and the effect size. The larger the sample size, the smaller the standard error, and looking for the larger effect sizes will similarly increase the power.

The general summary of these different decision combinations are included in the table below.

| | $H_0$ is TRUE, [Gene is NOT differentially expressed] | $H_1$ is TRUE, [Gene is differentially expressed] | |
|---|---|---|---|
| Accept $H_0$ (claim that the gene is not differentially expressed) | True Negatives (TN) | False Negatives (FN) ,type II error | $m_0$: number of truly null hypotheses |

|  | $H_0$ is TRUE,<br>[Gene is NOT<br>differentially<br>expressed] | $H_1$ is TRUE,<br>[Gene is<br>differentially<br>expressed] |  |
|---|---|---|---|
| reject $H_0$<br>(claim that<br>the gene is<br>differentially<br>expressed) | False Positives<br>(FP) ,type I error | True Positives<br>(TP) | $m - m_0$: number of<br>truly alternative<br>hypotheses |

We expect to make more type I errors as the number of tests increase, which means we will reject the null hypothesis by mistake. For example, if we perform a test at the 5% significance level, there is a 5% chance of incorrectly rejecting the null hypothesis if the null hypothesis is true. However, if we make 1000 tests where all null hypotheses are true for each of them, the average number of incorrect rejections is 50. And if we apply the rules of probability, there is almost a 100% chance that we will have at least one incorrect rejection. There are multiple statistical techniques to prevent this from happening. These techniques generally push the P-values obtained from multiple tests to higher values; if the individual P-value is low enough it survives this process. The simplest method is just to multiply the individual P-value ($p_i$) by the number of tests ($m$), $m \cdot p_i$. This is called "Bonferroni correction". However, this is too harsh if you have thousands of tests. Other methods are developed to remedy this. Those methods rely on ranking the P-values and dividing $m \cdot p_i$ by the rank, $i, : \frac{m \cdot p_i}{i}$, which is derived from the Benjamini–Hochberg procedure. This procedure is developed to control for "False Discovery Rate (FDR)" , which is the proportion of false positives among all significant tests. And in practical terms, we get the "FDR-adjusted P-value" from the procedure described above. This gives us an estimate of the proportion of false discoveries for a given test. To elaborate, p-value of 0.05 implies that 5% of all tests will be false positives. An FDR-adjusted p-value of 0.05 implies that 5% of significant tests will be false positives. The FDR-adjusted P-values will result in a lower number of false positives.

One final method that is also popular is called the "q-value" method and related to the method above. This procedure relies on estimating the proportion of true null hypotheses from the distribution of raw p-values and using that quantity to come up with what is called a "q-value", which is also an FDR-adjusted P-value (Storey and Tibshirani, 2003). That can be practically defined as "the proportion of significant features that turn out to be false leads." A q-value 0.01 would mean 1% of the tests called significant at this level will be truly null on average. Within the genomics community q-value and FDR adjusted P-value are synonymous although they can be calculated differently.

In R, the base function `p.adjust()` implements most of the p-value correction methods described above. For the q-value, we can use the `qvalue` package from Bioconductor. Below we demonstrate how to use them on a set of simulated p-values. The plot in Figure 3.10 shows that Bonferroni correction does a terrible job. FDR(BH) and q-value approach are better but, the q-value approach is more permissive than FDR(BH).

```
library(qvalue)
data(hedenfalk)

qvalues <- qvalue(hedenfalk$p)$q
bonf.pval=p.adjust(hedenfalk$p,method ="bonferroni")
fdr.adj.pval=p.adjust(hedenfalk$p,method ="fdr")

plot(hedenfalk$p,qvalues,pch=19,ylim=c(0,1),
     xlab="raw P-values",ylab="adjusted P-values")
points(hedenfalk$p,bonf.pval,pch=19,col="red")
points(hedenfalk$p,fdr.adj.pval,pch=19,col="blue")
legend("bottomright",legend=c("q-value","FDR (BH)","Bonferroni"),
       fill=c("black","blue","red"))
```

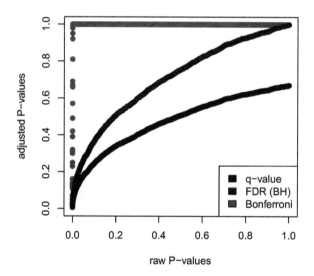

**FIGURE 3.10:** Adjusted P-values via different methods and their relationship to raw P-values.

### 3.2.4   Moderated t-tests: Using information from multiple comparisons

In genomics, we usually do not do one test but many, as described above. That means we may be able to use the information from the parameters obtained from all comparisons to influence the individual parameters. For example, if you have many variances calculated for thousands of genes across samples, you can force individual variance estimates to shrink toward the mean or the median of the distribution of variances. This usually creates better performance in individual variance estimates and therefore better performance in significance testing, which depends on variance estimates. How much the values are shrunk toward a common value depends on the exact method used. These tests in general are called moderated t-tests or shrinkage t-tests. One approach popularized by Limma software is to use so-called "Empirical Bayes methods". The main formulation in these methods is $\hat{V}_g = aV_0 + bV_g$, where $V_0$ is the background variability and $V_g$ is the individual variability. Then, these methods estimate $a$ and $b$ in various ways to come up with a "shrunk" version of the variability, $\hat{V}_g$. Bayesian inference can make use of prior knowledge to make inference about properties of the data. In a Bayesian viewpoint, the prior knowledge, in this case variability of other genes, can be used to calculate the variability of an individual gene. In our case, $V_0$ would be the prior knowledge we have on the variability of the genes and we use that knowledge to influence our estimate for the individual genes.

Below we are simulating a gene expression matrix with 1000 genes, and 3 test and 3 control groups. Each row is a gene, and in normal circumstances we would like to find differentially expressed genes. In this case, we are simulating them from the same distribution, so in reality we do not expect any differences. We then use the adjusted standard error estimates in empirical Bayesian spirit but, in a very crude way. We just shrink the gene-wise standard error estimates towards the median with equal $a$ and $b$ weights. That is to say, we add the individual estimate to the median of the standard error distribution from all genes and divide that quantity by 2. So if we plug that into the above formula, what we do is:

$$\hat{V}_g = (V_0 + V_g)/2$$

In the code below, we are avoiding for loops or apply family functions by using vectorized operations. The code below samples gene expression values from a hypothetical distribution. Since all the values come from the same distribution, we do not expect differences between groups. We then calculate moderated and unmoderated t-test statistics and plot the P-value distributions for tests. The results are shown in Figure 3.11.

```
set.seed(100)

#sample data matrix from normal distribution

gset=rnorm(3000,mean=200,sd=70)
data=matrix(gset,ncol=6)

# set groups
group1=1:3
group2=4:6
n1=3
n2=3
dx=rowMeans(data[,group1])-rowMeans(data[,group2])

require(matrixStats)

# get the esimate of pooled variance
stderr = sqrt( (rowVars(data[,group1])*(n1-1) +
        rowVars(data[,group2])*(n2-1)) / (n1+n2-2) * ( 1/n1 + 1/n2 ))

# do the shrinking towards median
mod.stderr = (stderr + median(stderr)) / 2 # moderation in variation

# esimate t statistic with moderated variance
t.mod <- dx / mod.stderr

# calculate P-value of rejecting null
p.mod = 2*pt( -abs(t.mod), n1+n2-2 )

# esimate t statistic without moderated variance
t = dx / stderr

# calculate P-value of rejecting null
p = 2*pt( -abs(t), n1+n2-2 )

par(mfrow=c(1,2))
hist(p,col="cornflowerblue",border="white",main="",xlab="P-values t-test")
mtext(paste("signifcant tests:",sum(p<0.05))  )
hist(p.mod,col="cornflowerblue",border="white",main="",
     xlab="P-values mod. t-test")
mtext(paste("signifcant tests:",sum(p.mod<0.05))  )
```

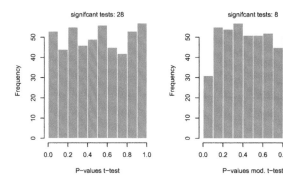

**FIGURE 3.11:** The distributions of P-values obtained by t-tests and moderated t-tests.

– **Want to know more ?**
* Basic statistical concepts
  · "Cartoon guide to statistics" by Gonick & Smith (Gonick and Smith, 2005). Provides central concepts depicted as cartoons in a funny but clear and accurate manner.
  · "OpenIntro Statistics" (Diez et al., 2015) (Free e-book `http://openintro.org`). This book provides fundamental statistical concepts in a clear and easy way. It includes R code.
* Hands-on statistics recipes with R
  · "The R book" (Crawley, 2012). This is the main R book for anyone interested in statistical concepts and their application in R. It requires some background in statistics since the main focus is applications in R.
* Moderated tests
  · Comparison of moderated tests for differential expression (De Hertogh et al., 2010) `http://bmcbioinformatics.biomedcentral.com/articles/10.1186/1471-2105-11-17`
  · Limma method developed for testing differential expression between genes using a moderated test (Smyth Gordon, 2004) `http://www.statsci.org/smyth/pubs/ebayes.pdf`

## 3.3 Relationship between variables: Linear models and correlation

In genomics, we would often need to measure or model the relationship between variables. We might want to know about expression of a particular gene in liver in relation to the dosage of a drug that patient receives. Or, we may want to know DNA methylation of a certain locus in the genome in relation to the age of the sample donor. Or, we might be interested in the relationship between histone modifications and gene expression. Is there a linear relationship, the more histone modification the more the gene is expressed ?

In these situations and many more, linear regression or linear models can be used to model the relationship with a "dependent" or "response" variable (expression or methylation in the above examples) and one or more "independent" or "explanatory" variables (age, drug dosage or histone modification in the above examples). Our simple linear model has the following components.

$$Y = \beta_0 + \beta_1 X + \epsilon$$

In the equation above, $Y$ is the response variable and $X$ is the explanatory variable. $\epsilon$ is the mean-zero error term. Since the line fit will not be able to precisely predict the $Y$ values, there will be some error associated with each prediction when we compare it to the original $Y$ values. This error is captured in the $\epsilon$ term. We can alternatively write the model as follows to emphasize that the model approximates $Y$, in this case notice that we removed the $\epsilon$ term: $Y \sim \beta_0 + \beta_1 X$.

The plot below in Figure 3.12 shows the relationship between histone modification (trimethylated forms of histone H3 at lysine 4, aka H3K4me3) and gene expression for 100 genes. The blue line is our model with estimated coefficients ($\hat{y} = \hat{\beta}_0 + \hat{\beta}_1 X$, where $\hat{\beta}_0$ and $\hat{\beta}_1$ are the estimated values of $\beta_0$ and $\beta_1$, and $\hat{y}$ indicates the prediction). The red lines indicate the individual errors per data point, indicated as $\epsilon$ in the formula above.

There could be more than one explanatory variable. We then simply add more $X$ and $\beta$ to our model. If there are two explanatory variables our model will look like this:

$$Y = \beta_0 + \beta_1 X_1 + \beta_2 X_2 + \epsilon$$

In this case, we will be fitting a plane rather than a line. However, the fitting process which we will describe in the later sections will not change for our gene expression problem. We can introduce one more histone modification, H3K27me3. We will then have a linear model with 2 explanatory variables and the fitted plane will look like the one in Figure 3.13. The gene expression values are shown as dots below and

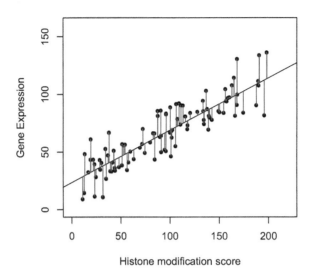

**FIGURE 3.12:** Relationship between histone modification score and gene expression. Increasing histone modification, H3K4me3, seems to be associated with increasing gene expression. Each dot is a gene.

above the fitted plane. Linear regression and its extensions which make use of other distributions (generalized linear models) are central in computational genomics for statistical tests. We will see more of how regression is used in statistical hypothesis testing for computational genomics in Chapters 8 and 10.

### 3.3.0.1 Matrix notation for linear models

We can naturally have more explanatory variables than just two. The formula below has $n$ explanatory variables.

$$Y = \beta_0 + \beta_1 X_1 + \beta_2 X_2 + \beta_3 X_3 + .. + \beta_n X_n + \epsilon$$

If there are many variables, it would be easier to write the model in matrix notation. The matrix form of linear model with two explanatory variables will look like the one below. The first matrix would be our data matrix. This contains our explanatory variables and a column of 1s. The second term is a column vector of $\beta$ values. We also add a vector of error terms, $\epsilon$s, to the matrix multiplication.

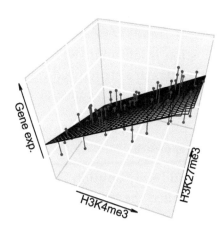

**FIGURE 3.13:** Association of gene expression with H3K4me3 and H3K27me3 histone modifications.

$$\mathbf{Y} = \begin{bmatrix} 1 & X_{1,1} & X_{1,2} \\ 1 & X_{2,1} & X_{2,2} \\ 1 & X_{3,1} & X_{3,2} \\ 1 & X_{4,1} & X_{4,2} \end{bmatrix} \begin{bmatrix} \beta_0 \\ \beta_1 \\ \beta_2 \end{bmatrix} + \begin{bmatrix} \epsilon_1 \\ \epsilon_2 \\ \epsilon_3 \\ \epsilon_0 \end{bmatrix}$$

The multiplication of the data matrix and $\beta$ vector and addition of the error terms simply results in the following set of equations per data point:

$$Y_1 = \beta_0 + \beta_1 X_{1,1} + \beta_2 X_{1,2} + \epsilon_1$$
$$Y_2 = \beta_0 + \beta_1 X_{2,1} + \beta_2 X_{2,2} + \epsilon_2$$
$$Y_3 = \beta_0 + \beta_1 X_{3,1} + \beta_2 X_{3,2} + \epsilon_3$$
$$Y_4 = \beta_0 + \beta_1 X_{4,1} + \beta_2 X_{4,2} + \epsilon_4$$

This expression involving the multiplication of the data matrix, the $\beta$ vector and vector of error terms ($\epsilon$) could be simply written as follows.

$$Y = X\beta + \epsilon$$

In the equation, above $Y$ is the vector of response variables, $X$ is the data matrix, and $\beta$ is the vector of coefficients. This notation is more concise and often used in scientific papers. However, this also means you need some understanding of linear algebra to follow the math laid out in such resources.

### 3.3.1   How to fit a line

At this point a major question is left unanswered: How did we fit this line? We basically need to define $\beta$ values in a structured way. There are multiple ways of understanding how to do this, all of which converge to the same end point. We will describe them one by one.

#### 3.3.1.1   The cost or loss function approach

This is the first approach and in my opinion is easiest to understand. We try to optimize a function, often called the "cost function" or "loss function". The cost function is the sum of squared differences between the predicted $\hat{Y}$ values from our model and the original $Y$ values. The optimization procedure tries to find $\beta$ values that minimize this difference between the reality and predicted values.

$$min \sum (y_i - (\beta_0 + \beta_1 x_i))^2$$

Note that this is related to the error term, $\epsilon$, we already mentioned above. We are trying to minimize the squared sum of $\epsilon_i$ for each data point. We can do this minimization by a bit of calculus. The rough algorithm is as follows:

1.   Pick a random starting point, random $\beta$ values.
2.   Take the partial derivatives of the cost function to see which direction is the way to go in the cost function.
3.   Take a step toward the direction that minimizes the cost function.
       •Step size is a parameter to choose, there are many variants.
4.   Repeat step 2,3 until convergence.

This is the basis of the "gradient descent" algorithm. With the help of partial derivatives we define a "gradient" on the cost function and follow that through multiple iterations until convergence, meaning until the results do not improve defined by a margin. The algorithm usually converges to optimum $\beta$ values. In Figure 3.14, we show the cost function over various $\beta_0$ and $\beta_1$ values for the histone modification and gene expression data set. The algorithm will pick a point on this graph and traverse it incrementally based on the derivatives and converge to the bottom of the cost function "well". Such optimization methods are the core of machine learning methods we will cover later in Chapters 4 and 5.

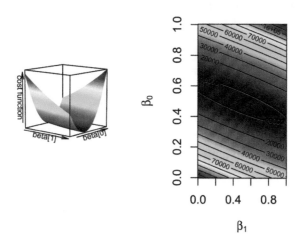

**FIGURE 3.14:** Cost function landscape for linear regression with changing beta values. The optimization process tries to find the lowest point in this landscape by implementing a strategy for updating beta values toward the lowest point in the landscape.

### 3.3.1.2 Not cost function but maximum likelihood function

We can also think of this problem from a more statistical point of view. In essence, we are looking for best statistical parameters, in this case $\beta$ values, for our model that are most likely to produce such a scatter of data points given the explanatory variables. This is called the "maximum likelihood" approach. The approach assumes that a given response variable $y_i$ follows a normal distribution with mean $\beta_0 + \beta_1 x_i$ and variance $s^2$. Therefore the probability of observing any given $y_i$ value is dependent on the $\beta_0$ and $\beta_1$ values. Since $x_i$, the explanatory variable, is fixed within our data set, we can maximize the probability of observing any given $y_i$ by varying $\beta_0$ and $\beta_1$ values. The trick is to find $\beta_0$ and $\beta_1$ values that maximize the probability of observing all the response variables in the dataset given the explanatory variables. The probability of observing a response variable $y_i$ with assumptions we described above is shown below. Note that this assumes variance is constant and $s^2 = \frac{\sum \epsilon_i}{n-2}$ is an unbiased estimation for population variance, $\sigma^2$.

$$P(y_i) = \frac{1}{s\sqrt{2\pi}} e^{-\frac{1}{2}\left(\frac{y_i - (\beta_0 + \beta_1 x_i)}{s}\right)^2}$$

Following from the probability equation above, the likelihood function (shown as $L$ below) for linear regression is multiplication of $P(y_i)$ for all data points.

$$L = P(y_1)P(y_2)P(y_3)..P(y_n) = \prod_{i=1}^{n} P_i$$

This can be simplified to the following equation by some algebra, assumption of normal distribution, and taking logs (since it is easier to add than multiply).

$$ln(L) = -nln(s\sqrt{2\pi}) - \frac{1}{2s^2}\sum_{i=1}^{n}(y_i - (\beta_0 + \beta_1 x_i))^2$$

As you can see, the right part of the function is the negative of the cost function defined above. If we wanted to optimize this function we would need to take the derivative of the function with respect to the $\beta$ parameters. That means we can ignore the first part since there are no $\beta$ terms there. This simply reduces to the negative of the cost function. Hence, this approach produces exactly the same result as the cost function approach. The difference is that we defined our problem within the domain of statistics. This particular function has still to be optimized. This can be done with some calculus without the need for an iterative approach.

The maximum likelihood approach also opens up other possibilities for regression. For the case above, we assumed that the points around the mean are distributed by normal distribution. However, there are other cases where this assumption may not hold. For example, for the count data the mean and variance relationship is not constant; the higher the mean counts, the higher the variance. In these cases, the regression framework with maximum likelihood estimation can still be used. We simply change the underlying assumptions about the distribution and calculate the likelihood with a new distribution in mind, and maximize the parameters for that likelihood. This gives way to "generalized linear model" approach where errors for the response variables can have other distributions than normal distribution. We will see examples of these generalized linear models in Chapter 8 and 10.

### 3.3.1.3   Linear algebra and closed-form solution to linear regression

The last approach we will describe is the minimization process using linear algebra. If you find this concept challenging, feel free to skip it, but scientific publications and other books frequently use matrix notation and linear algebra to define and solve regression problems. In this case, we do not use an iterative approach. Instead, we will minimize the cost function by explicitly taking its derivatives with respect to $\beta$'s and setting them to zero. This is doable by employing linear algebra and matrix calculus. This approach is also called "ordinary least squares". We will not show the whole derivation here, but the following expression is what we are trying to minimize in matrix notation, which is basically a different notation of the same minimization problem defined above. Remember $\epsilon_i = Y_i - (\beta_0 + \beta_1 x_i)$

$$\sum \epsilon_i^2 = \epsilon^T \epsilon = (Y - \beta X)^T (Y - \beta X)$$
$$= Y^T Y - 2\beta^T Y + \beta^T X^T X \beta$$

After rearranging the terms, we take the derivative of $\epsilon^T \epsilon$ with respect to $\beta$, and equalize that to zero. We then arrive at the following for estimated $\beta$ values, $\hat{\beta}$:

$$\hat{\beta} = (X^T X)^{-1} X^T Y$$

This requires you to calculate the inverse of the $X^T X$ term, which could be slow for large matrices. Using an iterative approach over the cost function derivatives will be faster for larger problems. The linear algebra notation is something you will see in the papers or other resources often. If you input the data matrix X and solve the $(X^T X)^{-1}$, you get the following values for $\beta_0$ and $\beta_1$ for simple regression. However, we should note that this simple linear regression case can easily be solved algebraically without the need for matrix operations. This can be done by taking the derivative of $\sum (y_i - (\beta_0 + \beta_1 x_i))^2$ with respect to $\beta_1$, rearranging the terms and equalizing the derivative to zero.

$$\hat{\beta}_1 = \frac{\sum (x_i - \overline{X})(y_i - \overline{Y})}{\sum (x_i - \overline{X})^2}$$
$$\hat{\beta}_0 = \overline{Y} - \hat{\beta}_1 \overline{X}$$

#### 3.3.1.4 Fitting lines in R

After all this theory, you will be surprised how easy it is to fit lines in R. This is achieved just by the lm() function, which stands for linear models. Let's do this for a simulated data set and plot the fit. The first step is to simulate the data. We will decide on $\beta_0$ and $\beta_1$ values. Then we will decide on the variance parameter, $\sigma$, to be used in simulation of error terms, $\epsilon$. We will first find $Y$ values, just using the linear equation $Y = \beta 0 + \beta_1 X$, for a set of $X$ values. Then, we will add the error terms to get our simulated values.

```
# set random number seed, so that the random numbers from the text
# is the same when you run the code.
set.seed(32)

# get 50 X values between 1 and 100
x = runif(50,1,100)

# set b0,b1 and variance (sigma)
b0 = 10
b1 = 2
sigma = 20
# simulate error terms from normal distribution
eps = rnorm(50,0,sigma)
```

```
# get y values from the linear equation and addition of error terms
y = b0 + b1*x+ eps
```

Now let us fit a line using the lm() function. The function requires a formula, and optionally a data frame. We need to pass the following expression within the lm() function, y~x, where y is the simulated $Y$ values and x is the explanatory variables $X$. We will then use the abline() function to draw the fit. The resulting plot is shown in Figure 3.15.

```
mod1=lm(y~x)
```

```
# plot the data points
plot(x,y,pch=20,
     ylab="Gene Expression",xlab="Histone modification score")
# plot the linear fit
abline(mod1,col="blue")
```

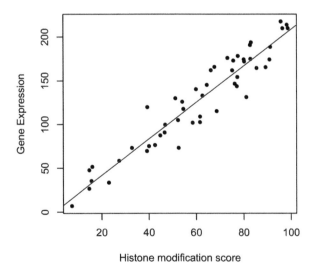

**FIGURE 3.15:** Gene expression and histone modification score modeled by linear regression.

### 3.3.2 How to estimate the error of the coefficients

Since we are using a sample to estimate the coefficients, they are not exact; with every random sample they will vary. In Figure 3.16, we take multiple samples from the population and fit lines to each sample; with each sample the lines slightly change. We are overlaying the points and the lines for each sample on top of the other samples. When we take 200 samples and fit lines for each of them, the line fits are variable. And, we get a normal-like distribution of $\beta$ values with a defined mean and standard deviation, which is called standard error of the coefficients.

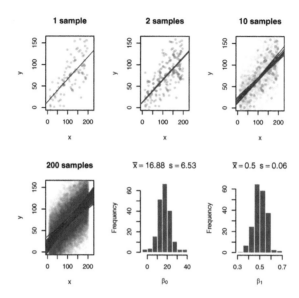

**FIGURE 3.16:** Regression coefficients vary with every random sample. The figure illustrates the variability of regression coefficients when regression is done using a sample of data points. Histograms depict this variability for $b_0$ and $b_1$ coefficients.

Normally, we will not have access to the population to do repeated sampling, model fitting, and estimation of the standard error for the coefficients. But there is statistical theory that helps us infer the population properties from the sample. When we assume that error terms have constant variance and mean zero , we can model the uncertainty in the regression coefficients, $\beta$s. The estimates for standard errors of $\beta$s for simple regression are as follows and shown without derivation.

$$s = RSE = \sqrt{\frac{\sum (y_i - (\beta_0 + \beta_1 x_i))^2}{n-2}} = \sqrt{\frac{\sum \epsilon^2}{n-2}}$$

$$SE(\hat{\beta}_1) = \frac{s}{\sqrt{\sum (x_i - \overline{X})^2}}$$

$$SE(\hat{\beta}_0) = s\sqrt{\frac{1}{n} + \frac{\overline{X}^2}{\sum (x_i - \overline{X})^2}}$$

Notice that that $SE(\beta_1)$ depends on the estimate of variance of residuals shown as $s$ or **Residual Standard Error (RSE)**. Notice also the standard error depends on the spread of $X$. If $X$ values have more variation, the standard error will be lower. This intuitively makes sense since if the spread of $X$ is low, the regression line will be able to wiggle more compared to a regression line that is fit to the same number of points but covers a greater range on the X-axis.

The standard error estimates can also be used to calculate confidence intervals and test hypotheses, since the following quantity, called t-score, approximately follows a t-distribution with $n - p$ degrees of freedom, where $n$ is the number of data points and $p$ is the number of coefficients estimated.

$$\frac{\hat{\beta}_i - \beta_t est}{SE(\hat{\beta}_i)}$$

Often, we would like to test the null hypothesis if a coefficient is equal to zero or not. For simple regression, this could mean if there is a relationship between the explanatory variable and the response variable. We would calculate the t-score as follows $\frac{\hat{\beta}_i - 0}{SE(\hat{\beta}_i)}$, and compare it to the t-distribution with $d.f. = n - p$ to get the p-value.

We can also calculate the uncertainty of the regression coefficients using confidence intervals, the range of values that are likely to contain $\beta_i$. The 95% confidence interval for $\hat{\beta}_i$ is $\hat{\beta}_i \pm t_{0.975} SE(\hat{\beta}_i)$. $t_{0.975}$ is the 97.5% percentile of the t-distribution with $d.f. = n-p$.

In R, the `summary()` function will test all the coefficients for the null hypothesis $\beta_i = 0$. The function takes the model output obtained from the `lm()` function. To demonstrate this, let us first get some data. The procedure below simulates data to be used in a regression setting and it is useful to examine what the linear model expects to model the data.

Since we have the data, we can build our model and call the `summary` function. We will then use the `confint()` function to get the confidence intervals on the coefficients and the `coef()` function to pull out the estimated coefficients from the model.

```
mod1=lm(y~x)
summary(mod1)
```

```
##
## Call:
## lm(formula = y ~ x)
##
## Residuals:
##     Min      1Q  Median      3Q     Max
## -77.11  -18.44    0.33   16.06   57.23
```

```
##
## Coefficients:
##              Estimate Std. Error t value Pr(>|t|)
## (Intercept) 13.24538    6.28869   2.106   0.0377 *
## x            0.49954    0.05131   9.736 4.54e-16 ***
## ---
## Signif. codes:  0 '***' 0.001 '**' 0.01 '*' 0.05 '.' 0.1 ' ' 1
##
## Residual standard error: 28.77 on 98 degrees of freedom
## Multiple R-squared:  0.4917, Adjusted R-squared:  0.4865
## F-statistic: 94.78 on 1 and 98 DF,  p-value: 4.537e-16
```

```
# get confidence intervals
confint(mod1)
```

```
##                   2.5 %      97.5 %
## (Intercept) 0.7656777 25.7250883
## x           0.3977129  0.6013594
```

```
# pull out coefficients from the model
coef(mod1)
```

```
## (Intercept)           x
##   13.2453830   0.4995361
```

The `summary()` function prints out an extensive list of values. The "Coefficients" section has the estimates, their standard error, t score, and the p-value from the hypothesis test $H_0 : \beta_i = 0$. As you can see, the estimate we get for the coefficients and their standard errors are close to the ones we get from repeatedly sampling and getting a distribution of coefficients. This is statistical inference at work, so we can estimate the population properties within a certain error using just a sample.

### 3.3.3  Accuracy of the model

If you have observed the table output of the `summary()` function, you must have noticed there are some other outputs, such as "Residual standard error", "Multiple R-squared" and "F-statistic". These are metrics that are useful for assessing the accuracy of the model. We will explain them one by one.

**RSE** is simply the square-root of the sum of squared error terms, divided by degrees of freedom, $n - p$. For the simple linear regression case, degrees of freedom is $n - 2$. Sum of the squares of the error terms is also called the **"Residual sum of squares"**, RSS. So the RSE is calculated as follows:

$$s = RSE = \sqrt{\frac{\sum (y_i - \hat{Y}_i)^2}{n - p}} = \sqrt{\frac{RSS}{n - p}}$$

The RSE is a way of assessing the model fit. The larger the RSE the worse the model is. However, this is an absolute measure in the units of $Y$ and we have nothing to compare against. One idea is that we divide it by the RSS of a simpler model for comparative purposes. That simpler model is in this case is the model with the intercept, $\beta_0$. A very bad model will have close to zero coefficients for explanatory variables, and the RSS of that model will be close to the RSS of the model with only the intercept. In such a model the intercept will be equal to $\overline{Y}$. As it turns out, the RSS of the model with just the intercept is called the *"Total Sum of Squares"* or TSS. A good model will have a low $RSS/TSS$. The metric $R^2$ uses these quantities to calculate a score between 0 and 1, and the closer to 1, the better the model. Here is how it is calculated:

$$R^2 = 1 - \frac{RSS}{TSS} = \frac{TSS - RSS}{TSS} = 1 - \frac{RSS}{TSS}$$

The $TSS - RSS$ part of the formula is often referred to as "explained variability" in the model. The bottom part is for "total variability". With this interpretation, the higher the "explained variability", the better the model. For simple linear regression with one explanatory variable, the square root of $R^2$ is a quantity known as the absolute value of the correlation coefficient, which can be calculated for any pair of variables, not only the response and the explanatory variables. *Correlation* is the general measure of linear relationship between two variables. One of the most popular flavors of correlation is the Pearson correlation coefficient. Formally, it is the *covariance* of X and Y divided by multiplication of standard deviations of X and Y. In R, it can be calculated with the cor() function.

$$r_{xy} = \frac{cov(X,Y)}{\sigma_x \sigma_y} = \frac{\sum_{i=1}^{n}(x_i - \bar{x})(y_i - \bar{y})}{\sqrt{\sum_{i=1}^{n}(x_i - \bar{x})^2 \sum_{i=1}^{n}(y_i - \bar{y})^2}}$$

In the equation above, *cov* is the covariance; this is again a measure of how much two variables change together, like correlation. If two variables show similar behavior, they will usually have a positive covariance value. If they have opposite behavior, the covariance will have a negative value. However, these values are boundless. A normalized way of looking at covariance is to divide covariance by the multiplication of standard errors of X and Y. This bounds the values to -1 and 1, and as mentioned above, is called Pearson correlation coefficient. The values that change in a similar manner will have a positive coefficient, the values that change in an opposite manner will have a negative coefficient, and pairs that do not have a linear relationship will have 0 or near 0 correlation. In Figure 3.17, we are showing $R^2$, the correlation coefficient, and covariance for different scatter plots.

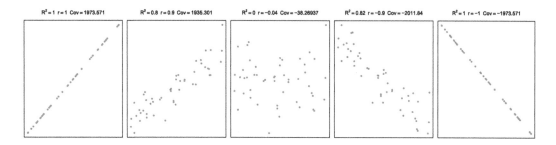

**FIGURE 3.17:** Correlation and covariance for different scatter plots.

For simple linear regression, correlation can be used to assess the model. However, this becomes useless as a measure of general accuracy if there is more than one explanatory variable as in multiple linear regression. In that case, $R^2$ is a measure of accuracy for the model. Interestingly, the square of the correlation of predicted values and original response variables $((cor(Y, \hat{Y}))^2$ ) equals $R^2$ for multiple linear regression.

The last accuracy measure, or the model fit in general we are going to explain is *F-statistic*. This is a quantity that depends on the RSS and TSS again. It can also answer one important question that other metrics cannot easily answer. That question is whether or not any of the explanatory variables have predictive value or in other words if all the explanatory variables are zero. We can write the null hypothesis as follows:

$$H_0 : \beta_1 = \beta_2 = \beta_3 = ... = \beta_p = 0$$

where the alternative is:

$$H_1 : \text{at least one } \beta_i \neq 0$$

Remember that $TSS - RSS$ is analogous to "explained variability" and the RSS is analogous to "unexplained variability". For the F-statistic, we divide explained variance by unexplained variance. Explained variance is just the $TSS - RSS$

divided by degrees of freedom, and unexplained variance is the RSE. The ratio will follow the F-distribution with two parameters, the degrees of freedom for the explained variance and the degrees of freedom for the unexplained variance. The F-statistic for a linear model is calculated as follows.

$$F = \frac{(TSS - RSS)/(p-1)}{RSS/(n-p)} = \frac{(TSS - RSS)/(p-1)}{RSE} \sim F(p-1, n-p)$$

If the variances are the same, the ratio will be 1, and when $H_0$ is true, then it can be shown that expected value of $(TSS - RSS)/(p-1)$ will be $\sigma^2$, which is estimated by the RSE. So, if the variances are significantly different, the ratio will need to be significantly bigger than 1. If the ratio is large enough we can reject the null hypothesis. To assess that, we need to use software or look up the tables for F statistics with calculated parameters. In R, function `qf()` can be used to calculate critical value of the ratio. Benefit of the F-test over looking at significance of coefficients one by one is that we circumvent multiple testing problem. If there are lots of explanatory variables at least 5% of the time (assuming we use 0.05 as P-value significance cut-off), p-values from coefficient t-tests will be wrong. In summary, F-test is a better choice for testing if there is any association between the explanatory variables and the response variable.

### 3.3.4 Regression with categorical variables

An important feature of linear regression is that categorical variables can be used as explanatory variables, this feature is very useful in genomics where explanatory variables can often be categorical. To put it in context, in our histone modification example we can also include if promoters have CpG islands or not as a variable. In addition, in differential gene expression, we usually test the difference between different conditions, which can be encoded as categorical variables in a linear regression. We can sure use the t-test for that as well if there are only 2 conditions, but if there are more conditions and other variables to control for, such as age or sex of the samples, we need to take those into account for our statistics, and the t-test alone cannot handle such complexity. In addition, when we have categorical variables we can also have numeric variables in the model and we certainly do not have to include only one type of variable in a model.

The simplest model with categorical variables includes two levels that can be encoded in 0 and 1. Below, we show linear regression with categorical variables. We then plot the fitted line. This plot is shown in Figure 3.18.

```
set.seed(100)
gene1=rnorm(30,mean=4,sd=2)
```

```
gene2=rnorm(30,mean=2,sd=2)
gene.df=data.frame(exp=c(gene1,gene2),
                   group=c( rep(1,30),rep(0,30) ) )

mod2=lm(exp~group,data=gene.df)
summary(mod2)
```

```
##
## Call:
## lm(formula = exp ~ group, data = gene.df)
##
## Residuals:
##     Min      1Q  Median      3Q     Max
## -4.7290 -1.0664  0.0122  1.3840  4.5629
##
## Coefficients:
##              Estimate Std. Error t value Pr(>|t|)
## (Intercept)    2.1851     0.3517   6.214 6.04e-08 ***
## group          1.8726     0.4973   3.765 0.000391 ***
## ---
## Signif. codes:  0 '***' 0.001 '**' 0.01 '*' 0.05 '.' 0.1 ' ' 1
##
## Residual standard error: 1.926 on 58 degrees of freedom
## Multiple R-squared:  0.1964, Adjusted R-squared:  0.1826
## F-statistic: 14.18 on 1 and 58 DF,  p-value: 0.0003905
```

```
require(mosaic)
plotModel(mod2)
```

We can even compare more levels, and we do not even have to encode them ourselves. We can pass categorical variables to the lm() function.

```
gene.df=data.frame(exp=c(gene1,gene2,gene2),
                   group=c( rep("A",30),rep("B",30),rep("C",30) )
                   )

mod3=lm(exp~group,data=gene.df)
summary(mod3)
```

```
##
## Call:
```

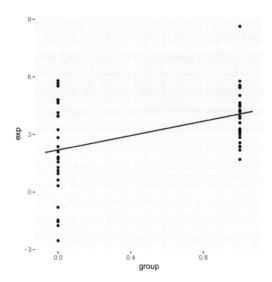

**FIGURE 3.18:** Linear model with a categorical variable coded as 0 and 1.

```
## lm(formula = exp ~ group, data = gene.df)
##
## Residuals:
##     Min      1Q  Median      3Q     Max
## -4.7290 -1.0793 -0.0976  1.4844  4.5629
##
## Coefficients:
##             Estimate Std. Error t value Pr(>|t|)
## (Intercept)   4.0577     0.3781  10.731  < 2e-16 ***
## groupB       -1.8726     0.5348  -3.502 0.000732 ***
## groupC       -1.8726     0.5348  -3.502 0.000732 ***
## ---
## Signif. codes:  0 '***' 0.001 '**' 0.01 '*' 0.05 '.' 0.1 ' ' 1
##
## Residual standard error: 2.071 on 87 degrees of freedom
## Multiple R-squared:  0.1582, Adjusted R-squared:  0.1388
## F-statistic: 8.174 on 2 and 87 DF,  p-value: 0.0005582
```

### 3.3.5 Regression pitfalls

In most cases one should look at the error terms (residuals) vs. the fitted values plot. Any structure in this plot indicates problems such as non-linearity, correlation of error terms, non-constant variance or unusual values driving the fit. Below we briefly explain the potential issues with the linear regression.

### 3.3.5.0.1 Non-linearity

If the true relationship is far from linearity, prediction accuracy is reduced and all the other conclusions are questionable. In some cases, transforming the data with $logX$, $\sqrt{X}$, and $X^2$ could resolve the issue.

### 3.3.5.0.2 Correlation of explanatory variables

If the explanatory variables are correlated that could lead to something known as multicolinearity. When this happens SE estimates of the coefficients will be too large. This is usually observed in time-course data.

### 3.3.5.0.3 Correlation of error terms

This assumes that the errors of the response variables are uncorrelated with each other. If they are, the confidence intervals of the coefficients might be too narrow.

### 3.3.5.0.4 Non-constant variance of error terms

This means that different response variables have the same variance in their errors, regardless of the values of the predictor variables. If the errors are not constant (ex: the errors grow as X values increase), this will result in unreliable estimates in standard errors as the model assumes constant variance. Transformation of data, such as $logX$ and $\sqrt{X}$, could help in some cases.

### 3.3.5.0.5 Outliers and high leverage points

Outliers are extreme values for Y and high leverage points are unusual X values. Both of these extremes have the power to affect the fitted line and the standard errors. In some cases (Ex: if there are measurement errors), they can be removed from the data for a better fit.

- **Want to know more ?**
    * Linear models and derivations of equations including matrix notation
        - *Applied Linear Statistical Models* by Kutner, Nachtsheim, et al. (Kutner et al., 2003)
        - *Elements of Statistical Learning* by Hastie & Tibshirani (Friedman et al., 2001)
        - *An Introduction to Statistical Learning* by James, Witten, et al. (James et al., 2013)

## 3.4   Exercises

### 3.4.1   How to summarize collection of data points: The idea behind statistical distributions

1.  Calculate the means and variances of the rows of the following simulated data set, and plot the distributions of means and variances using `hist()` and `boxplot()` functions. [Difficulty: **Beginner/Intermediate**]

```
set.seed(100)

#sample data matrix from normal distribution
gset=rnorm(600,mean=200,sd=70)
data=matrix(gset,ncol=6)
```

2.  Using the data generated above, calculate the standard deviation of the distribution of the means using the `sd()` function. Compare that to the expected standard error obtained from the central limit theorem keeping in mind the population parameters were $\sigma = 70$ and $n = 6$. How does the estimate from the random samples change if we simulate more data with `data=matrix(rnorm(6000,mean=200,sd=70),ncol=6)`? [Difficulty: **Beginner/Intermediate**]

3.  Simulate 30 random variables using the `rpois()` function. Do this 1000 times and calculate the mean of each sample. Plot the sampling distributions of the means using a histogram. Get the 2.5th and 97.5th percentiles of the distribution. [Difficulty: **Beginner/Intermediate**]

4.  Use the `t.test()` function to calculate confidence intervals of the mean on the first random sample `pois1` simulated from the `rpois()` function below. [Difficulty: **Intermediate**]

```
#HINT
set.seed(100)

#sample 30 values from poisson dist with lamda paramater =30
pois1=rpois(30,lambda=5)
```

5.  Use the bootstrap confidence interval for the mean on `pois1`. [Difficulty: **Intermediate/Advanced**]

6. Compare the theoretical confidence interval of the mean from the t.test and the bootstrap confidence interval. Are they similar? [Difficulty: **Intermediate/Advanced**]

7. Try to re-create the following figure, which demonstrates the CLT concept.[Difficulty: **Advanced**]

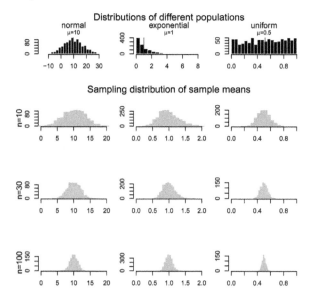

## 3.4.2 How to test for differences in samples

1. Test the difference of means of the following simulated genes using the randomization, t-test(), and wilcox.test() functions. Plot the distributions using histograms and boxplots. [Difficulty: **Intermediate/Advanced**]

```
set.seed(101)
gene1=rnorm(30,mean=4,sd=3)
gene2=rnorm(30,mean=3,sd=3)
```

2. Test the difference of the means of the following simulated genes using the randomization, t-test() and wilcox.test() functions. Plot the distributions using histograms and boxplots. [Difficulty: **Intermediate/Advanced**]

```
set.seed(100)
gene1=rnorm(30,mean=4,sd=2)
gene2=rnorm(30,mean=2,sd=2)
```

3. We need an extra data set for this exercise. Read the gene expression data set as follows: `gexpFile=system.file("extdata","geneExpMat.rds",package="compGenomRData")` `data=readRDS(gexpFile)`. The data has 100 differentially expressed genes. The first 3 columns are the test samples, and the last 3 are the control samples. Do a t-test for each gene (each row is a gene), and record the p-values. Then, do a moderated t-test, as shown in section "Moderated t-tests" in this chapter, and record the p-values. Make a p-value histogram and compare two approaches in terms of the number of significant tests with the 0.05 threshold. On the p-values use FDR (BH), Bonferroni and q-value adjustment methods. Calculate how many adjusted p-values are below 0.05 for each approach. [Difficulty: **Intermediate/Advanced**]

### 3.4.3   Relationship between variables: Linear models and correlation

Below we are going to simulate X and Y values that are needed for the rest of the exercise.

```
# set random number seed, so that the random numbers from the text
# is the same when you run the code.
set.seed(32)

# get 50 X values between 1 and 100
x = runif(50,1,100)

# set b0,b1 and variance (sigma)
b0 = 10
b1 = 2
sigma = 20
# simulate error terms from normal distribution
eps = rnorm(50,0,sigma)
# get y values from the linear equation and addition of error terms
y = b0 + b1*x+ eps
```

1. Run the code then fit a line to predict Y based on X. [Difficulty:**Intermediate**]

2. Plot the scatter plot and the fitted line. [Difficulty:**Intermediate**]

3. Calculate correlation and R^2. [Difficulty:**Intermediate**]

4. Run the `summary()` function and try to extract P-values for the model from the object returned by `summary`. See `?summary.lm`. [Difficulty:**Intermediate/Advanced**]

5. Plot the residuals vs. the fitted values plot, by calling the `plot()` function with `which=1` as the second argument. First argument is the model returned by `lm()`. [Difficulty:**Advanced**]

6. For the next exercises, read the data set histone modification data set. Use the following to get the path to the file:

```
hmodFile=system.file("extdata",
                "HistoneModeVSgeneExp.rds",
                package="compGenomRData")`
```

There are 3 columns in the dataset. These are measured levels of H3K4me3, H3K27me3 and gene expression per gene. Once you read in the data, plot the scatter plot for H3K4me3 vs. expression. [Difficulty:**Beginner**]

7. Plot the scatter plot for H3K27me3 vs. expression. [Difficulty:**Beginner**]

8. Fit the model for prediction of expression data using: 1) Only H3K4me3 as explanatory variable, 2) Only H3K27me3 as explanatory variable, and 3) Using both H3K4me3 and H3K27me3 as explanatory variables. Inspect the `summary()` function output in each case, which terms are significant. [Difficulty:**Beginner/Intermediate**]

9. Is using H3K4me3 and H3K27me3 better than the model with only H3K4me3? [Difficulty:**Intermediate**]

10. Plot H3k4me3 vs. H3k27me3. Inspect the points that do not follow a linear trend. Are they clustered at certain segments of the plot? Bonus: Is there any biological or technical interpretation for those points? [Difficulty:**Intermediate/Advanced**]

# 4

## *Exploratory Data Analysis with Unsupervised Machine Learning*

In this chapter, we will focus on using some of the machine learning techniques to explore genomics data. The goals of data exploration are usually many. Generally, we want to understand how the variables in our data set relate to each other and how the samples defined by those variables relate to each other. These points of information can be used to generate a hypothesis, find outliers in the samples or identify sample groups that need more data points. In this chapter, we will focus on two main classes of techniques: "clustering" and "dimension reduction". We will show how to use these techniques and how to visualize them using R. As these techniques are fundamental for data analysis, we will see more of their use cases in Chapters 8, 9, 10 and 11.

## 4.1 Clustering: Grouping samples based on their similarity

In genomics, we would very frequently want to assess how our samples relate to each other. Are our replicates similar to each other? Do the samples from the same treatment group have similar genome-wide signals? Do the patients with similar diseases have similar gene expression profiles? Take the last question for example. We need to define a distance or similarity metric between patients' expression profiles and use that metric to find groups of patients that are more similar to each other than the rest of the patients. This, in essence, is the general idea behind clustering. We need a distance metric and a method to utilize that distance metric to find self-similar groups. Clustering is a ubiquitous procedure in bioinformatics as well as any field that deals with high-dimensional data. It is very likely that every genomics paper containing multiple samples has some sort of clustering. Due to this ubiquity and general usefulness, it is an essential technique to learn.

### 4.1.1 Distance metrics

The first required step for clustering is the distance metric. This is simply a measurement of how similar gene expressions are to each other. There are many options for distance metrics and the choice of the metric is quite important for clustering.

**TABLE 4.1:** Gene expressions from patients

|          | IRX4 | OCT4 | PAX6 |
|----------|------|------|------|
| patient1 | 11   | 10   | 1    |
| patient2 | 13   | 13   | 3    |
| patient3 | 2    | 4    | 10   |
| patient4 | 1    | 3    | 9    |

Consider a simple example where we have four patients and expression of three genes measured in Table 4.1. Which patients look similar to each other based on their gene expression profiles ?

It may not be obvious from the table at first sight, but if we plot the gene expression profile for each patient (shown in Figure 4.1), we will see that expression profiles of patient 1 and patient 2 are more similar to each other than patient 3 or patient 4.

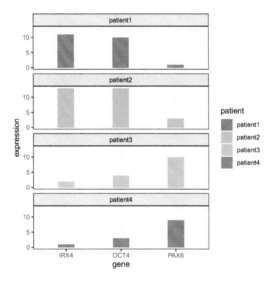

**FIGURE 4.1:** Gene expression values for different patients. Certain patients have gene expression values that are similar to each other.

But how can we quantify what we see? A simple metric for distance between gene expression vectors between a given patient pair is the sum of the absolute difference between gene expression values. This can be formulated as follows: $d_{AB} = \sum_{i=1}^{n} |e_{Ai} - e_{Bi}|$, where $d_{AB}$ is the distance between patients A and B, and the $e_{Ai}$ and $e_{Bi}$ are expression values of the $i$th gene for patients A and B. This distance metric is called the **"Manhattan distance"** or **"L1 norm"**.

Another distance metric uses the sum of squared distances and takes the square root of resulting value; this metric can be formulated as: $d_{AB} = \sqrt{\sum_{i=1}^{n} (e_{Ai} - e_{Bi})^2}$.

This distance is called **"Euclidean Distance"** or **"L2 norm"**. This is usually the default distance metric for many clustering algorithms. Due to the squaring operation, values that are very different get higher contribution to the distance. Due to this, compared to the Manhattan distance, it can be affected more by outliers. But, generally if the outliers are rare, this distance metric works well.

The last metric we will introduce is the **"correlation distance"**. This is simply $d_{AB} = 1 - \rho$, where $\rho$ is the Pearson correlation coefficient between two vectors; in our case those vectors are gene expression profiles of patients. Using this distance the gene expression vectors that have a similar pattern will have a small distance, whereas when the vectors have different patterns they will have a large distance. In this case, the linear correlation between vectors matters, although the scale of the vectors might be different.

Now let's see how we can calculate these distances in R. First, we have our gene expression per patient table.

```
df
```

```
##          IRX4 OCT4 PAX6
## patient1   11   10    1
## patient2   13   13    3
## patient3    2    4   10
## patient4    1    3    9
```

Next, we calculate the distance metrics using the `dist()` function and `1-cor()` expression.

```
dist(df,method="manhattan")
```

```
##          patient1 patient2 patient3
## patient2        7
## patient3       24       27
## patient4       25       28        3
```

```
dist(df,method="euclidean")
```

```
##            patient1  patient2  patient3
## patient2  4.123106
## patient3 14.071247 15.842980
## patient4 14.594520 16.733201  1.732051
```

```
as.dist(1-cor(t(df))) # correlation distance
```

```
##                  patient1     patient2     patient3
## patient2 0.004129405
## patient3 1.988522468  1.970725343
## patient4 1.988522468  1.970725343  0.000000000
```

#### 4.1.1.1  Scaling before calculating the distance

Before we proceed to the clustering, there is one more thing we need to take care of. Should we normalize our data? The scale of the vectors in our expression matrix can affect the distance calculation. Gene expression tables might have some sort of normalization, so the values are in comparable scales. But somehow, if a gene's expression values are on a much higher scale than the other genes, that gene will affect the distance more than others when using Euclidean or Manhattan distance. If that is the case we can scale the variables. The traditional way of scaling variables is to subtract their mean, and divide by their standard deviation, this operation is also called "standardization". If this is done on all genes, each gene will have the same effect on distance measures. The decision to apply scaling ultimately depends on our data and what you want to achieve. If the gene expression values are previously normalized between patients, having genes that dominate the distance metric could have a biological meaning and therefore it may not be desirable to further scale variables. In R, the standardization is done via the `scale()` function. Here we scale the gene expression values.

```
df
```

```
##            IRX4 OCT4 PAX6
## patient1    11   10    1
## patient2    13   13    3
## patient3     2    4   10
## patient4     1    3    9
```

```
scale(df)
```

```
##                  IRX4        OCT4        PAX6
## patient1   0.6932522   0.5212860  -1.0733721
## patient2   1.0194886   1.1468293  -0.6214260
## patient3  -0.7748113  -0.7298004   0.9603856
## patient4  -0.9379295  -0.9383149   0.7344125
## attr(,"scaled:center")
## IRX4 OCT4 PAX6
```

```
## 6.75 7.50 5.75
## attr(,"scaled:scale")
##       IRX4     OCT4     PAX6
## 6.130525 4.795832 4.425306
```

### 4.1.2   Hiearchical clustering

This is one of the most ubiquitous clustering algorithms. Using this algorithm you can see the relationship of individual data points and relationships of clusters. This is achieved by successively joining small clusters to each other based on the inter-cluster distance. Eventually, you get a tree structure or a dendrogram that shows the relationship between the individual data points and clusters. The height of the dendrogram is the distance between clusters. Here we can show how to use this on our toy data set from four patients. The base function in R to do hierarchical clustering in hclust(). Below, we apply that function on Euclidean distances between patients. The resulting clustering tree or dendrogram is shown in Figure 4.1.

```
d=dist(df)
hc=hclust(d,method="complete")
plot(hc)
```

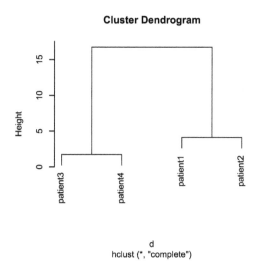

**FIGURE 4.2:** Dendrogram of distance matrix.

In the above code snippet, we have used the method="complete" argument without explaining it. The method argument defines the criteria that directs how the sub-clusters are merged. During clustering, starting with single-member clusters, the clusters are merged based on the distance between them. There are many different ways to define distance between clusters, and based on which definition you use, the hierarchical clustering results change. So the `method` argument controls that. There are a couple of values this argument can take; we list them and their description below:

- **"complete"** stands for "Complete Linkage" and the distance between two clusters is defined as the largest distance between any members of the two clusters.
- **"single"** stands for "Single Linkage" and the distance between two clusters is defined as the smallest distance between any members of the two clusters.
- **"average"** stands for "Average Linkage" or more precisely the UPGMA (Unweighted Pair Group Method with Arithmetic Mean) method. In this case, the distance between two clusters is defined as the average distance between any members of the two clusters.
- **"ward.D2"** and **"ward.D"** stands for different implementations of Ward's minimum variance method. This method aims to find compact, spherical clusters by selecting clusters to merge based on the change in the cluster variances. The clusters are merged if the increase in the combined variance over the sum of the cluster-specific variances is the minimum compared to alternative merging operations.

In real life, we would get expression profiles from thousands of genes and we will typically have many more patients than our toy example. One such data set is gene expression values from 60 bone marrow samples of patients with one of the four main types of leukemia (ALL, AML, CLL, CML) or no-leukemia controls. We trimmed that data set down to the top 1000 most variable genes to be able to work with it more easily, since genes that are not very variable do not contribute much to the distances between patients. We will now use this data set to cluster the patients and display the values as a heatmap and a dendrogram. The heatmap shows the expression values of genes across patients in a color coded manner. The heatmap function, `pheatmap()`, that we will use performs the clustering as well. The matrix that contains gene expressions has the genes in the rows and the patients in the columns. Therefore, we will also use a column-side color code to mark the patients based on their leukemia type. For the hierarchical clustering, we will use Ward's method designated by the `clustering_method` argument to the `pheatmap()` function. The resulting heatmap is shown in Figure 4.3.

```
library(pheatmap)
expFile=system.file("extdata","leukemiaExpressionSubset.rds",
                    package="compGenomRData")
mat=readRDS(expFile)

# set the leukemia type annotation for each sample
annotation_col = data.frame(
```

```
             LeukemiaType  =substr(colnames(mat),1,3))
rownames(annotation_col)=colnames(mat)
```

```
pheatmap(mat,show_rownames=FALSE,show_colnames=FALSE,
         annotation_col=annotation_col,
         scale = "none",clustering_method="ward.D2",
         clustering_distance_cols="euclidean")
```

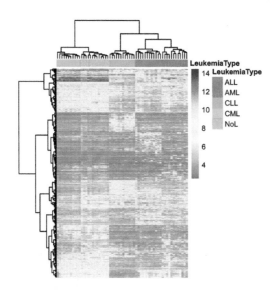

**FIGURE 4.3:** Heatmap of gene expression values from leukemia patients. Each column represents a patient. Columns are clustered using gene expression and color coded by disease type: ALL, AML, CLL, CML or no-leukemia.

As we can observe in the heatmap, each cluster has a distinct set of expression values. The main clusters almost perfectly distinguish the leukemia types. Only one CML patient is clustered as a non-leukemia sample. This could mean that gene expression profiles are enough to classify leukemia type. More detailed analysis and experiments are needed to verify that, but by looking at this exploratory analysis we can decide where to focus our efforts next.

#### 4.1.2.1 Where to cut the tree ?

The example above seems like a clear-cut example where we can pick clusters from the dendrogram by eye. This is mostly due to Ward's method, where compact clusters are preferred. However, as is usually the case, we do not have patient labels and it would be difficult to tell which leaves (patients) in the dendrogram we should consider as part of the same cluster. In other words, how deep we should cut the dendrogram so that every patient sample still connected via the remaining sub-dendrograms constitute clusters. The `cutree()` function provides the functionality to output either desired number of clusters or clusters obtained from cutting the dendrogram at a certain height. Below, we will cluster the patients with hierarchical clustering using the default method "complete linkage" and cut the dendrogram at a certain height. In this case, you will also observe that, changing from Ward's distance to complete linkage had an effect on clustering. Now the two clusters that are defined by Ward's distance are closer to each other and harder to separate from each other, shown in Figure 4.4.

```
hcl=hclust(dist(t(mat)))
plot(hcl,labels = FALSE, hang= -1)
rect.hclust(hcl, h = 80, border = "red")
```

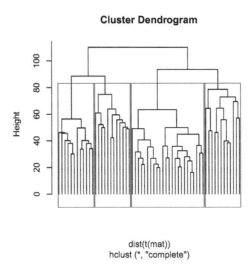

**FIGURE 4.4:** Dendrogram of Leukemia patients clustered by hierarchical clustering. Rectangles show the cluster we will get if we cut the tree at 'height=80'.

```
clu.k5=cutree(hcl,k=5) # cut tree so that there are 5 clusters

clu.h80=cutree(hcl,h=80) # cut tree/dendrogram from height 80
table(clu.k5) # number of samples for each cluster

## clu.k5
##  1  2  3  4  5
## 12  3  9 12 24
```

Apart from the arbitrary values for the height or the number of clusters, how can we define clusters more systematically? As this is a general question, we will show how to decide the optimal number of clusters later in this chapter.

### 4.1.3  K-means clustering

Another very common clustering algorithm is k-means. This method divides or partitions the data points, our working example patients, into a pre-determined, "k" number of clusters (Hartigan and Wong, 1979). Hence, these types of methods are generally called "partitioning" methods. The algorithm is initialized with randomly chosen $k$ centers or centroids. In a sense, a centroid is a data point with multiple values. In our working example, it is a hypothetical patient with gene expression values. But in the initialization phase, those gene expression values are chosen randomly within the boundaries of the gene expression distributions from real patients. As the next step in the algorithm, each patient is assigned to the closest centroid, and in the next iteration, centroids are set to the mean of values of the genes in the cluster. This process of setting centroids and assigning patients to the clusters repeats itself until the sum of squared distances to cluster centroids is minimized.

As you might see, the cluster algorithm starts with random initial centroids. This feature might yield different results for each run of the algorithm. We will now show how to use the k-means method on the gene expression data set. We will use set.seed() for reproducibility. In the wild, you might want to run this algorithm multiple times to see if your clustering results are stable.

```
set.seed(101)

# we have to transpore the matrix t()
# so that we calculate distances between patients
kclu=kmeans(t(mat),centers=5)

# number of data points in each cluster
table(kclu$cluster)

##
##  1  2  3  4  5
## 12 14 11 12 11
```

Now let us check the percentage of each leukemia type in each cluster. We can visualize this as a table. Looking at the table below, we see that each of the 5 clusters predominantly represents one of the 4 leukemia types or the control patients without leukemia.

```
type2kclu = data.frame(
                  LeukemiaType =substr(colnames(mat),1,3),
                  cluster=kclu$cluster)
```

```
table(type2kclu)
```

```
##                  cluster
## LeukemiaType  1  2  3  4  5
##          ALL 12  0  0  0  0
##          AML  0  1  0  0 11
##          CLL  0  0  0 12  0
##          CML  0  1 11  0  0
##          NoL  0 12  0  0  0
```

Another related and maybe more robust algorithm is called **"k-medoids"** clustering (Reynolds et al., 2006). The procedure is almost identical to k-means clustering with a couple of differences. In this case, centroids chosen are real data points in our case patients, and the metric we are trying to optimize in each iteration is based on the Manhattan distance to the centroid. In k-means this was based on the sum of squared distances, so Euclidean distance. Below we show how to use the k-medoids clustering function `pam()` from the `cluster` package.

```
kmclu=cluster::pam(t(mat),k=5) #  cluster using k-medoids
```

```
# make a data frame with Leukemia type and cluster id
type2kmclu = data.frame(
                  LeukemiaType =substr(colnames(mat),1,3),
                  cluster=kmclu$cluster)
```

```
table(type2kmclu)
```

```
##                  cluster
## LeukemiaType  1  2  3  4  5
##          ALL 12  0  0  0  0
##          AML  0 10  1  1  0
##          CLL  0  0  0  0 12
##          CML  0  0  0 12  0
##          NoL  0  0 12  0  0
```

We cannot visualize the clustering from partitioning methods with a tree like we did for hierarchical clustering. Even if we can get the distances between patients the algorithm does not return the distances between clusters out of the box. However, if we had a way to visualize the distances between patients in 2 dimensions we could see the how patients and clusters relate to each other. It turns out that there is a way to compress between patient distances to a 2-dimensional plot. There are many ways to do this, and we introduce these dimension-reduction methods including the one we will use later in this chapter. For now, we are going to use a method called "multi-dimensional scaling" and plot the patients in a 2D plot color coded by their cluster assignments shown in Figure 4.5. We will explain this method in more detail in the Multi-dimensional scaling section below.

```r
# Calculate distances
dists=dist(t(mat))

# calculate MDS
mds=cmdscale(dists)

# plot the patients in the 2D space
plot(mds,pch=19,col=rainbow(5)[kclu$cluster])

# set the legend for cluster colors
legend("bottomright",
       legend=paste("clu",unique(kclu$cluster)),
       fill=rainbow(5)[unique(kclu$cluster)],
       border=NA,box.col=NA)
```

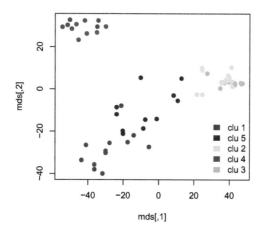

**FIGURE 4.5:** K-means cluster memberships are shown in a multi-dimensional scaling plot.

The plot we obtained shows the separation between clusters. However, it does not do a great job showing the separation between clusters 3 and 4, which represent CML and "no leukemia" patients. We might need another dimension to properly visualize that separation. In addition, those two clusters were closely related in the hierarchical clustering as well.

### 4.1.4    How to choose "k", the number of clusters

Up to this point, we have avoided the question of selecting optimal number clusters. How do we know where to cut our dendrogram or which k to choose ? First of all, this is a difficult question. Usually, clusters have different granularity. Some clusters are tight and compact and some are wide, and both these types of clusters can be in the same data set. When visualized, some large clusters may look like they may have sub-clusters. So should we consider the large cluster as one cluster or should we consider the sub-clusters as individual clusters? There are some metrics to help but there is no definite answer. We will show a couple of them below.

#### 4.1.4.1    Silhouette

One way to determine the quality of the clustering is to measure the expected self-similar nature of the points in a set of clusters. The silhouette value does just that and it is a measure of how similar a data point is to its own cluster compared to other clusters (Rousseeuw, 1987). The silhouette value ranges from -1 to +1, where values that are positive indicate that the data point is well matched to its own cluster, if the value is zero it is a borderline case, and if the value is minus it means that the data point might be mis-clustered because it is more similar to a neighboring cluster. If most data points have a high value, then the clustering is appropriate. Ideally, one can create many different clusterings with each with a different $k$ parameter indicating the number of clusters, and assess their appropriateness using the average silhouette values. In R, silhouette values are referred to as silhouette widths in the documentation.

A silhouette value is calculated for each data point. In our working example, each patient will get silhouette values showing how well they are matched to their assigned clusters. Formally this calculated as follows. For each data point $i$, we calculate $a(i)$, which denotes the average distance between $i$ and all other data points within the same cluster. This shows how well the point fits into that cluster. For the same data point, we also calculate $b(i)$, which denotes the lowest average distance of $i$ to all points in any other cluster, of which $i$ is not a member. The cluster with this lowest average $b(i)$ is the "neighboring cluster" of data point $i$ since it is the next best fit cluster for that data point. Then, the silhouette value for a given data point is
$s(i) = \frac{b(i)-a(i)}{\max\{a(i),b(i)\}}$.

As described, this quantity is positive when $b(i)$ is high and $a(i)$ is low, meaning that the data point $i$ is self-similar to its cluster. And the silhouette value, $s(i)$, is negative if it is more similar to its neighbors than its assigned cluster.

In R, we can calculate silhouette values using the `cluster::silhouette()` function. Below, we calculate the silhouette values for k-medoids clustering with the `pam()` function with k=5. The resulting silhouette values are shown in Figure 4.6.

```
library(cluster)
set.seed(101)
pamclu=cluster::pam(t(mat),k=5)
plot(silhouette(pamclu),main=NULL)
```

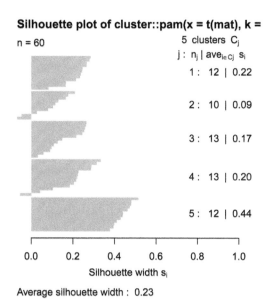

**FIGURE 4.6:** Silhouette values for k-medoids with 'k=5'.

Now, let us calculate the average silhouette value for different $k$ values and compare. We will use `sapply()` function to get average silhouette values across $k$ values between 2 and 7. Within `sapply()` there is an anonymous function that that does the clustering and calculates average silhouette values for each $k$. The plot showing average silhouette values for different $k$ values is shown in Figure 4.7.

```
Ks=sapply(2:7,
    function(i)
      summary(silhouette(pam(t(mat),k=i)))$avg.width)
plot(2:7,Ks,xlab="k",ylab="av. silhouette",type="b",
    pch=19)
```

In this case, it seems the best value for $k$ is 4. The k-medoids function `pam()` will

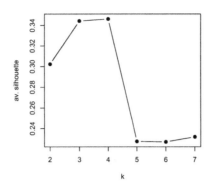

**FIGURE 4.7:** Average silhouette values for k-medoids clustering for 'k' values between 2 and 7.

usually cluster CML and "no Leukemia" cases together when k=4, which are also related clusters according to the hierarchical clustering we did earlier.

### 4.1.4.2 Gap statistic

As clustering aims to find self-similar data points, it would be reasonable to expect with the correct number of clusters the total within-cluster variation is minimized. Within-cluster variation for a single cluster can simply be defined as the sum of squares from the cluster mean, which in this case is the centroid we defined in the k-means algorithm. The total within-cluster variation is then the sum of within-cluster variations for each cluster. This can be formally defined as follows:

$$W_k = \sum_{k=1}^{K} \sum_{\mathrm{x}_i \in C_k} (\mathrm{x}_i - \mu_k)^2$$

where $\mathrm{x}_i$ is a data point in cluster $k$, and $\mu_k$ is the cluster mean, and $W_k$ is the total within-cluster variation quantity we described. However, the problem is that the variation quantity decreases with the number of clusters. The more centroids we have, the smaller the distances to the centroids become. A more reliable approach would be somehow calculating the expected variation from a reference null distribution and compare that to the observed variation for each $k$. In the gap statistic approach, the expected distribution is calculated via sampling points from the boundaries of the original data and calculating within-cluster variation quantity for multiple rounds of sampling (Tibshirani et al., 2001). This way we have an expectation about the variability when there is no clustering, and then compare that expected variation to the observed within-cluster variation. The expected variation should also go down with the increasing number of clusters, but for the optimal number of clusters, the expected variation will be furthest away from observed variation. This distance is called the **"gap statistic"** and defined as follows: $\mathrm{Gap}_n(k) = E_n^*\{\log W_k\} - \log W_k$, where $E_n^*\{\log W_k\}$ is the expected variation in log-scale under a sample size $n$ from the reference distribution and $\log W_k$ is the observed variation. Our aim is to choose the $k$ number of clusters that maximizes $\mathrm{Gap}_n(k)$.

We can easily calculate the gap statistic with the `cluster::clusGap()` function. We will now use that function to calculate the gap statistic for our patient gene expression data. The resulting gap statistics are shown in Figure 4.8.

```
library(cluster)
set.seed(101)
# define the clustering function
pam1 <- function(x,k)
  list(cluster = pam(x,k, cluster.only=TRUE))

# calculate the gap statistic
pam.gap= clusGap(t(mat), FUN = pam1, K.max = 8,B=50)

# plot the gap statistic accross k values
plot(pam.gap, main = "Gap statistic for the 'Leukemia' data")
```

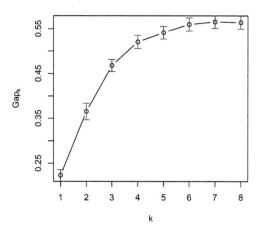

**FIGURE 4.8:** Gap statistic for clustering the leukemia dataset with k-medoids (pam) algorithm.

In this case, the gap statistic shows that $k = 7$ is the best if we take the maximum value as the best. However, after $k = 6$, the statistic has more or less a stable curve. This observation is incorporated into algorithms that can select the best $k$ value based on the gap statistic. A reasonable way is to take the simulation error (error bars in 4.8) into account, and take the smallest $k$ whose gap statistic is larger or equal to the one of $k + 1$ minus the simulation error. Formally written, we would pick the smallest $k$ satisfying the following condition: $\text{Gap}(k) \geq \text{Gap}(k + 1) - s_{k+1}$, where $s_{k+1}$ is the simulation error for $\text{Gap}(k + 1)$.

Using this procedure gives us $k = 6$ as the optimum number of clusters. Biologically, we know that there are 5 main patient categories but this does not mean there are no sub-categories or sub-types for the cancers we are looking at.

### 4.1.4.3 Other methods

There are several other methods that provide insight into how many clusters. In fact, the package NbClust provides 30 different ways to determine the number of optimal clusters and can offer a voting mechanism to pick the best number. Below, we show how to use this function for some of the optimal number of cluster detection methods.

```
library(NbClust)
nb = NbClust(data=t(mat),
            distance = "euclidean", min.nc = 2,
        max.nc = 7, method = "kmeans",
        index=c("kl","ch","cindex","db","silhouette",
                "duda","pseudot2","beale","ratkowsky",
                "gap","gamma","mcclain","gplus",
                "tau","sdindex","sdbw"))

table(nb$Best.nc[1,]) # consensus seems to be 3 clusters
```

However, readers should keep in mind that clustering is an exploratory technique. If you have solid labels for your data points, maybe clustering is just a sanity check, and you should just do predictive modeling instead. However, in biology there are rarely solid labels and things have different granularity. Take the leukemia patients case we have been using for example, it is known that leukemia types have subtypes and those sub-types that have different mutation profiles and consequently have different molecular signatures. Because of this, it is not surprising that some optimal cluster number techniques will find more clusters to be appropriate. On the other hand, CML (chronic myeloid leukemia) is a slow progressing disease and maybe their molecular signatures are closer to "no leukemia" patients, so clustering algorithms may confuse the two depending on what granularity they are operating with. It is always good to look at the heatmaps after clustering, if you have meaningful self-similar data points, even if the labels you have do not agree that there can be different clusters, you can perform downstream analysis to understand the sub-clusters better. As we have seen, we can estimate the optimal number of clusters but we cannot take that estimation as the absolute truth. Given more data points or a different set of expression signatures, you may have different optimal clusterings, or the supposed optimal clustering might overlook previously known sub-groups of your data.

---

## 4.2 Dimensionality reduction techniques: Visualizing complex data sets in 2D

In statistics, dimension reduction techniques are a set of processes for reducing the number of random variables by obtaining a set of principal variables. For example, in the context of a gene expression matrix across different patient samples, this might mean getting a set of new variables that cover the variation in sets of genes. This way samples can be represented by a couple of principal variables instead of thousands of genes. This is useful for visualization, clustering and predictive modeling.

### 4.2.1 Principal component analysis

Principal component analysis (PCA) is maybe the most popular technique to examine high-dimensional data. There are multiple interpretations of how PCA reduces dimensionality. We will first focus on geometrical interpretation, where this operation can be interpreted as rotating the original dimensions of the data. For this, we go back to our example gene expression data set. In this example, we will represent our patients with expression profiles of just two genes, CD33 (ENSG00000105383) and PYGL (ENSG00000100504). This way we can visualize them in a scatter plot (see Figure 4.9).

```
plot(mat[rownames(mat)=="ENSG00000100504",],
     mat[rownames(mat)=="ENSG00000105383",],pch=19,
     ylab="CD33 (ENSG00000105383)",
     xlab="PYGL (ENSG00000100504)")
```

PCA rotates the original data space such that the axes of the new coordinate system point to the directions of highest variance of the data. The axes or new variables are termed principal components (PCs) and are ordered by variance: The first component, PC 1, represents the direction of the highest variance of the data. The direction of the second component, PC 2, represents the highest of the remaining variance orthogonal to the first component. This can be naturally extended to obtain the required number of components, which together span a component space covering the desired amount of variance. In our toy example with only two

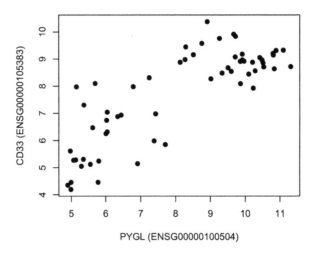

**FIGURE 4.9:** Gene expression values of CD33 and PYGL genes across leukemia patients.

genes, the principal components are drawn over the original scatter plot and in the next plot we show the new coordinate system based on the principal components. We will calculate the PCA with the `princomp()` function; this function returns the new coordinates as well. These new coordinates are simply a projection of data over the new coordinates. We will decorate the scatter plots with eigenvectors showing the direction of greatest variation. Then, we will plot the new coordinates (the resulting plot is shown in Figure 4.10). These are automatically calculated by the `princomp()` function. Notice that we are using the `scale()` function when plotting coordinates and also before calculating the PCA. This function centers the data, meaning it subtracts the mean of each column vector from the elements in the vector. This essentially gives the columns a zero mean. It also divides the data by the standard deviation of the centered columns. These two operations help bring the data to a common scale, which is important for PCA not to be affected by different scales in the data.

```
par(mfrow=c(1,2))

# create the subset of the data with two genes only
# notice that we transpose the matrix so samples are
# on the columns
sub.mat=t(mat[rownames(mat) %in% c("ENSG00000100504","ENSG00000105383"),])
```

```r
# ploting our genes of interest as scatter plots
plot(scale(mat[rownames(mat)=="ENSG00000100504",]),
     scale(mat[rownames(mat)=="ENSG00000105383",]),
     pch=19,
     ylab="CD33 (ENSG00000105383)",
     xlab="PYGL (ENSG00000100504)",
     col=as.factor(annotation_col$LeukemiaType),
     xlim=c(-2,2),ylim=c(-2,2))

# create the legend for the Leukemia types
legend("bottomright",
       legend=unique(annotation_col$LeukemiaType),
       fill =palette("default"),
       border=NA,box.col=NA)

# calculate the PCA only for our genes and all the samples
pr=princomp(scale(sub.mat))


# plot the direction of eigenvectors
# pr$loadings returned by princomp has the eigenvectors
arrows(x0=0, y0=0, x1 = pr$loadings[1,1],
       y1 = pr$loadings[2,1],col="pink",lwd=3)
arrows(x0=0, y0=0, x1 = pr$loadings[1,2],
       y1 = pr$loadings[2,2],col="gray",lwd=3)


# plot the samples in the new coordinate system
plot(-pr$scores,pch=19,
     col=as.factor(annotation_col$LeukemiaType),
     ylim=c(-2,2),xlim=c(-4,4))

# plot the new coordinate basis vectors
arrows(x0=0, y0=0, x1 =-2,
       y1 = 0,col="pink",lwd=3)
arrows(x0=0, y0=0, x1 = 0,
       y1 = -1,col="gray",lwd=3)
```

As you can see, the new coordinate system is useful by itself. The X-axis, which represents the first component, separates the data along the lymphoblastic and myeloid leukemias.

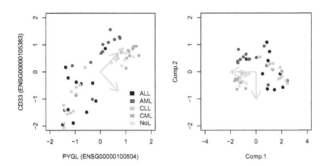

**FIGURE 4.10:** Geometric interpretation of PCA finding eigenvectors that point to the direction of highest variance. Eigenvectors can be used as a new coordinate system.

PCA in this case, is obtained by calculating eigenvectors of the covariance matrix via an operation called eigen decomposition. The covariance matrix is obtained by covariance of pairwise variables of our expression matrix, which is simply $\text{cov}(X, Y) = \frac{1}{n} \sum_{i=1}^{n} (x_i - \mu_X)(y_i - \mu_Y)$, where $X$ and $Y$ are expression values of genes in a sample in our example. This is a measure of how things vary together, if highly expressed genes in sample A are also highly expressed in sample B and lowly expressed in sample A are also lowly expressed in sample B, then sample A and B will have positive covariance. If the opposite is true, then they will have negative covariance. This quantity is related to correlation, and as we saw in the previous chapter, correlation is standardized covariance. Covariance of variables can be obtained with the `cov()` function, and eigen decomposition of such a matrix will produce a set of orthogonal vectors that span the directions of highest variation. In 2D, you can think of this operation as rotating two perpendicular lines together until they point to the directions where most of the variation in the data lies, similar to Figure 4.10. An important intuition is that, after the rotation prescribed by eigenvectors is complete, the covariance between variables in this rotated dataset will be zero. There is a proper mathematical relationship between covariances of the rotated dataset and the original dataset. That's why operating on the covariance matrix is related to the rotation of the original dataset.

```
cov.mat=cov(sub.mat) # calculate covariance matrix
cov.mat
eigen(cov.mat) # obtain eigen decomposition for eigen values and vectors
```

Eigenvectors and eigenvalues of the covariance matrix indicate the direction and the magnitude of variation of the data. In our visual example, the eigenvectors are so-called principal components. The eigenvector indicates the direction and the eigenvalues indicate the variation in that direction. Eigenvectors and values exist in pairs: every eigenvector has a corresponding eigenvalue and the eigenvectors are linearly independent from each other, which means they are orthogonal or uncorrelated as in our working example above. The eigenvectors are ranked by their corresponding eigenvalue, the higher the eigenvalue the more important the eigenvector is, because it explains more of the variation compared to the other eigenvectors. This feature of PCA makes the dimension reduction possible. We can sometimes display data sets that have many variables only in 2D or 3D because these top eigenvectors are sometimes enough to capture most of variation in the data. The `screeplot()` function takes the output of the `princomp()` or `prcomp()` functions as input and plots the variance explained by eigenvectors.

#### 4.2.1.1 Singular value decomposition and principal component analysis

A more common way to calculate PCA is through something called singular value decomposition (SVD). This results in another interpretation of PCA, which is called "latent factor" or "latent component" interpretation. In a moment, it will be clearer what we mean by "latent factors". SVD is a matrix factorization or decomposition algorithm that decomposes an input matrix, $X$, to three matrices as follows: $X = USV^T$. In essence, many matrices can be decomposed as a product of multiple matrices and we will come to other techniques later in this chapter. Singular value decomposition is shown in Figure 4.11. $U$ is the matrix with eigenarrays on the columns and this has the same dimensions as the input matrix; you might see elsewhere the columns are called eigenassays. $S$ is the matrix that contains the singular values on the diagonal. The singular values are also known as eigenvalues and their square is proportional to explained variation by each eigenvector. Finally, the matrix $V^T$ contains the eigenvectors on its rows. Its interpretation is still the same. Geometrically, eigenvectors point to the direction of highest variance in the data. They are uncorrelated or geometrically orthogonal to each other. These interpretations are identical to the ones we made before. The slight difference is that the decomposition seems to output $V^T$, which is just the transpose of the matrix $V$. However, the SVD algorithms in R usually return the matrix $V$. If you want the eigenvectors, you either simply use the columns of matrix $V$ or rows of $V^T$.

One thing that is new in Figure 4.11 is the concept of eigenarrays. The eigenarrays, sometimes called eigenassays, represent the sample space and can be used to plot the relationship between samples rather than genes. In this way, SVD offers additional information than the PCA using the covariance matrix. It offers us a way to summarize both genes and samples. As we can project the gene expression profiles over the top two eigengenes and get a 2D representation of genes, but with the SVD, we can also project the samples over the top two eigenarrays and get a representation of samples in 2D scatter plot. The eigenvector could represent independent expression programs across samples, such as cell-cycle, if we had time-

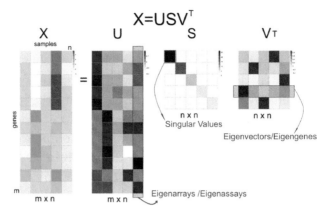

**FIGURE 4.11:** Singular value decomposition (SVD) explained in a diagram.

based expression profiles. However, there is no guarantee that each eigenvector will be biologically meaningful. Similarly each eigenarray represents samples with specific expression characteristics. For example, the samples that have a particular pathway activated might be correlated to an eigenarray returned by SVD.

Previously, in order to map samples to the reduced 2D space we had to transpose the genes-by-samples matrix before using the `princomp()` function. We will now first use SVD on the genes-by-samples matrix to get eigenarrays and use that to plot samples on the reduced dimensions. We will project the columns in our original expression data on eigenarrays and use the first two dimensions in the scatter plot. If you look at the code you will see that for the projection we use $U^T X$ operation, which is just $SV^T$ if you follow the linear algebra. We will also perform the PCA this time with the `prcomp()` function on the transposed genes-by-samples matrix to get similar information, and plot the samples on the reduced coordinates.

```
par(mfrow=c(1,2))
d=svd(scale(mat)) # apply SVD
assays=t(d$u) %*% scale(mat) # projection on eigenassays
plot(assays[1,],assays[2,],pch=19,
     col=as.factor(annotation_col$LeukemiaType))
#plot(d$v[,1],d$v[,2],pch=19,
#     col=annotation_col$LeukemiaType)
pr=prcomp(t(mat),center=TRUE,scale=TRUE) # apply PCA on transposed matrix

# plot new coordinates from PCA, projections on eigenvectors
# since the matrix is transposed eigenvectors represent
plot(pr$x[,1],pr$x[,2],col=as.factor(annotation_col$LeukemiaType))
```

As you can see in Figure 4.12, the two approaches yield separation of samples,

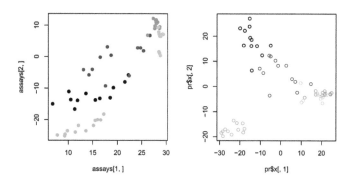

**FIGURE 4.12:** SVD on the matrix and its transpose.

although they are slightly different. The difference comes from the centering and scaling. In the first case, we scale and center columns and in the second case we scale and center rows since the matrix is transposed. If we do not do any scaling or centering we would get identical plots.

#### 4.2.1.1.1  *Eigenvectors as latent factors/variables*

Finally, we can introduce the latent factor interpretation of PCA via SVD. As we have already mentioned, eigenvectors can also be interpreted as expression programs that are shared by several genes such as cell cycle expression program when measuring gene expression across samples taken in different time points. In this interpretation, linear combination of expression programs makes up the expression profile of the genes. Linear combination simply means multiplying the expression program with a weight and adding them up. Our $USV^T$ matrix multiplication can be rearranged to yield such an understanding, we can multiply eigenarrays $U$ with the diagonal eigenvalues $S$, to produce an m-by-n weights matrix called $W$, so $W = US$ and we can re-write the equation as just weights by eigenvectors matrix, $X = WV^T$ as shown in Figure 4.13.

This simple transformation now makes it clear that indeed, if eigenvectors represent expression programs, their linear combination makes up individual gene expression profiles. As an example, we can show the linear combination of the first two eigenvectors can approximate the expression profile of a hypothetical gene in the gene expression matrix. Figure 4.14 shows eigenvector 1 and eigenvector 2 combined with certain weights in $W$ matrix can approximate gene expression pattern our example gene.

However, SVD does not care about biology. The eigenvectors are just obtained from the data with constraints of orthogonality and the direction of variation. There are examples of eigenvectors representing real expression programs but that

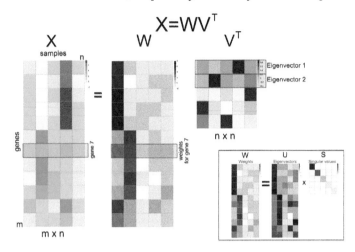

**FIGURE 4.13:** Singular value decomposition (SVD) reorganized as multiplication of m-by-n weights matrix and eigenvectors.

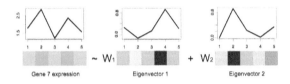

**FIGURE 4.14:** Gene expression of a gene can be regarded as a linear combination of eigenvectors.

does not mean eigenvectors will always be biologically meaningful. Sometimes a combination of them might make more sense in biology than single eigenvectors. This is also the same for the other matrix factorization techniques we describe below.

### 4.2.2 Other matrix factorization methods for dimensionality reduction

We must mention a few other techniques that are similar to SVD in spirit. Remember, we mentioned that every matrix can be decomposed to other matrices where matrix multiplication operations reconstruct the original matrix, which is in general called "matrix factorization". In the case of SVD/PCA, the constraint is that eigenvectors/arrays are orthogonal, however, there are other decomposition algorithms with other constraints.

#### 4.2.2.1 Independent component analysis (ICA)

We will first start with independent component analysis (ICA) which is an extension of PCA. ICA algorithm decomposes a given matrix $X$ as follows: $X = SA$

(Hyvärinen, 2013). The rows of $A$ could be interpreted similar to the eigengenes and columns of $S$ could be interpreted as eigenarrays. These components are sometimes called metagenes and metasamples in the literature. Traditionally, $S$ is called the source matrix and $A$ is called mixing matrix. ICA is developed for a problem called "blind-source separation". In this problem, multiple microphones record sound from multiple instruments, and the task is to disentangle sounds from original instruments since each microphone is recording a combination of sounds. In this respect, the matrix $S$ contains the original signals (sounds from different instruments) and their linear combinations identified by the weights in $A$, and the product of $A$ and $S$ makes up the matrix $X$, which is the observed signal from different microphones. With this interpretation in mind, if the interest is strictly expression patterns that represent the hidden expression programs, we see that the genes-by-samples matrix is transposed to a samples-by-genes matrix, so that the columns of $S$ represent these expression patterns, here referred to as "metagenes", hopefully representing distinct expression programs (Figure 4.15 ).

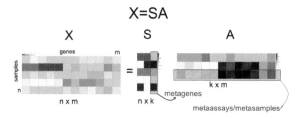

**FIGURE 4.15:** Independent Component Analysis (ICA).

ICA requires that the columns of the $S$ matrix, the "metagenes" in our example above, are statistically independent. This is a stronger constraint than uncorrelatedness. In this case, there should be no relationship between non-linear transformation of the data either. There are different ways of ensuring this statistical indepedence and this is the main constraint when finding the optimal $A$ and $S$ matrices. The various ICA algorithms use different proxies for statistical independence, and the definition of that proxy is the main difference between many ICA algorithms. The algorithm we are going to use requires that metagenes or sources in the $S$ matrix are non-Gaussian (non-normal) as possible. Non-Gaussianity is shown to be related to statistical independence (Hyvärinen, 2013). Below, we are using the `fastICA::fastICA()` function to extract 2 components and plot the rows of matrix $A$ which represents metagenes shown in Figure 4.16. This way, we can visualize samples in a 2D plot. If we wanted to plot the relationship between genes we would use the columns of matrix $S$.

```
library(fastICA)
ica.res=fastICA(t(mat),n.comp=2) # apply ICA
```

```
# plot reduced dimensions
plot(ica.res$S[,1],ica.res$S[,2],col=as.factor(annotation_col$LeukemiaType))
```

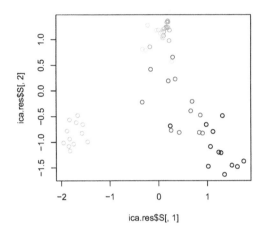

**FIGURE 4.16:** Leukemia gene expression values per patient on reduced dimensions by ICA.

#### 4.2.2.2 Non-negative matrix factorization (NMF)

Non-negative matrix factorization algorithms are series of algorithms that aim to decompose the matrix $X$ into the product of matrices $W$ and $H$, $X = WH$ (Figure 4.17) (Lee and Seung, 2001). The constraint is that $W$ and $H$ must contain non-negative values, so must $X$. This is well suited for data sets that cannot contain negative values such as gene expression. This also implies additivity of components or latent factors. This is in line with the idea that expression pattern of a gene across samples is the weighted sum of multiple metagenes. Unlike ICA and SVD/PCA, the metagenes can never be combined in a subtractive way. In this sense, expression programs potentially captured by metagenes are combined additively.

The algorithms that compute NMF try to minimize the cost function $D(X, WH)$, which is the distance between $X$ and $WH$. The early algorithms just use the Euclidean distance, which translates to $\sum(X - WH)^2$; this is also known as the Frobenius norm and you will see in the literature it is written as : $\|X - WH\|_F$. However, this is not the only distance metric; other distance metrics are also used in NMF algorithms. In addition, there could be other parameters to optimize that relates to sparseness of the $W$ and $H$ matrices. With sparse $W$ and $H$, each entry in the $X$ matrix is expressed as the sum of a small number of components. This

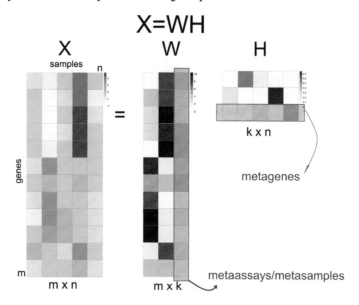

**FIGURE 4.17**: Non-negative matrix factorization summary.

makes the interpretation easier, if the weights are 0 then there is no contribution from the corresponding factors.

Below, we are plotting the values of metagenes (rows of $H$) for components 1 and 3, shown in Figure 4.18. In this context, these values can also be interpreted as the relationship between samples. If we wanted to plot the relationship between genes we would plot the columns of the $W$ matrix.

```
library(NMF)
res=NMF::nmf(mat,rank=3,seed="nndsvd") # nmf with 3 components/factors
w <- basis(res) # get W
h <- coef(res)  # get H


# plot 1st factor against 3rd factor
plot(h[1,],h[3,],col=as.factor(annotation_col$LeukemiaType),pch=19)
```

We should add the note that, due to random starting points of the optimization algorithm, NMF is usually run multiple times and a consensus clustering approach is used when clustering samples. This simply means that samples are clustered together if they cluster together in multiple runs of the NMF. The NMF package we used above has built-in ways to achieve this. In addition, NMF is a family of algorithms. The choice of cost function to optimize the difference between $X$ and $WH$, and the methods used for optimization create multiple variants of NMF. The "method" parameter in the above nmf() function controls the algorithm choice for NMF.

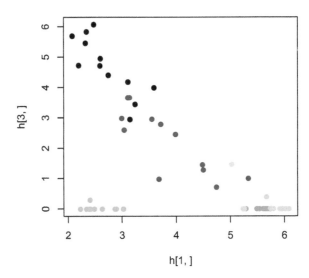

**FIGURE 4.18:** Leukemia gene expression values per patient on reduced dimensions by NMF. Components 1 and 3 is used for the plot.

#### 4.2.2.3 Choosing the number of components and ranking components in importance

In both ICA and NMF, there is no well-defined way to rank components or to select the number of components. There are a couple of approaches that might suit both ICA and NMF for ranking components. One can use the norms of columns/rows in mixing matrices. This could simply mean take the sum of absolute values in mixing matrices. For our ICA example above, we would take the sum of the absolute values of the rows of $A$ since we transposed the input matrix $X$ before ICA. And for the NMF, we would use the columns of $W$. These ideas assume that the larger coefficients in the weight or mixing matrices indicate more important components.

For selecting the optimal number of components, the NMF package provides different strategies. One way is to calculate the RSS for each $k$, the number of components, and take the $k$ where the RSS curve starts to stabilize. However, these strategies require that you run the algorithm with multiple possible component numbers. The `nmf` function will run these automatically when the `rank` argument is a vector of numbers. For ICA there is no straightforward way to choose the right number of components. A common strategy is to start with as many components as variables and try to rank them by their usefulness.

**– Want to know more ?**

The NMF package vignette has extensive information on how to run NMF to get stable results and an estimate of components: `https://cran.r-project.org/web/packages/NMF/vignettes/NMF-vignette.pdf`

### 4.2.3  Multi-dimensional scaling

MDS is a set of data analysis techniques that displays the structure of distance data in a high-dimensional space into a lower dimensional space without much loss of information (Cox and Cox, 2000). The overall goal of MDS is to faithfully represent these distances with the lowest possible dimensions. The so-called "classical multi-dimensional scaling" algorithm, tries to minimize the following function:

$$Stress_D(z_1, z_2, ..., z_N) = \left( \frac{\sum_{i,j} (d_{ij} - \|z_i - z_j\|)^2}{\sum_{i,j} d_{ij}^2} \right)^{1/2}$$

Here the function compares the new data points on the lower dimension $(z_1, z_2, ..., z_N)$ to the input distances between data points or distance between samples in our gene expression example. It turns out, this problem can be efficiently solved with SVD/PCA on the scaled distance matrix, the projection on eigenvectors will be the most optimal solution for the equation above. Therefore, classical MDS is sometimes called Principal Coordinates Analysis in the literature. However, later variants improve on classical MDS by using this as a starting point and optimize a slightly different cost function that again measures how well the low-dimensional distances correspond to high-dimensional distances. This variant is called non-metric MDS and due to the nature of the cost function, it assumes a less stringent relationship between the low-dimensional distances $|z_{i}-z_{j}|$ and input distances $d_{ij}$. Formally, this procedure tries to optimize the following function.

$$Stress_D(z_1, z_2, ..., z_N) = \left( \frac{\sum_{i,j} (\|z_i - z_j\| - \theta(d_{ij}))^2}{\sum_{i,j} \|z_i - z_j\|^2} \right)^{1/2}$$

The core of a non-metric MDS algorithm is a two-fold optimization process. First the optimal monotonic transformation of the distances has to be found, which is shown in the above formula as $\theta(d_{ij})$. Secondly, the points on a low dimension configuration have to be optimally arranged, so that their distances match the scaled distances as closely as possible. These two steps are repeated until some convergence criteria is reached. This usually means that the cost function does not improve much after certain number of iterations. The basic steps in a non-metric MDS algorithm are:

1) Find a random low-dimensional configuration of points, or in the variant we will be using below we start with the configuration returned by classical MDS.

2) Calculate the distances between the points in the low dimension $\|z_i - z_j\|$, $z_i$ and $z_j$ are vector of positions for samples $i$ and $j$.

3) Find the optimal monotonic transformation of the input distance, $\theta(d_{ij})$, to approximate input distances to low-dimensional distances. This is achieved by isotonic regression, where a monotonically increasing free-form function is fit. This step practically ensures that ranking of low-dimensional distances are similar to rankings of input distances.

4) Minimize the stress function by re-configuring low-dimensional space and keeping $\theta$ function constant.

5) Repeat from Step 2 until convergence.

We will now demonstrate both classical MDS and Kruskal's isometric MDS.

```
mds=cmdscale(dist(t(mat)))
isomds=MASS::isoMDS(dist(t(mat)))


## initial  value 15.907414
## final  value 13.462986
## converged


# plot the patients in the 2D space
par(mfrow=c(1,2))
plot(mds,pch=19,col=as.factor(annotation_col$LeukemiaType),
     main="classical MDS")
plot(isomds$points,pch=19,col=as.factor(annotation_col$LeukemiaType),
     main="isotonic MDS")
```

The resulting plot is shown in Figure 4.19. In this example, there is not much difference between isotonic MDS and classical MDS. However, there might be cases where different MDS methods provide visible changes in the scatter plots.

### 4.2.4 t-Distributed Stochastic Neighbor Embedding (t-SNE)

t-SNE maps the distances in high-dimensional space to lower dimensions and it is similar to the MDS method in this respect. But the benefit of this particular method is that it tries to preserve the local structure of the data so the distances

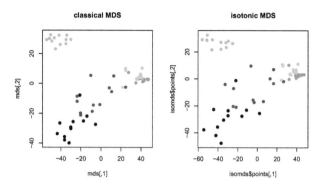

**FIGURE 4.19:** Leukemia gene expression values per patient on reduced dimensions by classical MDS and isometric MDS.

and grouping of the points we observe in lower dimensions such as a 2D scatter plot is as close as possible to the distances we observe in the high-dimensional space (Maaten and Hinton, 2008). As with other dimension reduction methods, you can choose how many lower dimensions you need. The main difference of t-SNE, as mentiones above, is that it tries to preserve the local structure of the data. This kind of local structure embedding is missing in the MDS algorithm, which also has a similar goal. MDS tries to optimize the distances as a whole, whereas t-SNE optimizes the distances with the local structure in mind. This is defined by the "perplexity" parameter in the arguments. This parameter controls how much the local structure influences the distance calculation. The lower the value, the more the local structure is taken into account. Similar to MDS, the process is an optimization algorithm. Here, we also try to minimize the divergence between observed distances and lower dimension distances. However, in the case of t-SNE, the observed distances and lower dimensional distances are transformed using a probabilistic framework with their local variance in mind.

From here on, we will provide a bit more detail on how the algorithm works in case the conceptual description above is too shallow. In t-SNE the Euclidean distances between data points are transformed into a conditional similarity between points. This is done by assuming a normal distribution on each data point with a variance calculated ultimately by the use of the "perplexity" parameter. The perplexity parameter is, in a sense, a guess about the number of the closest neighbors each point has. Setting it to higher values gives more weight to global structure. Given $d_{ij}$ is the Euclidean distance between point $i$ and $j$, the similarity score $p_{ij}$ is calculated as shown below.

$$p_{j|i} = \frac{\exp(-\|d_{ij}\|^2 / 2\sigma_i^2)}{\sum_{k \neq i} \exp(-\|d_{ik}\|^2 / 2\sigma_i^2)}$$

This distance is symmetrized by incorporating $p_{i|j}$ as shown below.

$$p_{ij} = \frac{p_{j|i} + p_{i|j}}{2n}$$

For the distances in the reduced dimension, we use t-distribution with one degree of freedom. In the formula below, $\|y_i - y_j\|^2$ is Euclidean distance between points $i$ and $j$ in the reduced dimensions.

$$q_{ij} = \frac{(1 + \|y_i - y_j\|^2)^{-1}}{(\sum_{k \neq l} 1 + \|y_k - y_l\|^2)^{-1}}$$

As most of the algorithms we have seen in this section, t-SNE is an optimization process in essence. In every iteration the points along lower dimensions are re-arranged to minimize the formulated difference between the observed joint probabilities ($p_{ij}$) and low-dimensional joint probabilities ($q_{ij}$). Here we are trying to compare probability distributions. In this case, this is done using a method called Kullback-Leibler divergence, or KL-divergence. In the formula below, since the $p_{ij}$ is pre-defined using original distances, the only way to optimize is to play with $q_{ij}$ because it depends on the configuration of points in the lower dimensional space. This configuration is optimized to minimize the KL-divergence between $p_{ij}$ and $q_{ij}$.

$$KL(P\|Q) = \sum_{i,j} p_{ij} \log \frac{p_{ij}}{q_{ij}}.$$

Strictly speaking, KL-divergence measures how well the distribution $P$ which is observed using the original data points can be approximated by distribution $Q$, which is modeled using points on the lower dimension. If the distributions are identical, KL-divergence would be 0. Naturally, the more divergent the distributions are, the higher the KL-divergence will be.

We will now show how to use t-SNE on our gene expression data set using the Rtsne package . We are setting the random seed because again, the t-SNE optimization algorithm has random starting points and this might create non-identical results in every run. After calculating the t-SNE lower dimension embeddings we plot the points in a 2D scatter plot, shown in Figure 4.20.

```
library("Rtsne")
set.seed(42) # Set a seed if you want reproducible results
tsne_out <- Rtsne(t(mat),perplexity = 10) # Run TSNE
 #image(t(as.matrix(dist(tsne_out$Y))))
# Show the objects in the 2D tsne representation
```

```
plot(tsne_out$Y,col=as.factor(annotation_col$LeukemiaType),
     pch=19)

# create the legend for the Leukemia types
legend("bottomleft",
       legend=unique(annotation_col$LeukemiaType),
       fill =palette("default"),
       border=NA,box.col=NA)
```

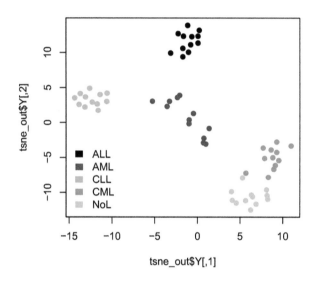

**FIGURE 4.20:** t-SNE of leukemia expression dataset.

As you might have noticed, we set again a random seed with the `set.seed()` function. The optimization algorithm starts with random configuration of points in the lower dimension space, and in each iteration it tries to improve on the previous lower dimension conflagration, which is why starting points can result in different final outcomes.

**– Want to know more ?**

* How perplexity affects t-sne, interactive examples: `https://distill.pub/2016/misread-tsne/`
* More on perplexity: `https://blog.paperspace.com/dimension-reduction-with-t-sne/`
* Intro to t-SNE: `https://www.oreilly.com/learning/an-illustrated-introduction-to-the-t-sne-algorithm`

## 4.3  Exercises

For this set of exercises we will be using the expression data shown below:

```
expFile=system.file("extdata",
                    "leukemiaExpressionSubset.rds",
                    package="compGenomRData")
mat=readRDS(expFile)
```

### 4.3.1  Clustering

1.  We want to observe the effect of data transformation in this exercise. Scale the expression matrix with the `scale()` function. In addition, try taking the logarithm of the data with the `log2()` function prior to scaling. Make box plots of the unscaled and scaled data sets using the `boxplot()` function. [Difficulty: **Beginner/Intermediate**]

2.  For the same problem above using the unscaled data and different data transformation strategies, use the `ward.d` distance in hierarchical clustering and plot multiple heatmaps. You can try to use the `pheatmap` library or any other library that can plot a heatmap with a dendrogram. Which data-scaling strategy provides more homogeneous clusters with respect to disease types? [Difficulty: **Beginner/Intermediate**]

3.  For the transformed and untransformed data sets used in the exercise above, use the silhouette for deciding number of clusters using hierarchical clustering. [Difficulty: **Intermediate/Advanced**]

4.  Now, use the Gap Statistic for deciding the number of clusters in hier-

archical clustering. Is it the same number of clusters identified by two methods? Is it similar to the number of clusters obtained using the k-means algorithm in the chapter. [Difficulty: **Intermediate/Advanced**]

### 4.3.2 Dimension reduction

We will be using the leukemia expression data set again. You can use it as shown in the clustering exercises.

1. Do PCA on the expression matrix using the `princomp()` function and then use the `screeplot()` function to visualize the explained variation by eigenvectors. How many top components explain 95% of the variation? [Difficulty: **Beginner**]

2. Our next tasks are to remove eigenvectors and reconstruct the matrix using SVD, then calculate the reconstruction error as the difference between original and reconstructed matrix. HINT: You have to use the `svd()` function and equalize eigenvalue to 0 for the component you want to remove. [Difficulty: **Intermediate/Advanced**]

3. Produce a 10-component ICA from the expression data set. Remove each component and measure the reconstruction error without that component. Rank the components by decreasing reconstruction-error. [Difficulty: **Advanced**]

4. In this exercise we use the `Rtsne()` function on the leukemia expression data set. Try to increase and decrease perplexity t-sne, and describe the observed changes in 2D plots. [Difficulty: **Beginner**]

# 5

## Predictive Modeling with Supervised Machine Learning

In this chapter we will introduce supervised machine learning applications for predictive modeling. In genomics, we are often faced with biological questions to answer using lots of data. Some of those questions can easily fit in the domain of machine learning, where algorithms will learn a mathematical model of the input data in order to make decisions about similar data, previously unseen by the model. Often we are trying to predict a medical or biological variable of interest using molecular signatures obtained via genomics methods. To give you a better idea, we listed some of the machine learning applications in genomics:

- Predicting gene expression from epigenetic modifications (Dong et al., 2012).
- Predicting gene locations (Mathe et al., 2002).
- Predicting enhancer or other regulatory regions (Fernandez and Miranda-Saavedra, 2012).
- Predicting drug response based on genomics (Wang et al., 2011).
- Predicting healthy/disease status or disease subtypes based on genomics (Kourou et al., 2015).
- Predicting the effect of SNPs on gene regulation (Zhou and Troyanskaya, 2015).
- Calling SNPs (Poplin et al., 2018).

Apart from prediction of an outcome, machine learning can be used to understand which predictor variables are the most important for prediction performance. This often gives insights into the biology as well. Many machine learning algorithms have either built-in variable importance assessment or can be wrapped around a model-agnostic variable importance method. For example, we may want to find which epigenetic modifications are most important for gene expression prediction. Although decades of molecular biology gives a pretty good idea for this, we could arrive at similar conclusions by building a machine learning model to predict gene expression using histone modifications H3K27ac, H3K27me, H3K4me1, H3K4me3, and DNA methylation. We can then check which of these are most important for gene expression prediction using variable importance metrics.

In this chapter, we will show how to use supervised machine learning models to solve problems in genomics. We will go over general steps in machine learning applications. In addition, we will introduce how to use some of the most popular supervised machine learning models in practice.

## 5.1   How are machine learning models fit?

We already have quite an insight on how machine learning models are fit. We have previously seen clustering methods, which are unsupervised machine learning models, and we have seen linear regression which is a simple machine learning model if we disregard its objectives for statistical inference.

Machine learning models are optimization methods at their core. They all depend on defining a "cost" or "loss" function to minimize. For example, in linear regression the difference between the predicted and the original values are being minimized. When we have a data set with the correct answer such as original values or class labels, this is called supervised learning. We use the structure in the data to predict a value, and optimization methods help us use the right structure or patterns in the data. The supervised machine learning methods use predictor variables such as gene expression values or other genomic scores to build a mathematical function, or a mapping method if you will. This function maps a predictor variable vector or matrix from a given sample to the response variable: labels/classes or numeric values. The response variable is also called the "dependent variable". Then, the predictions are simply output of mathematical functions, $f(X)$. These functions take predictor variables, $X$, as input. The variables in $X$ are also called "independent variables", "explanatory variables" or "features". The functions also have internal parameters that help map $X$ to the predicted values. The optimization works on the parameters of $f(X)$ and tries to minimize the difference between the function output and original response variables $(Y)$: $\sum(Y - f(X))^2$. Now, this is just a simplification of the actual "cost" or "loss" function. Especially in classification problems, cost functions can take different forms, but the idea is the same. You have a mathematical expression you can minimize by searching for the optimal parameter values. The core ingredients of a machine learning algorithm are the same and they are listed as follows:

1) Define a prediction function or method $f(X)$.
2) Devise a function (called the loss or cost function) to optimize the difference between your predictions and observed values, such as $\sum(Y - f(X))^2$.
3) Apply mathematical optimization methods to find the best parameter values for $f(X)$ in relation to the cost/loss function.

Similarly, clustering and dimension reduction techniques can use optimization methods, but they do so without having a correct answer to predict or train with. In this case, they find patterns or structure in the data without trying to estimate a correct answer. These patterns are groupings of samples or variables, such as common gene expression patterns, that can be obtained from dimension reduction techniques such as PCA. In general, dimension reduction algorithms can be thought of as optimization procedures that are trying to minimize $X - WH$. Here, $X$ is our original data set and $WH$ is the product of potentially two lower dimension matrices, $W$ and $H$. In this case, the optimization procedure hopefully gives us the lower-dimensional space so that we can represent our data without losing too much information.

### 5.1.1 Machine learning vs. statistics

Machine learning and statistics are related and sometimes overlapping fields. Statistical inference is the main purpose of statistics. The aim of inference is to find statistical properties of the underlying data and to estimate the uncertainty about those properties. However, while doing so, the field of statistics developed dimension reduction and regression techniques that are the cornerstone of machine learning applications.

Both machine learning and statistics share the same overarching goal, which is learning from the data. The difference between the two is that machine learning emphasizes optimization and performance over statistical inference. Statistics is also concerned about performance but would like to calculate the uncertainty associated with parameters of the model. It will try to model the population statistics from the sample data points to assess that uncertainty. Having said that, many machine learning algorithms, including a couple we will introduce below, are developed by scientists who will define themselves as statisticians, and work at statistics departments of universities.

## 5.2 Steps in supervised machine learning

There are many methods to use for supervised learning problems. However, there are similar steps that you will need to follow whatever machine learning method you choose to train. These steps are briefly described below and we will get back to these in detail later in the chapter:

- Pre-processing data: We might have to use normalization and data transformation procedures.
- Training and test data split: Decide which strategy you want to use for evaluation purposes. You need to use a test set to evaluate your model later on.
- Training the model: This is where your choice of supervised learning algorithm becomes relevant. "Training" generally means your data set is used in optimization of the loss function to find parameters for $f(x)$.

- Estimating performance of the model: This is about which metrics to use to evaluate performance and how to calculate those metrics.
- Model tuning and selection: We try different parameters and select the best model.

Many of these steps are identical for different supervised learning methods. Therefore, we will use the caret[1] package to perform these steps, which streamlines the steps and provides a similar interface for different supervised learning methods. There are other similar packages, such as mlr[2], that can provide similar functionality. For now, we will focus on classification models, which is a subset of supervised learning models. In these types of models, we try to predict a categorical response variable, such as if a patient has the disease or not, or what type of disease the patient has based on predictor variables.

---

## 5.3  Use case: Disease subtype from genomics data

We will start our illustration of machine learning using a real dataset from tumor biopsies. We will use the gene expression data of glioblastoma tumor samples from The Cancer Genome Atlas project. We will try to predict the subtype of this disease using molecular markers. This subtype is characterized by large-scale epigenetic alterations called the "CpG island methylator phenotype" or "CIMP" (Noushmehr et al., 2010); half of the patients in our data set have this subtype and the rest do not, and we will try to predict which ones have the CIMP subtype. There two data objects we need for this exercise, one for gene expression values per tumor sample and the other one is subtype annotation per patient. In the expression data set, every row is a patient and every column is a gene expression value. There are 184 tumor samples. This data set might be a bit small for real-world applications, however it is very relevant for the genomics focus of this book and the small datasets take less time to train, which is useful for reproducibility purposes. We will read these data sets from the **compGenomRData** package now with the readRDS() function.

```
# get file paths
fileLGGexp=system.file("extdata",
                       "LGGrnaseq.rds",
                       package="compGenomRData")
fileLGGann=system.file("extdata",
                       "patient2LGGsubtypes.rds",
```

---

[1]http://topepo.github.io/caret/index.html
[2]https://mlr.mlr-org.com/

```
                                package="compGenomRData")
# gene expression values
gexp=readRDS(fileLGGexp)
head(gexp[,1:5])
```

```
##          TCGA-CS-4941 TCGA-CS-4944 TCGA-CS-5393 TCGA-CS-5394 TCGA-CS-5395
## A1BG         72.2326       24.7132       46.3789       37.9659       19.5162
## A1CF          0.0000        0.0000        0.0000        0.0000        0.0000
## A2BP1       524.4997      105.4092      323.5828       19.7390      299.5375
## A2LD1       144.0856       18.0154       29.0942        7.5945      202.1231
## A2ML1       521.3941      159.3746      164.6157       63.5664      953.4106
## A2M       17944.7205    10894.9590    16480.1130     9217.7919    10801.8461
```

```
dim(gexp)
```

```
## [1] 20501    184
```

```
# patient annotation
patient=readRDS(fileLGGann)
head(patient)
```

```
##               subtype
## TCGA-FG-8185     CIMP
## TCGA-DB-5276     CIMP
## TCGA-P5-A77X     CIMP
## TCGA-IK-8125     CIMP
## TCGA-DU-A5TR     CIMP
## TCGA-E1-5311     CIMP
```

```
dim(patient)
```

```
## [1] 184    1
```

---

## 5.4 Data preprocessing

We will have to preprocess the data before we start training. This might include exploratory data analysis to see how variables and samples relate to each other.

For example, we might want to check the correlation between predictor variables and keep only one variable from that group. In addition, some training algorithms might be sensitive to data scales or outliers. We should deal with those issues in this step. In some cases, the data might have missing values. We can choose to remove the samples that have missing values or try to impute them. Many machine learning algorithms will not be able to deal with missing values.

We will show how to do this in practice using the `caret::preProcess()` function and base R functions. Please note that there are more preprocessing options available than we will show here. There are more possibilities in `caret::preProcess()` function and base R functions, we are just going to cover a few basics in this section.

### 5.4.1   Data transformation

The first thing we will do is data normalization and transformation. We have to take care of data scale issues that might come from how the experiments are performed and the potential problems that might occur during data collection. Ideally, each tumor sample has a similar distribution of gene expression values. Systematic differences between tumor samples must be corrected. We check if there are such differences using box plots. We will only plot the first 50 tumor samples so that the figure is not too squished. The resulting boxplot is shown in Figure 5.1.

```
boxplot(gexp[,1:50],outline=FALSE,col="cornflowerblue")
```

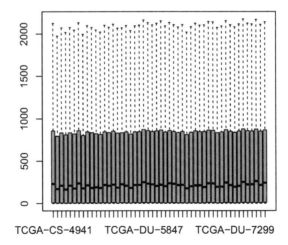

**FIGURE 5.1:** Boxplots for gene expression values.

It seems there was some normalization done on this data. Gene expression values per sample seem to have the same scale. However, it looks like they have long-tailed distributions, so a log transformation may fix that. These long-tailed distributions have outliers and this might adversely affect the models. Below, we show the effect of log transformation on the gene expression profile of a patient. We add a pseudo count of 1 to avoid log(0). The resulting histograms are shown in Figure 5.2.

```
par(mfrow=c(1,2))
hist(gexp[,5],xlab="gene expression",main="",border="blue4",
    col="cornflowerblue")
hist(log10(gexp+1)[,5], xlab="gene expression log scale",main="",
    border="blue4",col="cornflowerblue")
```

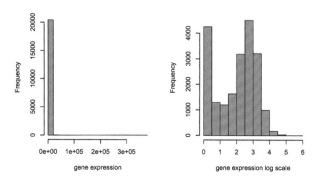

**FIGURE 5.2:** Gene expression distribution for the 5th patient (left). Log transformed gene expression distribution for the same patient (right).

Since taking a log seems to work to tame the extreme values, we do that below and also add 1 pseudo-count to be able to deal with 0 values:

```
gexp=log10(gexp+1)
```

Another thing we can do in combination with this is to winsorize the data, which caps extreme values to the 1st and 99th percentiles or to other user-defined percentiles. But before we go forward, we should transpose our data. In this case, the predictor variables are gene expression values and they should be on the column side. It was OK to leave them on the row side, to check systematic errors with box plots, but machine learning algorithms require that predictor variables are on the column side.

```
# transpose the data set
tgexp <- t(gexp)
```

### 5.4.2 Filtering data and scaling

We can filter predictor variables which have low variation. They are not likely to have any predictive importance since there is not much variation and they will just slow our algorithms. The more variables, the slower the algorithms will be generally. The `caret::preProcess()` function can help filter the predictor variables with near zero variance.

```
library(caret)
# remove near zero variation for the columns at least
# 85% of the values are the same
# this function creates the filter but doesn't apply it yet
nzv=preProcess(tgexp,method="nzv",uniqueCut = 15)

# apply the filter using "predict" function
# return the filtered dataset and assign it to nzv_tgexp
# variable
nzv_tgexp=predict(nzv,tgexp)
```

In addition, we can also choose arbitrary cutoffs for variability. For example, we can choose to take the top 1000 variable predictors.

```
SDs=apply(tgexp,2,sd )
topPreds=order(SDs,decreasing = TRUE)[1:1000]
tgexp=tgexp[,topPreds]
```

We can also center the data, which as we have seen in Chapter 4, is subtracting the mean. Following this, the predictor variables will have zero means. In addition, we can scale the data. When we scale, each value of the predictor variable is divided by its standard deviation. Therefore predictor variables will have the same standard deviation. These manipulations are generally used to improve the numerical stability of some calculations. In distance-based metrics, it could be beneficial to at least center the data. We will now center the data using the `preProcess()` function. This is more practical than the `scale()` function because when we get a new data point, we can use the `predict()` function and `processCenter` object to process it just like we did for the training samples.

```
library(caret)
processCenter=preProcess(tgexp, method = c("center"))
tgexp=predict(processCenter,tgexp)
```

We will next filter the predictor variables that are highly correlated. You may choose not to do this as some methods can handle correlation between predictor variables. However, the fewer predictor variables we have, the faster the model fitting can be done.

```
# create a filter for removing higly correlated variables
# if two variables are highly correlated only one of them
# is removed
corrFilt=preProcess(tgexp, method = "corr",cutoff = 0.9)
tgexp=predict(corrFilt,tgexp)
```

### 5.4.3 Dealing with missing values

In real-life situations, there will be missing values in our data. In genomics, we might not have values for certain genes or genomic locations due to technical problems during experiments. We have to be able to deal with these missing values. For demonstration purposes, we will now introduce NA values in our data, the "NA" value is normally used to encode missing values in R. We then show how to check and deal with those. One way is to impute them; here, we again use a machine learning algorithm to guess the missing values. Another option is to discard the samples with missing values or discard the predictor variables with missing values. First, we replace one of the values as NA and check if it is there.

```
missing_tgexp=tgexp
missing_tgexp[1,1]=NA
anyNA(missing_tgexp) # check if there are NA values
```

```
## [1] TRUE
```

Next, we will try to remove that gene from the set. Removing genes or samples both have downsides. You might be removing a predictor variable that could be important for the prediction. Removing samples with missing values will decrease the number of samples in the training set. The code below checks which values are NA in the matrix, then runs a column sum and keeps everything where the column sum is equal to 0. The column sums where there are NA values will be higher than 0 depending on how many NA values there are in a column.

```
gexpnoNA=missing_tgexp[ , colSums(is.na(missing_tgexp)) == 0]
```

We will next try to impute the missing value(s). Imputation can be as simple as assigning missing values to the mean or median value of the variable, or assigning the mean/median of values from nearest neighbors of the sample having the missing value. We will show both using the `caret::preProcess()` function. First, let us run the median imputation.

```
library(caret)
mImpute=preProcess(missing_tgexp,method="medianImpute")
imputedGexp=predict(mImpute,missing_tgexp)
```

Another imputation method that is more precise than the median imputation is to impute the missing values based on the nearest neighbors of the samples. In this case, the algorithm finds samples that are most similar to the sample vector with NA values. Next, the algorithm averages the non-missing values from those neighbors and replaces the missing value with that value.

```
library(RANN)
knnImpute=preProcess(missing_tgexp,method="knnImpute")
knnimputedGexp=predict(knnImpute,missing_tgexp)
```

## 5.5  Splitting the data

At this point we might choose to split the data into the test and the training partitions. The reason for this is that we need an independent test we did not train on. This will become clearer in the following sections, but without having a separate test set, we cannot assess the performance of our model or tune it properly.

### 5.5.1  Holdout test dataset

There are multiple data split strategies. For starters, we will split 30% of the data as the test. This method is the gold standard for testing performance of our model. By doing this, we have a separate data set that the model has never seen. First, we create a single data frame with predictors and response variables.

```
tgexp=merge(patient,tgexp,by="row.names")

# push sample ids back to the row names
rownames(tgexp)=tgexp[,1]
tgexp=tgexp[,-1]
```

Now that the response variable or the class label is merged with our dataset, we can split it into test and training sets with the `caret::createPartition()` function.

```
set.seed(3031) # set the random number seed for reproducibility

# get indices for 70% of the data set
intrain <- createDataPartition(y = tgexp[,1], p= 0.7)[[1]]

# seperate test and training sets
training <- tgexp[intrain,]
testing <- tgexp[-intrain,]
```

### 5.5.2   Cross-validation

In some cases, we might have too few data points and it might be too costly to set aside a significant portion of the data set as a holdout test set. In these cases a resampling-based technique such as cross-validation may be useful.

Cross-validation works by splitting the data into randomly sampled $k$ subsets, called k-folds. So, for example, in the case of 5-fold cross-validation with 100 data points, we would create 5 folds, each containing 20 data points. We would then build models and estimate errors 5 times. Each time, four of the groups are combined (resulting in 80 data points) and used to train your model. Then the 5th group of 20 points that was not used to construct the model is used to estimate the test error. In the case of 5-fold cross-validation, we would have 5 error estimates that could be averaged to obtain a more robust estimate of the test error.

An extreme case of k-fold cross-validation, is to equalize the $k$ to the number of data points or in our case, the number of tumor samples. This is called leave-one-out cross-validation (LOOCV). This could be better than k-fold cross-validation but it takes too much time to train that many models if the number of data points is large.

The `caret` package has built-in cross-validation functionality for all the machine learning methods and we will be using that in the later sections.

### 5.5.3   Bootstrap resampling

Another method to estimate the prediction error is to use bootstrap resampling. This is a general method we have already introduced in Chapter 3. It can be used to estimate variability of any statistical parameter. In this case, that parameter is the test error or test accuracy.

The training set is drawn from the original set with replacement (same size as the original set), then we build a model with this bootstrap resampled set. Next, we take the data points that are not selected for the random sample and predict labels for them. These data points are called the "out-of-the-bag (OOB) sample". We repeat this process many times and record the error for the OOB samples. We can take the average of the OOB errors to estimate the real test error. This is a powerful method that is not only used to estimate test error but incorporated into the training part of some machine learning methods such as random forests. Normally, we should repeat the process hundreds or up to a thousand times to get good estimates. However, the limiting factor would be the time it takes to construct and test that many models. Twenty to 30 repetitions might be enough if the time cost of training is too high. Again, the `caret` package provides the bootstrap interface for many machine learning models for sampling before training and estimating the error on OOB samples.

## 5.6   Predicting the subtype with k-nearest neighbors

One of the easiest things to wrap our heads around when we are trying to predict a label such as disease subtype is to look for similar samples and assign the labels of those similar samples to our sample.

Conceptually, k-nearest neighbors (k-NN) is very similar to clustering algorithms we have seen earlier. If we have a measure of distance between the samples, we can find the nearest $k$ samples to our new sample and use a voting method to decide on the label of our new sample.

Let us run the k-NN algorithm with our cancer data. For illustrative purposes, we provide the same data set for training and test data. Providing the training data as the test data shows us the training error or accuracy, which is how the model is doing on the data it is trained with. Below we are running k-NN with the `caret:knn3()` function. The most important argument is `k`, which is the number of nearest neighbors to consider. In this case, we set it to 5. We will later discuss how to find the best `k`.

```
library(caret)
knnFit=knn3(x=training[,-1], # training set
            y=training[,1], # training set class labels
            k=5)
# predictions on the test set
trainPred=predict(knnFit,training[,-1])
```

## 5.7 Assessing the performance of our model

We have to define some metrics to see if our model worked. The algorithm is trying to reduce the classification error, or in other words it is trying to increase the training accuracy. For the assessment of performance, there are other different metrics to consider. All the metrics for 2-class classification depend on the table below, which shows the number of true positives (TP), false positives (FP), true negatives (TN) and false negatives (FN), similar to a table we used in the hypothesis testing section in the statistics chapter previously.

|  | Actual CIMP | Actual noCIMP |
| --- | --- | --- |
| Predicted as CIMP | True Positives (TP) | False Positive (FP) |
| Predicted as noCIMP | False Positives (FN) | True negatives (TN) |

Accuracy is the first metric to look at. This metric is simply $(TP + TN)/(TP + TN + FP + FN)$ and shows the proportion of times we were right. There are other accuracy metrics that are important and output by caret functions. We will go over some of them here.

Precision, $TP/(TP + FP)$, is about the confidence we have on our CIMP calls. If our method is very precise, we will have low false positives. That means every time we call a CIMP event, we would be relatively certain it is not a false positive.

Sensitivity, $TP/(TP + FN)$, is how often we miss CIMP cases and call them as noCIMP. Making fewer mistakes in noCIMP cases will increase our sensitivity. You can think of sensitivity also in a sick/healthy context. A highly sensitive method will be good at classifying sick people when they are indeed sick.

Specificity, $TN/(TN + FP)$, is about how sure we are when we call something noCIMP. If our method is not very specific, we would call many patients CIMP, while in fact, they did not have the subtype. In the sick/healthy context, a highly specific method will be good at not calling healthy people sick.

An alternative to accuracy we showed earlier is "balanced accuracy". Accuracy does not perform well when classes have very different numbers of samples (class imbalance). For example, if you have 90 CIMP cases and 10 noCIMP cases, classifying all the samples as CIMP gives 0.9 accuracy score by default. Using the "balanced accuracy" metric can help in such situations. This is simply $(Precision + Sensitivity)/2$. In this case above with the class imbalance scenario, the "balanced accuracy" would be 0.5. Another metric that takes into account accuracy that could be generated by chance is the "Kappa statistic" or "Cohen's Kappa". This metric includes expected accuracy, which is affected by class imbalance in the training set and provides a metric corrected by that.

In the k-NN example above, we trained and tested on the same data. The model returned the predicted labels for our training. We can calculate the accuracy metrics using the `caret::confusionMatrix()` function. This is sometimes called training accuracy. If you take $1 - accuracy$, it will be the "training error".

```
# get k-NN prediction on the training data itself, with k=5
knnFit=knn3(x=training[,-1], # training set
            y=training[,1], # training set class labels
            k=5)


# predictions on the training set
trainPred=predict(knnFit,training[,-1],type="class")


# compare the predicted labels to real labels
# get different performance metrics
confusionMatrix(data=training[,1],reference=trainPred)


## Confusion Matrix and Statistics
##
##           Reference
## Prediction CIMP noCIMP
##     CIMP    65     0
##     noCIMP   2    63
##
##                Accuracy : 0.9846
##                  95% CI : (0.9455, 0.9981)
##     No Information Rate : 0.5154
##     P-Value [Acc > NIR] : <2e-16
##
```

```
##                          Kappa : 0.9692
##
##     Mcnemar's Test P-Value : 0.4795
##
##                    Sensitivity : 0.9701
##                    Specificity : 1.0000
##                 Pos Pred Value : 1.0000
##                 Neg Pred Value : 0.9692
##                     Prevalence : 0.5154
##                 Detection Rate : 0.5000
##        Detection Prevalence : 0.5000
##           Balanced Accuracy : 0.9851
##
##               'Positive' Class : CIMP
##
```

Now, let us see what our test set accuracy looks like again using the knn function and the confusionMatrix() function on the predicted and real classes.

```
# predictions on the test set, return class labels
testPred=predict(knnFit,testing[,-1],type="class")

# compare the predicted labels to real labels
# get different performance metrics
confusionMatrix(data=testing[,1],reference=testPred)
```

```
## Confusion Matrix and Statistics
##
##               Reference
## Prediction CIMP noCIMP
##      CIMP      27       0
##      noCIMP     2      25
##
##                    Accuracy : 0.963
##                      95% CI : (0.8725, 0.9955)
##        No Information Rate : 0.537
##        P-Value [Acc > NIR] : 2.924e-12
##
##                       Kappa : 0.9259
##
##     Mcnemar's Test P-Value : 0.4795
##
```

```
##             Sensitivity : 0.9310
##             Specificity : 1.0000
##          Pos Pred Value : 1.0000
##          Neg Pred Value : 0.9259
##              Prevalence : 0.5370
##          Detection Rate : 0.5000
##    Detection Prevalence : 0.5000
##        Balanced Accuracy : 0.9655
##
##         'Positive' Class : CIMP
##
```

Test set accuracy is not as good as the training accuracy, which is usually the case. That is why the best way to evaluate performance is to use test data that is not used by the model for training. That gives you an idea about real-world performance where the model will be used to predict data that is not previously seen.

### 5.7.1   Receiver Operating Characteristic (ROC) curves

One important and popular metric when evaluating performance is looking at receiver operating characteristic (ROC) curves. The ROC curve is created by evaluating the class probabilities for the model across a continuum of thresholds. Typically, in the case of two-class classification, the methods return a probability for one of the classes. If that probability is higher than $0.5$, you call the label, for example, class A. If less than $0.5$, we call the label class B. However, we can move that threshold and change what we call class A or B. For each candidate threshold, the resulting sensitivity and 1-specificity are plotted against each other. The best possible prediction would result in a point in the upper left corner, representing 100% sensitivity (no false negatives) and 100% specificity (no false positives). For the best model, the curve will be almost like a square. Since this is important information, area under the curve (AUC) is calculated. This is a quantity between 0 and 1, and the closer to 1, the better the performance of your classifier in terms of sensitivity and specificity. For an uninformative classification model, AUC will be $0.5$. Although, ROC curves are initially designed for two-class problems, later extensions made it possible to use ROC curves for multi-class problems.

ROC curves can also be used to determine alternate cutoffs for class probabilities for two-class problems. However, this will always result in a trade-off between sensitivity and specificity. Sometimes it might be desirable to limit the number of false positives because making such mistakes would be too costly for the individual cases. For example, if predicted with a certain disease, you might be recommended to have surgery. However, if your classifier has a relatively high false positive rate, low specificity, you might have surgery for no reason. Typically, you want your classification model to have high specificity and sensitivity, which may not always be possible in the real world. You might have to choose what is more important for a specific problem and try to increase that.

Next, we will show how to use ROC curves for our k-NN application. The method requires classification probabilities in the format where 0 probability denotes class "noCIMP" and probability 1 denotes class "CIMP". This way the ROC curve can be drawn by varying the probability cutoff for calling class a "noCIMP" or "CIMP". Below we are getting a similar probability from k-NN, but we have to transform it to the format we described above. Then, we feed those class probabilities to the pROC::roc() function to calculate the ROC curve and the area-under-the-curve. The resulting ROC curve is shown in Figure 5.3.

```r
library(pROC)

# get k-NN class probabilities
# prediction probabilities on the test set
testProbs=predict(knnFit,testing[,-1])

# get the roc curve
rocCurve <- pROC::roc(response = testing[,1],
              predictor = testProbs[,1],
           ## This function assumes that the second
           ## class is the class of interest, so we
           ## reverse the labels.
           levels = rev(levels(testing[,1])))
# plot the curve
plot(rocCurve, legacy.axes = TRUE)


# return area under the curve
pROC::auc(rocCurve)

## Area under the curve: 0.976
```

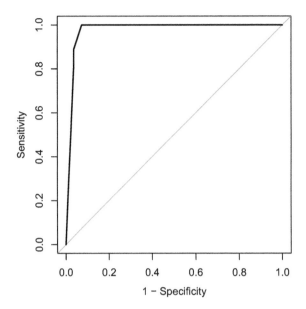

**FIGURE 5.3:** ROC curve for k-NN.

## 5.8    Model tuning and avoiding overfitting

How can we know that we picked the best $k$? One straightforward way is that we can try many different $k$ values and check the accuracy of our model. We will first check the effect of different $k$ values on training accuracy. Below, we will go through many $k$ values and calculate the training accuracy for each.

```
set.seed(101)
k=1:12 # set k values
trainErr=c() # set vector for training errors
for( i in k){

  knnFit=knn3(x=training[,-1], # training set
             y=training[,1], # training set class labels
             k=i)

  # predictions on the training set
  class.res=predict(knnFit,training[,-1],type="class")
```

```
# training error
err=1-confusionMatrix(training[,1],class.res)$overall[1]
trainErr[i]=err
}

# plot training error vs k
plot(k,trainErr,type="p",col="#CC0000",pch=20)

# add a smooth line for the trend
lines(loess.smooth(x=k, trainErr,degree=2),col="#CC0000")
```

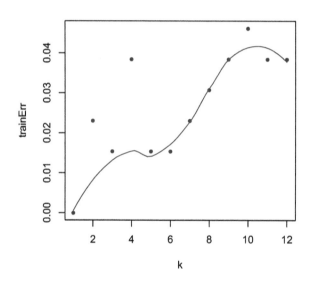

**FIGURE 5.4:** Training error for k-NN classification of glioma tumor samples.

The resulting training error plot is shown in Figure 5.4. We can see the effect of $k$ in the training error; as $k$ increases the model tends to be a bit worse on training. This makes sense because with large $k$ we take into account more and more neighbors, and at some point we start considering data points from the other classes as well and that decreases our accuracy.

However, looking at the training accuracy is not the right way to test the model as we have mentioned. The models are generally tested on the datasets that are not used when building model. There are different strategies to do this. We have already split part of our dataset as test set, so let us see how we do it on the test data using the code below. The resulting plot is shown in Figure 5.5.

```r
set.seed(31)
k=1:12
testErr=c()
for( i in k){

  knnFit=knn3(x=training[,-1], # training set
              y=training[,1], # training set class labels
              k=i)

  # predictions on the training set
  class.res=predict(knnFit,testing[,-1],type="class")
  testErr[i]=1-confusionMatrix(testing[,1],
                               class.res)$overall[1]

}


# plot training error
plot(k,trainErr,type="p",col="#CC0000",
     ylim=c(0.000,0.08),
     ylab="prediction error (1-accuracy)",pch=19)
# add a smooth line for the trend
lines(loess.smooth(x=k, trainErr,degree=2), col="#CC0000")

# plot test error
points(k,testErr,col="#00CC66",pch=19)
lines(loess.smooth(x=k,testErr,degree=2), col="#00CC66")
# add legend
legend("bottomright",fill=c("#CC0000","#00CC66"),
       legend=c("training","test"),bty="n")
```

The test data show a different thing, of course. It is not the best strategy to increase the $k$ indefinitely. The test error rate increases after a while. Increasing $k$ results in too many data points influencing the decision about the class of the new sample, this may not be desirable since this strategy might include points from other classes eventually. On the other hand, if we set $k$ too low, we are restricting the model to only look for a few neighbors.

In addition, $k$ values that give the best performance for the training set are not the best $k$ for the test set. In fact, if we stick with $k = 1$ as the best $k$ obtained from the training set, we would obtain a worse performance on the test set. In this case, we can talk about the concept of overfitting. This happens when our models fit the data in the training set extremely well but cannot perform well in the test data;

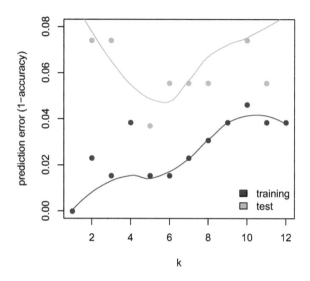

**FIGURE 5.5:** Training and test error for k-NN classification of glioma tumor samples.

in other words, they cannot generalize. Similarly, underfitting could occur when our models do not learn well from the training data and they are overly simplistic. Ideally, we should use methods that help us estimate the real test error when tuning the models such as cross-validation, bootstrap or holdout test set.

### 5.8.1 Model complexity and bias variance trade-off

The case of over- and underfitting is closely related to the model complexity and the related bias-variance trade-off. We will introduce these concepts now. First, let us point out that prediction error depends on the real value of the class label of the test case and predicted value. The test case label or value is not dependent on the prediction; the only thing that is variable here is the model. Therefore, if we could train multiple models with different data sets for the same problem, our predictions for the test set would vary. That means our prediction error would also vary. Now, with this setting we can talk about expected prediction error for a given machine learning model. This is the average error you would get for a test set if you were able to train multiple models. This expected prediction error can largely be decomposed into the variability of the predictions due to the model variability (variance) and the difference between the expected prediction values and the correct value of the response (bias). Formally, the expected prediction error, $E[Error]$ is decomposed as follows:

$$E[Error] = Bias^2 + Variance + \sigma_e^2$$

Note that in the above equation $\sigma_e^2$ is the irreducible error. This is the noise term that cannot fundamentally be accounted for by any model. The bias is formally the difference between the expected prediction value and the correct response value, $Y$: $Bias = (Y - E[PredictedValue])$. The variance is simply the variability of the prediction values when we construct models multiple times with different training sets for the same problem: $Variance = E[(PredictedValue - E[PredictedValue])^2]$. Note that this value of the variance does not depend of the correct value of the test cases.

The models that have high variance are generally more complex models that have many knobs or parameters than can fit the training data well. These models, due to their flexibility, can fit training data too much that it creates poor prediction performance in a new data set. On the other hand, simple, less complex models do not have the flexibility to fit every data set that well, so they can avoid overfitting. However, they can underfit if they are not flexible enough to model or at least approximate the true relationship between predictors and the response variable. The bias term is mostly about the general model performance (expected or average value of predictions) that can be attributed to approximating a real-life problem with simpler models. These simple models can have less variability in their predictions, so the prediction error will be mostly composed of the bias term.

In reality, there is always a trade-off between bias and variance (See Figure 5.6). Increasing the variance with complex models will decrease the bias, but that might overfit. Conversely, simple models will increase the bias at the expense of the model variance, and that might underfit. There is an optimal point for model complexity, a balance between overfitting and underfitting. In practice, there is no analytical way to find this optimal complexity. Instead we must use an accurate measure of prediction error and explore different levels of model complexity and choose the complexity level that minimizes the overall error. Another approach to this is to use "the one standard error rule". Instead of choosing the parameter that minimizes the error estimate, we can choose the simplest model whose error estimate is within one standard error of the best model (see Chapter 7 of (Friedman et al., 2001)). The rationale behind that is to choose a simple model with the hope that it would perform better in the unseen data since its performance is not different from the best model in a statistically significant way. You might see the option to choose the "one-standard-error" model in some machine learning packages.

In our k-NN example, lower $k$ values create a more flexible model. This might be counterintuitive, but as we have explained before having small $k$ values will fit the data in a very data-specific manner. It will probably not generalize well. Therefore in this respect, lower $k$ values will result in more complex models with high variance. On the other hand, higher $k$ values will result in less variance but higher bias. Figure

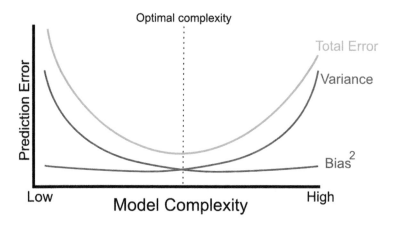

**FIGURE 5.6:** Variance-bias trade-off visualized as components of total prediction error in relation to model complexity.

5.7 shows the decision boundary for two different k-NN models with $k = 2$ and $k = 12$. To be able to plot this in 2D we ran the model on principal component 1 and 2 of the training data set, and predicted the class label of many points in this 2D space. As you can see, $k = 2$ creates a more variable model which tries aggressively to include all training samples in the correct class. This creates a high-variance model because the model could change drastically from dataset to dataset. On the other hand, setting $k = 12$ creates a model with a smoother decision boundary. This model will have less variance since it considers many points for a decision, and therefore the decision boundary is smoother.

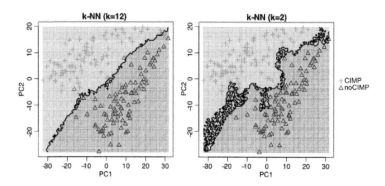

**FIGURE 5.7:** Decision boundary for different k values in k-NN models. k=12 creates a smooth decision boundary and ignores certain data points on either side of the boundary. k=2 is less smooth and more variable.

### 5.8.2    Data split strategies for model tuning and testing

The data split strategy is essential for accurate prediction of the test error. As we have seen in the model complexity/bias-variance discussion, estimating the prediction error is central for model tuning in order to find the model with the right complexity. Therefore, we will revisit this and show how to build and test models, and measure their prediction error in practice.

#### 5.8.2.1    Training-validation-test

This data split strategy creates three partitions of the dataset, training, validation, and test sets. In this strategy, the training set is used to train the data and the validation set is used to tune the model to the best possible model. The final partition, "test", is only used for the final test and should not be used to tune the model. This is regarded as the real-world prediction error for your model. This strategy works when you have a lot of data to do a three-way split. The test set we used above is most likely too small to measure the prediction error with just using a test set. In such cases, bootstrap or cross-validation should yield more stable results.

#### 5.8.2.2    Cross-validation

A more realistic approach when you do not have a lot of data to do the three-way split is cross-validation. You can use cross-validation in the model-tuning phase as well, instead of going with a single train-validation split. As with the three-way split, the final prediction error could be estimated with the test set. In other words, we can separate 80% of the data for model building with cross-validation, and the final model performance will be measured on the test set.

We have already split our glioma dataset into training and test sets. Now, we will show how to run a k-NN model with cross-validation using the `caret::train()` function. This function will use cross-validation to train models for different $k$ values. Every $k$ value will be trained and tested with cross-validation to estimate prediction performance for each $k$. We will then plot the cross-validation error and the resulting plot is shown in Figure 5.8.

```
set.seed(17)
# this method controls everything about training
# we will just set up 10-fold cross validation
trctrl <- trainControl(method = "cv",number=10)

# we will now train k-NN model
knn_fit <- train(subtype~., data = training,
                 method = "knn",
                 trControl=trctrl,
```

```
          tuneGrid = data.frame(k=1:12))

# best k value by cross-validation accuracy
knn_fit$bestTune

##   k
## 4 4

# plot k vs prediction error
plot(x=1:12,1-knn_fit$results[,2],pch=19,
     ylab="prediction error",xlab="k")
lines(loess.smooth(x=1:12,1-knn_fit$results[,2],degree=2),
      col="#CC0000")
```

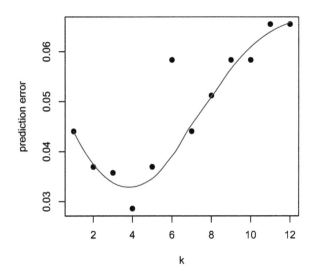

**FIGURE 5.8:** Cross-validated estimate of prediction error of k in k-NN models.

Based on Figure 5.8 the cross-validation accuracy reveals that $k = 5$ is the best $k$ value. On the other hand, we can also try bootstrap resampling and check the prediction error that way. We will again use the caret::trainControl() function to do the bootstrap sampling and estimate OOB-based error. However, for a small number of samples like we have in our example, the difference between the estimated and the true value of the prediction error can be large. Below we show how to use bootstrapping for the k-NN model.

```
set.seed(17)
# this method controls everything about training
# we will just set up 100 bootstrap samples and for each
# bootstrap OOB samples to test the error
trctrl <- trainControl(method = "boot",number=20,
                         returnResamp="all")


# we will now train k-NN model
knn_fit <- train(subtype~., data = training,
              method = "knn",
              trControl=trctrl,
              tuneGrid = data.frame(k=1:12))
```

## 5.9   Variable importance

Another important purpose of machine learning models could be to learn which variables are more important for the prediction. This information could lead to potential biological insights or could help design better data collection methods or experiments.

Variable importance metrics can be separated into two groups: those that are model dependent and those that are not. Many machine-learning methods come with built-in variable importance measures. These may be able to incorporate the correlation structure between the predictors into the importance calculation. Model-independent methods are not able to use any internal model data. We will go over some model-independent strategies below. The model-dependent importance measures will be mentioned when we introduce machine learning methods that have built-in variable importance measures.

One simple method for variable importance is to correlate or apply statistical tests to test the association of the predictor variable with the response variable. Variables can be ranked based on the strength of those associations. For classification problems, ROC curves can be computed by thresholding the predictor variable, and for each variable an AUC can be computed. The variables can be ranked based on these values. However, these methods completely ignore how variables would behave in the presence of other variables. The caret::filterVarImp() function implements some of these strategies.

If a variable is important for prediction, removing that variable before model training will cause a drop in performance. With this understanding, we can remove the variables one by one and train models without them and rank them by the loss of performance. The most important variables must cause the largest loss of performance. This strategy requires training and testing models as many times as the number of predictor variables. This will consume a lot of time. A related but more practical approach has been put forward to measure variable importance in a model-independent manner but without re-training (Biecek, 2018; Fisher et al., 2018). In this case, instead of removing the variables at training, variables are permuted at the test phase. The loss in prediction performance is calculated by comparing the labels/values from the original response variable to the labels/values obtained by running the permuted test data through the model. This is called "variable dropout loss". In this case, we are not really dropping out variables, but by permuting them, we destroy their relationship to the response variable. The dropout loss is compared to the "worst case" scenario where the response variable is permuted and compared against the original response variables, which is called "baseline loss". The algorithm ranks the variables by their variable dropout loss or by their ratio of variable dropout to baseline loss. Both quantities are proportional but the second one contains information about the baseline loss. Below, we run the DALEX::explain() function to do the permutation drop-out strategy for the variables. The function needs the machine learning model, and new data and its labels to do the permutation-based dropout strategy. In this case, we are feeding the function with the data we used for training. For visualization we can use the DALEX::feature_importance() function which plots the loss. Although, in this case we are not plotting the results. In the following sections, we will discuss method-specific variable importance measures.

```
library(DALEX)
set.seed(102)
# do permutation drop-out
explainer_knn<- DALEX::explain(knn_fit,
                      label="knn",
                      data =training[,-1],
                      y = as.numeric(training[,1]))

viknn=feature_importance(explainer_knn,n_sample=50,type="difference")
plot(viknn)
```

Although the variable drop-out strategy will still be slow if you have a lot of variables, the upside is that you can use any black-box model as long as you have access to the model to run new predictions. Later sections in this chapter will show methods with built-in variable importance metrics, since these are calculated during training it comes with less of an additional compute cost.

## 5.10   How to deal with class imbalance

A common hurdle in many applications of machine learning on genomic data is the large class imbalance. The imbalance refers to relative difference in the sizes of the groups being classified. For example, if we had class imbalance in our example data set we could have much more CIMP samples in the training than noCIMP samples, or the other way around. Another example with severe class imbalance would be enhancer prediction (Libbrecht and Noble, 2015). Depending on which training data set you use, you can have a couple of hundred to thousands of positive examples for enhancer locations in the human genome. In either case, the negative set, "not enhancer", will overwhelm the training, because the human genome is 3 billion base-pairs long and most of that does not overlap with an enhancer annotation. In whatever strategy you pick to build a negative set, it will contain many more data points than the positive set. As we have mentioned in the model performance section above, if we have a severe imbalance in the class sizes, the training algorithm may get better accuracy just by calling everything one class. This will be evident in specificity and sensitivity metrics, and the related balanced accuracy metric. Below, we will discuss a couple of techniques that might help when the training set has class imbalance.

### 5.10.1   Sampling for class balance

If we think class imbalance is a problem based on looking at the relative sizes of the classes and relevant accuracy metrics of a model, there are a couple of things that might help. First, we can try sampling or "stratified" sampling when we are constructing our training set. This simply means that before training we can we build the classification model with samples of the data so we have the same size classes. This could be down-sampling the classes with too many data points. For this purpose, you can simply use the `sample()` or `caret::downSample()` function and create your training set prior to modeling. In addition, the minority class could be up-sampled for the missing number of data points using sampling with replacement similar to bootstrap sampling with the `caret::upSample()` function. There are more advanced up-sampling methods such as the synthetic up-sampling method SMOTE (Chawla et al., 2002). In this method, each data point from the minority class is up-sampled synthetically by adding variability to the predictor variable vector from one of the k-nearest neighbors of the data point. Specifically, one neighbor is randomly chosen and the difference between predictor variables of the neighbor and the original data point is added to the original predictor variables after multiplying the difference values with a random number between 0 and 1. This creates synthetic data points that are similar to original data points but not identical. This method and other similar methods of synthetic sampling are available at smotefamily[3] package in CRAN.

In addition to the strategies above, some methods can do sampling during training to cope with the effects of class imbalance. For example, random forests has a sampling step during training, and this step can be altered to do stratified sampling. We will be introducing random forests later in the chapter.

However, even if we are doing the sampling on the training set to avoid problems, the test set proportions should have original class label proportions to evaluate the performance in a real-world situation.

### 5.10.2 Altering case weights

For some methods, we can use different case weights proportional to the imbalance suffered by the minority class. This means cases from the minority class will have higher case weights, which causes an effect as if we are up-sampling the minority class. Logistic regression-based methods and boosting methods are examples of algorithms that can utilize case weights, both of which will be introduced later.

### 5.10.3 Selecting different classification score cutoffs

Another simple approach for dealing with class imbalance is to select a prediction score cutoff that minimizes the excess true positives or false positives depending on the direction of the class imbalance. This can simply be done using ROC curves. For example, the classical prediction cutoff for a 2-class classification problems is 0.5. We can alter this cutoff to optimize sensitivity and specificity.

---

## 5.11 Dealing with correlated predictors

Highly correlated predictors can lead to collinearity issues and this can greatly increase the model variance, especially in the context of regression. In some cases, there could be relationships between multiple predictor variables and this is called multicollinearity. Having correlated variables will result in unnecessarily complex models with more than necessary predictor variables. From a data collection point of view, spending time and money for collecting correlated variables could be a waste of effort. In terms of linear regression or the models that are based on regression, the collinearity problem is more severe because it creates unstable models where statistical inference becomes difficult or unreliable. On the other hand, correlation between variables may not be a problem for the predictive performance if the

---

[3]https://cran.r-project.org/web/packages/smotefamily/index.html

correlation structure in the training and the future tests data sets are the same. However, more often, correlated structures within the training set might lead to overfitting.

Here are couple of things to do if collinearity is a problem:

- We can do PCA on the training data, which creates new variables removing the collinearity between them. We can then train models on these new dimensions. The downside is that it is harder to interpret these variables. They are now linear combinations of original variables. The variable importance would be harder to interpret.

- As we have already shown in the data preprocessing section, we can try variable filtering and reduce the number of correlated variables. However, this may not address the multicollinearity issue where linear combinations of variables might be correlated while they are not directly correlated themselves.

- Method-specific techniques such as regularization can decrease the effects of collinearity. Regularization, as we will see in the later chapter, is a technique that is used to prevent overfitting and it can also dampen the effects of collinearity. In addition, decision-tree-based methods could suffer less from the effects of collinearity.

## 5.12   Trees and forests: Random forests in action

### 5.12.1   Decision trees

Decision trees are a popular method for various machine learning tasks mostly because their interpretability is very high. A decision tree is a series of filters on the predictor variables. The series of filters end up in a class prediction. Each filter is a binary yes/no question, which creates bifurcations in the series of filters thus leading to a treelike structure. The filters are dependent on the type of predictor variables. If the variables are categorical, such as gender, then the filters could be "is gender female" type of questions. If the variables are continuous, such as gene expression, the filter could be "is PIGX expression larger than 210?". Every point where we filter samples based on these questions are called "decision nodes". The tree-fitting algorithm finds the best variables at decision nodes depending on how well they split the samples into classes after the application of the decision node. Decision trees handle both categorical and numeric predictor variables, they are easy to interpret, and they can deal with missing variables. Despite their advantages, decision trees tend to overfit if they are grown very deep and can learn irregular patterns. There are many variants of tree-based machine learning algorithms. However, most algorithms construct decision nodes in a top down manner. They select the best variables to use in decision nodes based on how homogeneous the sample sets are after the split. One measure of homogeneity is "Gini impurity". This measure is calculated for each subset after the split and later summed up as a weighted average. For a decision node that splits the data perfectly in a two-class problem, the gini impurity will be 0, and for a node that splits the data into a subset that has 50% class A and 50% class B the impurity will be 0.5. Formally, the gini impurity, $I_G(p)$, of a set of samples with known class labels for $K$ classes is the following, where $p_i$ is the probability of observing class $i$ in the subset:

$$I_G(p) = \sum_{i=1}^{K} p_i(1 - p_i) = \sum_{i=1}^{K} p_i - \sum_{i=1}^{K} p_i^2 = 1 - \sum_{i=1}^{K} p_i^2$$

For example, if a subset of data after split has 75% class A and 25% class B for that subset, the impurity would be $1 - (0.75^2 + 0.25^2) = 0.375$. If the other subset had 5% class A and 95% class B, its impurity would be $1 - (0.95^2 + 0.05^2) = 0.095$. If the subset sizes after the split were equal, total weighted impurity would be $0.5 * 0.375 + 0.5 * 0.095 = 0.235$. These calculations will be done for each potential variable and the split, and every node will be constructed based on gini impurity decrease. If the variable is continuous, the cutoff value will be decided based on the best impurity. For example, gene expression values will have splits such as "PIGX expression < 2.1". Here $2.1$ is the cutoff value that produces the best impurity. There are other homogeneity measures, however gini impurity is the one that is used for random forests, which we will introduce next.

### 5.12.2 Trees to forests

Random forests are devised to counter the shortcomings of decision trees. They are simply ensembles of decision trees. Each tree is trained with a different randomly selected part of the data with randomly selected predictor variables. The goal of introducing randomness is to reduce the variance of the model so it does not overfit, at the expense of a small increase in the bias and some loss of interpretability. This strategy generally boosts the performance of the final model.

The random forests algorithm tries to decorrelate the trees so that they learn different things about the data. It does this by selecting a random subset of variables. If one or a few predictor variables are very strong predictors for the response variable, these features will be selected in many of the trees, causing them to become correlated. Random subsampling of predictor variables ensures that not always the best predictors overall are selected for every tree, and, the model does have a chance to learn other features of the data.

Another sampling method introduced when building random forest models is bootstrap resampling before constructing each tree. This brings the advantage of out-of-the-bag (OOB) error prediction. In this case, the prediction error can be estimated for training samples that were OOB, meaning they were not used in the training, for some percentage of the trees. The prediction error for each sample can be estimated from the trees where that sample was OOB. OOB estimates claimed to be a good alternative to cross-validation estimated errors (Breiman, 2001).

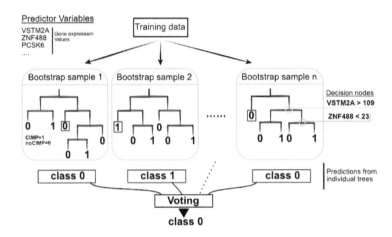

**FIGURE 5.9:** Random forest concept. Individual decision trees are built with sampling strategies. Votes from each tree define the final class.

For demonstration purposes, we will use the `caret` package interface to the `ranger` random forest package. This is a fast implementation of the original random forest algorithm. For random forests, we have two critical arguments. One of the most critical arguments for random forest is the number of predictor variables to sample in each split of the tree. This parameter controls the independence between the trees, and as explained before, this limits overfitting. Below, we are going to fit a random forest model to our tumor subtype problem. We will set `mtry=100` and not perform the training procedure to find the best `mtry` value for simplicity. However, it is good practice to run the model with cross-validation and let it pick the best parameters based on the cross-validation performance. It defaults to the square root of number of predictor variables. Another variable we can tune is the minimum node size of terminal nodes in the trees (`min.node.size`). This controls the depth of the trees grown. Setting this to larger numbers might cost a small loss in accuracy but the algorithm will run faster.

```
set.seed(17)

# we will do no resampling based prediction error
```

```
# although it is advised to do so even for random forests
trctrl <- trainControl(method = "none")

# we will now train random forest model
rfFit <- train(subtype~.,
               data = training,
               method = "ranger",
               trControl=trctrl,
               importance="permutation", # calculate importance
               tuneGrid = data.frame(mtry=100,
                                     min.node.size = 1,
                                     splitrule="gini")
               )
# print OOB error
rfFit$finalModel$prediction.error
```

```
## [1] 0.01538462
```

### 5.12.3  Variable importance

Random forests come with built-in variable importance metrics. One of the metrics is similar to the "variable dropout metric" where the predictor variables are permuted. In this case, OOB samples are used and the variables are permuted one at a time. Every time, the samples with the permuted variables are fed to the network and the decrease in accuracy is measured. Using this quantity, the variables can be ranked.

A less costly method with similar performance is to use gini impurity. Every time a variable is used in a tree to make a split, the gini impurity is less than the parent node. This method adds up these gini impurity decreases for each individual variable across the trees and divides it by the number of the trees in the forest. This metric is often consistent with the permutation importance measure (Breiman, 2001). Below, we are going to plot the permutation-based importance metric. This metric has been calculated during the run of the model above. We will use the `caret::varImp()` function to access the importance values and plot them using the `plot()` function; the result is shown in Figure 5.10.

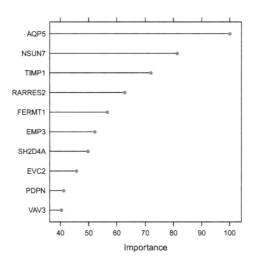

**FIGURE 5.10:** Top 10 important variables based on permutation-based method for the random forest classification.

```
plot(varImp(rfFit),top=10)
```

## 5.13  Logistic regression and regularization

Logistic regression is a statistical method that is used to model a binary response variable based on predictor variables. Although initially devised for two-class or binary response problems, this method can be generalized to multiclass problems. However, our example tumor sample data is a binary response or two-class problem, therefore we will not go into the multiclass case in this chapter.

Logistic regression is very similar to linear regression as a concept and it can be thought of as a "maximum likelihood estimation" problem where we are trying to find statistical parameters that maximize the likelihood of the observed data being sampled from the statistical distribution of interest. This is also very related to the general cost/loss function approach we see in supervised machine learning algorithms. In the case of binary response variables, the simple linear regression model, such as $y_i \sim \beta_0 + \beta_1 x_i$, would be a poor choice because it can easily generate values outside of the 0 to 1 boundary. What we need is a model that restricts the lower bound of the prediction to zero and an upper bound to 1. The first thing towards this requirement is to formulate the problem differently. If $y_i$ can only be 0 or 1, we can formulate $y_i$ as a realization of a random variable that can take the values one and zero with probabilities $p_i$ and $1 - p_i$, respectively. This random variable follows the Bernoulli distribution, and instead of predicting the binary variable we can formulate the problem as $p_i \sim \beta_0 + \beta_1 x_i$. However, our initial problem still stands, simple linear regression will still result in values that are beyond 0 and 1 boundaries. A model that satisfies the boundary requirement is the logistic equation shown below.

$$p_i = \frac{e^{(\beta_0 + \beta_1 x_i)}}{1 + e^{(\beta_0 + \beta_1 x_i)}}$$

This equation can be linearized by the following transformation

$$\text{logit}(p_i) = \ln\left(\frac{p_i}{1 - p_i}\right) = \beta_0 + \beta_1 x_i$$

The left-hand side is termed the logit, which stands for "logistic unit". It is also known as the log odds. In this case, our model will produce values on the log scale and with the logistic equation above, we can transform the values to the $0 - 1$ range. Now, the question remains: "What are the best parameter estimates for our training set". Within the maximum likelihood framework we have touched upon in Chapter 3, the best parameter estimates are the ones that maximize the likelihood of the statistical model actually producing the observed data. You can think of this fitting as a probability distribution to an observed data set. The parameters of the probability distribution should maximize the likelihood that the observed data came from the distribution in question. If we were using a Gaussian distribution we would change the mean and variance parameters until the observed data was more plausible to be drawn from that specific Gaussian distribution.

In logistic regression, the response variable is modeled with a binomial distribution or its special case Bernoulli distribution. The value of each response variable, $y_i$, is 0 or 1, and we need to figure out parameter $p_i$ values that could generate such a distribution of 0s and 1s. If we can find the best $p_i$ values for each tumor sample $i$, we would be maximizing the log-likelihood function of the model over the observed data. The maximum log-likelihood function for our binary response variable case is shown as Equation (5.1).

$$\ln(L) = \sum_{i=1}^{N}\left[\ln(1 - p_i) + y_i \ln\left(\frac{p_i}{1 - p_i}\right)\right] \tag{5.1}$$

In order to maximize this equation we have to find optimum $p_i$ values which are dependent on parameters $\beta_0$ and $\beta_1$, and also dependent on the values of predictor variables $x_i$. We can rearrange the equation replacing $p_i$ with the logistic equation.

In addition, many optimization functions minimize rather than maximize. Therefore, we will be using negative log likelihood, which is also called the "log loss" or
"logistic loss" function. The function below is the "log loss" function. We substituted
$p_i$ with the logistic equation and simplified the expression.

$$\mathrm{L}_{log} = -\ln(L) = -\sum_{i=1}^{N} \left[ -\ln(1 + e^{(\beta_0 + \beta_1 x_i)}) + y_i \left( \beta_0 + \beta_1 x_i \right) \right] \quad (5.2)$$

Now, let us see how this works in practice. First, as in the example above we will
use one predictor variable, the expression of one gene to classify tumor samples to
"CIMP" and "noCIMP" subtypes. We will be using PDPN gene expression, which
was one of the most important variables in our random forest model. We will use
the formula interface in `caret`, where we will supply the names of the response and
predictor variables in a formula. In this case, we will be using a core R function,
`glm()`, from the `stats` package. "glm" stands for generalized linear models, and it is
the main interface for different types of regression in R.

```
# fit logistic regression model
# method and family defines the type of regression
# in this case these arguments mean that we are doing logistic
# regression
lrFit = train(subtype ~ PDPN,
              data=training, trControl=trainControl("none"),
              method="glm", family="binomial")

# create data to plot the sigmoid curve
newdat <- data.frame(PDPN=seq(min(training$PDPN),
                      max(training$PDPN),len=100))

# predict probabilities for the simulated data
newdat$subtype = predict(lrFit, newdata=newdat, type="prob")[,1]

# plot the sigmoid curve and the training data
plot(ifelse(subtype=="CIMP",1,0) ~ PDPN,
     data=training, col="red4",
     ylab="subtype as 0 or 1", xlab="PDPN expression")
lines(subtype ~ PDPN, newdat, col="green4", lwd=2)
```

Figure 5.11 shows the sigmoidal curve that is fitted by the logistic regression.
"noCIMP" subtype has higher expression of the PDPN gene than the "CIMP" subtype. In other words, the higher the values of PDPN, the more likely that the tumor
sample will be classified as "noCIMP". We can also assess the performance of our

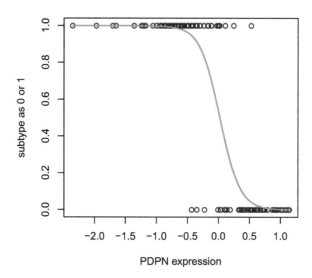

**FIGURE 5.11:** Sigmoid curve for prediction of subtype based on one predictor variable.

model with the test set and the training set. Let us try to do that again with the `caret::predict()` and `caret::confusionMatrix()` functions.

```
# training accuracy
class.res=predict(lrFit,training[,-1])
confusionMatrix(training[,1],class.res)$overall[1]
```

```
##  Accuracy
## 0.9461538
```

```
# test accuracy
class.res=predict(lrFit,testing[,-1])
confusionMatrix(testing[,1],class.res)$overall[1]
```

```
##  Accuracy
## 0.9259259
```

The test accuracy is slightly worse than the training accuracy. Overall this is not as good as k-NN, but remember we used only one predictor variable. We have thousands of genes as predictor variables. Now we will try to use all of them in the classification problem. After fitting the model, we will check training and test accuracy. We fit the model again with the `caret::train()` function.

```
lrFit2 = train(subtype ~ .,
                data=training,
                # no model tuning with sampling
                trControl=trainControl("none"),
                method="glm", family="binomial")
```

```
# training accuracy
class.res=predict(lrFit2,training[,-1])
confusionMatrix(training[,1],class.res)$overall[1]
```

```
## Accuracy
##         1
```

```
# test accuracy
class.res=predict(lrFit2,testing[,-1])
confusionMatrix(testing[,1],class.res)$overall[1]
```

```
##   Accuracy
## 0.4259259
```

Training accuracy is 1, so training error is 0, and nothing is misclassified in the training set. However, test accuracy/error is close to terrible. It does only little better than a random guess. If we randomly assigned class labels we would get 0.5 accuracy. The test set accuracy is 0.55 despite the 100% training accuracy. This is because the model overfits to the training data. There are too many variables in the model. The number of predictor variables is ~6.5 times more than the number of samples. The excess of predictor variables makes the model very flexible (high variance), and this leads to overfitting.

### 5.13.1 Regularization in order to avoid overfitting

If we can limit the flexibility of the model, this might help with performance on the unseen, new data sets. Generally, any modification of the learning method to improve performance on the unseen datasets is called regularization. We need regularization to introduce bias to the model and to decrease the variance. This can be achieved by modifying the loss function with a penalty term which effectively shrinks the estimates of the coefficients. Therefore these types of methods within the framework of regression are also called "shrinkage" methods or "penalized regression" methods.

One way to ensure shrinkage is to add the penalty term, $\lambda \sum \beta_j^2$, to the loss function. This penalty term is also known as the L2 norm or L2 penalty. It is calculated as the square root of the sum of the squared vector values. This term will help shrink the coefficients in the regression towards zero. The new loss function is as follows, where $j$ is the number of parameters/coefficients in the model and $L_{log}$ is the log loss function in Eq. (5.2).

$$L_{log} + \lambda \sum_{j=1}^{p} \beta_j^2 \qquad (5.3)$$

This penalized loss function is called "ridge regression" (Hoerl and Kennard, 1970). When we add the penalty, the only way the optimization procedure keeps the overall loss function minimum is to assign smaller values to the coefficients. The $\lambda$ parameter controls how much emphasis is given to the penalty term. The higher the $\lambda$ value, the more coefficients in the regression will be pushed towards zero. However, they will never be exactly zero. This is not desirable if we want the model to select important variables. A small modification to the penalty is to use the absolute values of $B_j$ instead of squared values. This penalty is called the "L1 norm" or "L1 penalty". The regression method that uses the L1 penalty is known as "Lasso regression" (Tibshirani, 1996).

$$L_{log} + \lambda \sum_{j=1}^{p} |\beta_j|$$

However, the L1 penalty tends to pick one variable at random when predictor variables are correlated. In this case, it looks like one of the variables is not important although it might still have predictive power. The Ridge regression on the other hand shrinks coefficients of correlated variables towards each other, keeping all of them. It has been shown that both Lasso and Ridge regression have their drawbacks and advantages (Friedman et al., 2010). More recently, a method called "elastic net" was proposed to include the best of both worlds (Zou and Hastie, 2005). This method uses both L1 and L2 penalties. The equation below shows the modified loss function by this penalty. As you can see the $\lambda$ parameter still controls the weight that is given to the penalty. This time the additional parameter $\alpha$ controls the weight given to L1 or L2 penalty and it is a value between 0 and 1.

$$L_{log} + \lambda \sum_{j=1}^{p} (\alpha \beta_j^2 + (1 - \alpha)|\beta_j|)$$

We have now got the concept behind regularization and we can see how it works in practice. We are going to use elastic net on our tumor subtype prediction problem. We will let cross-validation select the best $\lambda$ and we will fix the $\alpha$ parameter at 0.5.

```
set.seed(17)
library(glmnet)

# this method controls everything about training
# we will just set up 10-fold cross validation
trctrl <- trainControl(method = "cv",number=10)

# we will now train elastic net model
# it will try
enetFit <- train(subtype~., data = training,
                method = "glmnet",
                trControl=trctrl,
                # alpha and lambda paramters to try
                tuneGrid = data.frame(alpha=0.5,
                                    lambda=seq(0.1,0.7,0.05)))

# best alpha and lambda values by cross-validation accuracy
enetFit$bestTune

##   alpha lambda
## 1   0.5    0.1

# test accuracy
class.res=predict(enetFit,testing[,-1])
confusionMatrix(testing[,1],class.res)$overall[1]

##  Accuracy
## 0.9814815
```

As you can see regularization worked, the tuning step selected $\lambda = 1$, and we were able to get a satisfactory test set accuracy with the best model.

### 5.13.2  Variable importance

The variable importance of the penalized regression, especially for lasso and elastic net, is more or less out of the box. As discussed, these methods will set regression coefficients for irrelevant variables to zero. This provides a system for selecting important variables but it does not necessarily provide a way to rank them. Using the size of the regression coefficients is a way to rank predictor variables, however if the data is not normalized, you will get different scales for different variables. In our case, we normalized the data and we know that the variables have the same scale before they went into the training. We can use this fact and rank them based on the regression coefficients. The `caret::varImp()` function uses the coefficients to rank the variables from the elastic net model. Below, were going to plot the top 10 important variables which are normalized to the importance of the most important variable.

```
plot(varImp(enetFit),top=10)
```

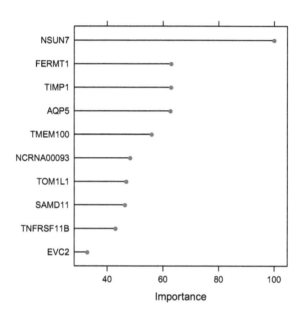

**FIGURE 5.12:** Variable importance metric for elastic net. This metric uses regression coefficients as importance.

– **Want to know more ?**

  * Lecture by Trevor Hastie on regularized regression. You probably need to understand the basics of regression and its terminology to follow this. However, the lecture is not very heavy on math. `https://youtu.be/BU2gjoLPfDc`

## 5.14   Other supervised algorithms

We will next introduce a couple of other supervised algorithms for completeness but in less detail. These algorithms are also as popular as the others we introduced above and people who are interested in computational genomics see them used in the field for different problems. These algorithms also fit to the general framework of optimization of a cost/loss function. However, the approaches to the construction of the cost function and the cost function itself are different in each case.

### 5.14.1   Gradient boosting

Gradient boosting is a prediction model that uses an ensemble of decision trees similar to random forest. However, the decision trees are added sequentially, which is why these models are also called "Multiple Additive Regression Trees (MART)" (Friedman and Meulman, 2003). Apart from this, you will see similar methods called "Gradient boosting machines (GBM)"(Friedman, 2001) or "Boosted regression trees (BRT)" (Elith et al., 2008) in the literature.

Generally, "boosting" refers to an iterative learning approach where each new model tries to focus on data points where the previous ensemble of simple models did not predict well. Gradient boosting is an improvement over that, where each new model tries to focus on the residual errors (prediction error for the current ensemble of models) of the previous model. Specifically in gradient boosting, the simple models are trees. As in random forests, many trees are grown but in this case, trees are sequentially grown and each tree focuses on fixing the shortcomings of the previous trees. Figure 5.13 shows this concept. One of the most widely used algorithms for gradient boosting is xGboost which stands for "extreme gradient boosting" (Chen and Guestrin, 2016). Below we will demonstrate how to use this on our problem. xGboost as well as other gradient boosting methods has many parameters to regularize and optimize the complexity of the model. Finding the best parameters for your problem might take some time. However, this flexibility comes with benefits; methods depending on xGboost have won many machine learning competitions (Chen and Guestrin, 2016).

The most important parameters are number of trees (nrounds), tree depth (max_depth), and learning rate or shrinkage (eta). Generally, the more trees we have, the better the algorithm will learn because each tree tries to fix classification errors that the previous tree ensemble could not perform. Having too many trees might cause overfitting. However, the learning rate parameter, eta, combats that by shrinking the contribution of each new tree. This can be set to lower values if you have many trees. You can either set a large number of trees and then tune the model with the learning rate parameter or set the learning rate low, say to 0.01 or

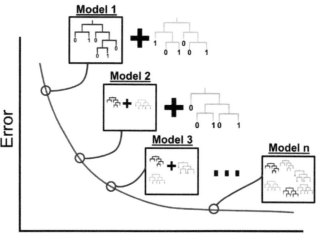

**FIGURE 5.13:** Gradient boosting machines concept. Individual decision trees are built sequentially in order to fix the errors from the previous trees.

0.1 and tune the number of trees. Similarly, tree depth also controls for overfitting. The deeper the tree, the more usually it will overfit. This has to be tuned as well; the default is at 6. You can try to explore a range around the default. Apart from these, as in random forests, you can subsample the training data and/or the predictive variables. These strategies can also help you counter overfitting.

We are now going to use xGboost with the caret package on our cancer subtype classification problem. We are going to try different learning rate parameters. In this instance, we also subsample the dataset before we train each tree. The "subsample" parameter controls this and we set this to be 0.5, which means that before we train a tree we will sample 50% of the data and use only that portion to train the tree.

```
library(xgboost)
set.seed(17)

# we will just set up 5-fold cross validation
trctrl <- trainControl(method = "cv",number=5)

# we will now train elastic net model
# it will try
gbFit <- train(subtype~., data = training,
                method = "xgbTree",
                trControl=trctrl,
```

```
# paramters to try
tuneGrid = data.frame(nrounds=200,
                      eta=c(0.05,0.1,0.3),
                      max_depth=4,
                      gamma=0,
                      colsample_bytree=1,
                      subsample=0.5,
                      min_child_weight=1))
```

```
# best parameters by cross-validation accuracy
gbFit$bestTune
```

```
##    nrounds max_depth eta gamma colsample_bytree min_child_weight subsample
## 2      200         4 0.1     0                1                1       0.5
```

Similar to random forests, we can estimate the variable importance for gradient boosting using the improvement in gini impurity or other performance-related metrics every time a variable is selected in a tree. Again, the `caret::varImp()` function can be used to plot the importance metrics.

– **Want to know more ?**

* More background on gradient boosting and XGboost: (`https://xgboost.readthedocs.io/en/latest/tutorials/model.html`). This explains the cost/loss function and regularization in more detail.
* Lecture on Gradient boosting and random forests by Trevor Hastie: (`https://youtu.be/wPqtzj5VZus`)

### 5.14.2   Support Vector Machines (SVM)

Support vector machines (SVM) were popularized in the 90s due the efficiency and the performance of the algorithm (Boser et al., 1992). The algorithm works by identifying the optimal decision boundary that separates the data points into different groups (or classes), and then predicts the class of new observations based on this separation boundary. Depending on the situation, the different groups might be separable by a linear straight line or by a non-linear boundary line or plane. If you review k-NN decision boundaries in Figure 5.7, you can see that the decision boundary is not linear. SVM can deal with linear or non-linear decision boundaries.

First, SVM can map the data to higher dimensions where the decision boundary can be linear. This is achieved by applying certain mathematical functions, called "kernel functions", to the predictor variable space. For example, a second-degree polynomial can be applied to predictor variables which creates new variables and in this new space the problem is linearly separable. Figure 5.14 demonstrates this concept where points in feature space are mapped to quadratic space where linear separation is possible.

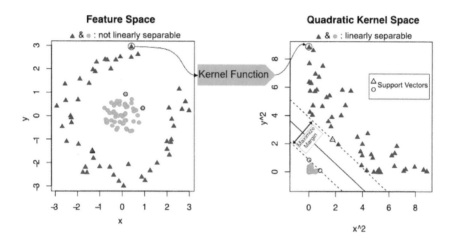

**FIGURE 5.14:** Support vector machine concept. With the help of a kernel function, points in feature space are mapped to higher dimensions where linear separation is possible.

Second, SVM not only tries to find a decision boundary, but tries to find the boundary with the largest buffer zone on the sides of the boundary. Having a boundary with a large buffer or "margin", as it is formally called, will perform better for the new data points not used in the model training (margin is marked in Figure 5.14 ). In addition, SVM calculates the decision boundary with some error toleration. As we have seen it may not always be possible to find a linear boundary that perfectly separates the classes. SVM tolerates some degree of error, as in data points on the wrong side of the decision boundary.

Another important feature of the algorithm is that SVM decides on the decision boundary by only relying on the "landmark" data points, formally known as "support vectors". These are points that are closest to the decision boundary and harder to classify. By keeping track of such points only for decision boundary creation, the computational complexity of the algorithm is reduced. However, this depends on the margin or the buffer zone. If we have a large margin then there are many landmark points. The extent of the margin is also related to the variance-bias trade-off. If the allowed margin is small the classification will try to find a boundary that makes fewer errors in the training set therefore might overfit. If the margin is larger, it will tolerate more errors in the training set and might generalize better. Practically, this is controlled by the "C" or "Cost" parameter in the SVM example we will show below. Another important choice we will make is the kernel function. Below we use the radial basis kernel function. This function provides an extra predictor dimension where the problem is linearly separable. The model we will use has only one parameter, which is "C". It is recommended that $C$ is in the form of $2^k$ where $k$ is in the range of -5 and 15 (Hsu et al., 2003). Another parameter that can be tuned is related to the radial basis function called "sigma". A smaller sigma means less bias and more variance, while a larger sigma means less variance and more bias. Again, exponential sequences are recommended for tuning that (Hsu et al., 2003). We will set it to 1 for demonstration purposes below.

```r
#svm code here
library(kernlab)
set.seed(17)


# we will just set up 5-fold cross validation
trctrl <- trainControl(method = "cv",number=5)


# we will now train elastic net model
# it will try
svmFit <- train(subtype~., data = training,
                # this SVM used radial basis function
                 method = "svmRadial",
                 trControl=trctrl,
                 tuneGrid=data.frame(C=c(0.25,0.5,1),
                                    sigma=1))
```

– **Want to know more ?**

* MIT lecture by Patrick Winston on SVM: https://youtu.be/_PwhiWxHK8o. This lecture explains the concept with some mathematical background. It is not hard to follow. You should be able to follow this if you know what vectors are and if you have some knowledge on derivatives and basic algebra.
* Online demo for SVM: (https://cs.stanford.edu/people/karpathy/svmjs/demo/). You can play with sigma and C parameters for radial basis SVM and see how they affect the decision boundary.

### 5.14.3 Neural networks and deep versions of it

Neural networks are another popular machine learning method which is recently regaining popularity. The earlier versions of the algorithm were popularized in the 80s and 90s. The advantage of neural networks is like SVM, they can model non-linear decision boundaries. The basic idea of neural networks is to combine the predictor variables in order to model the response variable as a non-linear function. In a neural network, input variables pass through several layers that combine the variables and transform those combinations and recombine outputs depending on how many layers the network has. In the conceptual example in Figure 5.15 the input nodes receive predictor variables and make linear combinations of them in the form of $\sum(w_i x i + b)$. Simply put, the variables are multiplied with weights and summed up. This is what we call "linear combination". These quantities are further fed into another layer called the hidden layer where an activation function is applied on the sums. And these results are further fed into an output node which outputs class probabilities assuming we are working on a classification algorithm. There could be many more hidden layers that will even further combine the output from hidden layers before them. The algorithm in the end also has a cost function similar to the logistic regression cost function, but it now has to estimate all the weight parameters: $w_i$. This is a more complicated problem than logistic regression because of the number of parameters to be estimated but neural networks are able to fit complex functions due their parameter space flexibility as well.

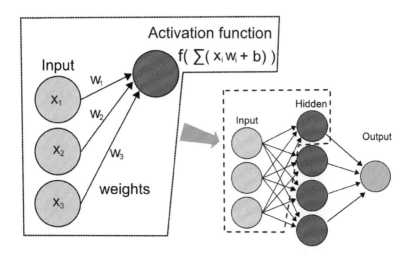

**FIGURE 5.15:** Diagram for a simple neural network, their combinations pass through hidden layers and are combined again for the output. Predictor variables are fed to the network and weights are adjusted to optimize the cost function.

In a practical sense, the number of nodes in the hidden layer (size) and some regularization on the weights can be applied to control for overfitting. This is called the calculated (decay) parameter controls for overfitting.

We will train a simple neural network on our cancer data set. In this simple example, the network architecture is somewhat fixed. We can only the choose number of nodes (denoted by "size") in the hidden layer and a regularization parameter (denoted by "decay"). Increasing the number of nodes in the hidden layer or in other implementations increasing the number of the hidden layers, will help model non-linear relationships but can overfit. One way to combat that is to limit the number of nodes in the hidden layer; another way is to regularize the weights. The decay parameter does just that, it penalizes the loss function by $decay(weigths^2)$. In the example below, we try 1 or 2 nodes in the hidden layer in the interest of simplicity and run-time. In addition, we set decay=0, which will correspond to not doing any regularization.

```
#svm code here
library(nnet)
set.seed(17)


# we will just set up 5-fold cross validation
trctrl <- trainControl(method = "cv",number=5)


# we will now train neural net model
# it will try
nnetFit <- train(subtype~., data = training,
                method = "nnet",
                trControl=trctrl,
                tuneGrid=data.frame(size=1:2,decay=0
                              ),
                # this is maximum number of weights
                # needed for the nnet method
                MaxNWts=2000)
```

The example we used above is a bit outdated. The modern "deep" neural networks provide much more flexibility in the number of nodes, number of layers and regularization options. In many areas, especially computer vision deep neural networks are the state-of-the-art (LeCun et al., 2015). These modern implementations of neural networks are available in R via the keras package and can also be trained via the caret package with the similar interface we have shown until now.

**– Want to know more ?**

      \* Deep neural networks in R: (https://keras.rstudio.com/). There are examples and background information on deep neural networks.

      \* Online demo for neural networks: (https://cs.stanford.edu/~karpathy/svmjs/demo/demonn.html). You can see the effect of the number of hidden layers and number of nodes on the decision boundary.

### 5.14.4 Ensemble learning

Ensemble learning models are simply combinations of different machine learning models. By now, we already introduced the concept of ensemble learning in random forests and gradient boosting. However, this concept can be generalized to combining all kinds of different models. "Random forests" is an ensemble of the same type of models, decision trees. We can also have ensembles of different types of models. For example, we can combine random forest, k-NN and elastic net models, and make class predictions based on the votes from those different models. Below, we are showing how to do this. We are going to get predictions for three different models on the test set, use majority voting to decide on the class label, and then check performance using `caret::confusionMatrix()`.

```
# predict with k-NN model
knnPred=as.character(predict(knnFit,testing[,-1],type="class"))
# predict with elastic Net model
enetPred=as.character(predict(enetFit,testing[,-1]))
# predict with random forest model
rfPred=as.character(predict(rfFit,testing[,-1]))

# do voting for class labels
# code finds the most frequent class label per row
votingPred=apply(cbind(knnPred,enetPred,rfPred),1,
                 function(x) names(which.max(table(x))))

# check accuracy
confusionMatrix(data=testing[,1],
                reference=as.factor(votingPred))$overall[1]
```

```
##   Accuracy
## 0.9814815
```

In the test set, we were able to obtain perfect accuracy after voting. More complicated and accurate ways to build ensembles exist. We could also use the mean of class probabilities instead of voting for final class predictions. We can even combine models in a regression-based scheme to assign weights to the votes or to the predicted class probabilities of each model. In these cases, the prediction performance of the ensembles can also be tested with sampling techniques such as cross-validation. You can think of this as another layer of optimization or modeling for combining results from different models. We will not pursue this further in this chapter but packages such as `caretEnsemble`[4], `SuperLearner`[5] or `mlr`[6] can combine models in various ways described above.

## 5.15  Predicting continuous variables: Regression with machine learning

Until now, we only considered methods that can help us predict class labels. However, all the methods we have shown can also be used to predict continuous variables. In this case, the methods will try to optimize the prediction in error which is usually in the form of the sum of squared errors (SSE): $SSE = \sum(Y - f(X))^2$, where $Y$ is the continuous response variable and $f(X)$ is the outcome of the machine learning model.

In this section, we are going to show how to use a supervised learning method for regression. All the methods we have introduced previously in the context of classification can also do regression. Technically, this is just a simple change in the cost function format and the optimization step still tries to optimize the parameters of the cost function. In many cases, if your response variable is numeric, methods in the `caret` package will automatically apply regression.

### 5.15.1  Use case: Predicting age from DNA methylation

We will demonstrate random forest regression using a different data set which has a continuous response variable. This time we are going to try to predict the age of individuals from their DNA methylation levels. Methylation is a DNA modification which has implications in gene regulation and cell state. We have introduced DNA methylation in depth in Chapters 1 and 10, however for now, what we need to know is that there are about 24 million CpG dinucleotides in the human genome. Their methylation status can be measured with quantitative assays and the value is between 0 and 1. If it is 0, the CpG is not methylated in any of the cells in the

---

[4]https://cran.r-project.org/web/packages/caretEnsemble/
[5]https://cran.r-project.org/web/packages/SuperLearner/index.html
[6]https://mlr.mlr-org.com/

sample, and if it is 1, the CpG is methylated in all the cells of the sample. It has been shown that methylation is predictive of the age of the individual that the sample is taken from (Numata et al., 2012; Horvath, 2013). Now, we will try to test that with a data set containing hundreds of individuals, their age, and methylation values for ~27000 CpGs. We first read in the files and construct a training set.

### 5.15.2 Reading and processing the data

Let us first read in the data. When we run the summary and histogram we see that the methylation values are between 0 and 1 and there are 108 samples (see Figure 5.16 ). Typically, methylation values have bimodal distribution. In this case many of them have values around 0 and the second-most frequent value bracket is around 0.9.

```
# file path for CpG methylation and age
fileMethAge=system.file("extdata",
                        "CpGmeth2Age.rds",
                        package="compGenomRData")

# read methylation-age table
ameth=readRDS(fileMethAge)
dim(ameth)
```

```
## [1]   108 27579
```

```
summary(ameth[,1:3])
```

```
##       Age            cg26211698          cg03790787
##   Min.   :-0.4986   Min.   :0.01223   Min.   :0.05001
##   1st Qu.:-0.4027   1st Qu.:0.01885   1st Qu.:0.07818
##   Median :18.8466   Median :0.02269   Median :0.08964
##   Mean   :25.9083   Mean   :0.02483   Mean   :0.09300
##   3rd Qu.:49.6110   3rd Qu.:0.02888   3rd Qu.:0.10423
##   Max.   :83.6411   Max.   :0.04883   Max.   :0.16271
```

```
# plot histogram of methylation values
hist(unlist(ameth[,-1]),border="white",
     col="cornflowerblue",main="",xlab="methylation values")
```

There are 27000 predictor variables. We can remove the ones that have low variation across samples. In this case, the methylation values are between 0 and 1. The CpGs that have low variation are not likely to have any association with age; they

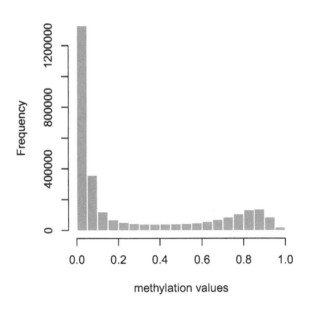

**FIGURE 5.16:** Histogram of methylation values in the training set for age prediction.

could simply be technical variation of the experiment. We will remove CpGs that have less than 0.1 standard deviation.

```
ameth=ameth[,c(TRUE,matrixStats::colSds(as.matrix(ameth[,-1]))>0.1)]
dim(ameth)
```

```
## [1]  108 2290
```

### 5.15.3 Running random forest regression

Now we can use random forest regression to predict the age from methylation values. We are then going to plot the predicted vs. observed ages and see how good our predictions are. The resulting plots are shown in Figure 5.17.

```
set.seed(18)
```

```
par(mfrow=c(1,2))
```

```
# we are not going to do any cross-validatin
# and rely on OOB error
```

```
trctrl <- trainControl(method = "none")

# we will now train random forest model
rfregFit <- train(Age~.,
                  data = ameth,
                  method = "ranger",
                  trControl=trctrl,
                  # calculate importance
                  importance="permutation",
                  tuneGrid = data.frame(mtry=50,
                                        min.node.size = 5,
                                        splitrule="variance")
                  )
# plot Observed vs OOB predicted values from the model
plot(ameth$Age,rfregFit$finalModel$predictions,
     pch=19,xlab="observed Age",
     ylab="OOB predicted Age")
mtext(paste("R-squared",
            format(rfregFit$finalModel$r.squared,digits=2)))

# plot residuals
plot(ameth$Age,(rfregFit$finalModel$predictions-ameth$Age),
     pch=18,ylab="residuals (predicted-observed)",
     xlab="observed Age",col="blue3")
abline(h=0,col="red4",lty=2)
```

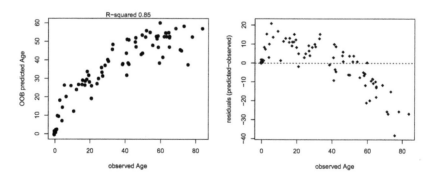

**FIGURE 5.17:** Observed vs. predicted age (Left). Residual plot showing that for older people the error increases (Right).

In this instance, we are using OOB errors and $R^2$ value which shows how the model performs on OOB samples. The model can capture the general trend and it has acceptable OOB performance. It is not perfect as it makes errors on average close to 10 years when predicting the age, and the errors are more severe for older people (Figure 5.17). This could be due to having fewer older people to model or missing/in-adequate predictor variables. However, everything we discussed in classification applies here. We had even fewer data points than the classification problem, so we did not do a split for a test data set. However, this should also be done for regression problems, especially when we are going to compare the performance of different models or want to have a better idea of the real-world performance of our model. We might also be interested in which variables are most important as in the classification problem; we can use the `caret:varImp()` function to get access to random-forest-specific variable importance metrics.

## 5.16   Exercises

### 5.16.1   Classification

For this set of exercises we will be using the gene expression and patient annotation data from the glioblastoma patient. You can read the data as shown below:

```
library(compGenomRData)
# get file paths
fileLGGexp=system.file("extdata",
                       "LGGrnaseq.rds",
                       package="compGenomRData")
fileLGGann=system.file("extdata",
                       "patient2LGGsubtypes.rds",
                       package="compGenomRData")
# gene expression values
gexp=readRDS(fileLGGexp)

# patient annotation
patient=readRDS(fileLGGann)
```

1.  Our first task is to not use any data transformation and do classification. Run the k-NN classifier on the data without any transformation or scaling. What is the effect on classification accuracy for k-NN predicting the CIMP and noCIMP status of the patient? [Difficulty: **Beginner**]

2. Bootstrap resampling can be used to measure the variability of the prediction error. Use bootstrap resampling with k-NN for the prediction accuracy. How different is it from cross-validation for different $k$s? [Difficulty: **Intermediate**]

3. There are a number of ways to get variable importance for a classification problem. Run random forests on the classification problem above. Compare the variable importance metrics from random forest and the one obtained from DALEX. How many variables are the same in the top 10? [Difficulty: **Advanced**]

4. Come up with a unified importance score by normalizing importance scores from random forests and DALEX, followed by taking the average of those scores. [Difficulty: **Advanced**]

### 5.16.2 Regression

For this set of problems we will use the regression data set where we tried to predict the age of the sample from the methylation values. The data can be loaded as shown below:

```
# file path for CpG methylation and age
fileMethAge=system.file("extdata",
                        "CpGmeth2Age.rds",
                        package="compGenomRData")

# read methylation-age table
ameth=readRDS(fileMethAge)
```

1. Run random forest regression and plot the importance metrics. [Difficulty: **Beginner**]

2. Split 20% of the methylation-age data as test data and run elastic net regression on the training portion to tune parameters and test it on the test portion. [Difficulty: **Intermediate**]

3. Run an ensemble model for regression using the **caretEnsemble** or **mlr** package and compare the results with the elastic net and random forest model. Did the test accuracy increase? **HINT:** You need to install these extra packages and learn how to use them in the context of ensemble models. [Difficulty: **Advanced**]

# 6

## Operations on Genomic Intervals and Genome Arithmetic

Considerable time in computational genomics is spent on overlapping different features of the genome. Each feature can be represented with a genomic interval within the chromosomal coordinate system. In addition, each interval can carry different sorts of information. An interval may for instance represent exon coordinates or a transcription factor binding site. On the other hand, you can have base-pair resolution, continuous scores over the genome such as read coverage, or scores that could be associated with only certain bases such as in the case of CpG methylation (see Figure 6.1). Typically, you will need to overlap intervals of interest with other features of the genome, again represented as intervals. For example, you may want to overlap transcription factor binding sites with CpG islands or promoters to quantify what percentage of binding sites overlap with your regions of interest. Overlapping mapped reads from high-throughput sequencing experiments with genomic features such as exons, promoters, and enhancers can also be classified as operations on genomic intervals. You can think of a million other ways that involve overlapping two sets of different features on the genome. This chapter aims to show how to do analysis involving operations on genomic intervals.

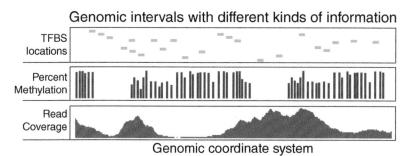

**FIGURE 6.1:** Summary of genomic intervals with different kinds of information.

## 6.1 Operations on genomic intervals with GenomicRanges package

The Bioconductor[1] project has a dedicated package called GenomicRanges[2] to deal with genomic intervals. In this section, we will provide use cases involving operations on genomic intervals. The main reason we will stick to this package is that it provides tools to do overlap operations. However, the package requires that users operate on specific data types that are conceptually similar to a tabular data structure implemented in a way that makes overlapping and related operations easier. The main object we will be using is called the GRanges object and we will also see some other related objects from the GenomicRanges package.

### 6.1.1 How to create and manipulate a GRanges object

GRanges (from GenomicRanges package) is the main object that holds the genomic intervals and extra information about those intervals. Here we will show how to create one. Conceptually, it is similar to a data frame and some operations such as using [ ] notation to subset the table will also work on GRanges, but keep in mind that not everything that works for data frames will work on GRanges objects.

```
library(GenomicRanges)
gr=GRanges(seqnames=c("chr1","chr2","chr2"),
          ranges=IRanges(start=c(50,150,200),
                         end=c(100,200,300)),
          strand=c("+","-","-")
)
gr
```

```
## GRanges object with 3 ranges and 0 metadata columns:
##       seqnames    ranges strand.
##          <Rle> <IRanges>  <Rle>
##   [1]     chr1    50-100      +
##   [2]     chr2   150-200      -
##   [3]     chr2   200-300      -
##   -------
##   seqinfo: 2 sequences from an unspecified genome; no seqlengths
```

```
# subset like a data frame
gr[1:2,]
```

---

[1]http://bioconductor.org
[2]http://www.bioconductor.org/packages/release/bioc/html/GenomicRanges.html

```
## GRanges object with 2 ranges and 0 metadata columns:
##         seqnames      ranges strand
##            <Rle> <IRanges>  <Rle>
##   [1]      chr1     50-100      +
##   [2]      chr2    150-200      -
##   -------
##   seqinfo: 2 sequences from an unspecified genome; no seqlengths
```

As you can see, it looks a bit like a data frame. Also, note that the peculiar second argument "ranges" basically contains the start and end positions of the genomic intervals. However, you cannot just give start and end positions, you actually have to provide another object of IRanges. Do not let this confuse you; GRanges actually depends on another object that is very similar to itself called IRanges and you have to provide the "ranges" argument as an IRanges object. In its simplest form, an IRanges object can be constructed by providing start and end positions to the IRanges() function. Think of it as something you just have to provide in order to construct the GRanges object.

GRanges can also contain other information about the genomic interval such as scores, names, etc. You can provide extra information at the time of the construction or you can add it later. Here is how you can do that:

```
gr=GRanges(seqnames=c("chr1","chr2","chr2"),
           ranges=IRanges(start=c(50,150,200),
                          end=c(100,200,300)),
           names=c("id1","id3","id2"),
           scores=c(100,90,50)
)
# or add it later (replaces the existing meta data)
mcols(gr)=DataFrame(name2=c("pax6","meis1","zic4"),
                    score2=c(1,2,3))

gr=GRanges(seqnames=c("chr1","chr2","chr2"),
           ranges=IRanges(start=c(50,150,200),
                          end=c(100,200,300)),
           names=c("id1","id3","id2"),
           scores=c(100,90,50)
)

# or appends to existing meta data
```

```
mcols(gr)=cbind(mcols(gr),
                      DataFrame(name2=c("pax6","meis1","zic4")) )
gr
```

```
## GRanges object with 3 ranges and 3 metadata columns:
##         seqnames     ranges strand |       names     scores        name2
##            <Rle>  <IRanges>  <Rle> | <character>  <numeric>  <character>
##   [1]       chr1     50-100      * |         id1        100         pax6
##   [2]       chr2    150-200      * |         id3         90        meis1
##   [3]       chr2    200-300      * |         id2         50         zic4
##   -------
##   seqinfo: 2 sequences from an unspecified genome; no seqlengths
```

```
# elementMetadata() and values() do the same things
elementMetadata(gr)
```

```
## DataFrame with 3 rows and 3 columns
##          names     scores        name2
##    <character>  <numeric>  <character>
## 1          id1        100         pax6
## 2          id3         90        meis1
## 3          id2         50         zic4
```

```
values(gr)
```

```
## DataFrame with 3 rows and 3 columns
##          names     scores        name2
##    <character>  <numeric>  <character>
## 1          id1        100         pax6
## 2          id3         90        meis1
## 3          id2         50         zic4
```

```
# you may also add metadata using the $ operator, as for data frames
gr$name3 = c("A","C", "B")
gr
```

```
## GRanges object with 3 ranges and 4 metadata columns:
##         seqnames     ranges strand |       names     scores        name2        name3
##            <Rle>  <IRanges>  <Rle> | <character>  <numeric>  <character>  <character>
##   [1]       chr1     50-100      * |         id1        100         pax6            A
##   [2]       chr2    150-200      * |         id3         90        meis1            C
```

```
##    [3]    chr2    200-300      * |          id2       50        zic4        B
##    -------
##    seqinfo: 2 sequences from an unspecified genome; no seqlengths
```

### 6.1.2   Getting genomic regions into R as GRanges objects

There are multiple ways you can read your genomic features into R and create a GRanges object. Most genomic interval data comes in a tabular format that has the basic information about the location of the interval and some other information. We already showed how to read BED files as a data frame in Chapter 2. Now we will show how to convert it to the GRanges object. This is one way of doing it, but there are more convenient ways described further in the text.

```
# read CpGi data set
filePath=system.file("extdata",
                     "cpgi.hg19.chr21.bed",
                     package="compGenomRData")
cpgi.df = read.table(filePath, header = FALSE,
                     stringsAsFactors=FALSE)
# remove chr names with "_"
cpgi.df =cpgi.df [grep("_",cpgi.df[,1],invert=TRUE),]

cpgi.gr=GRanges(seqnames=cpgi.df[,1],
                ranges=IRanges(start=cpgi.df[,2],
                               end=cpgi.df[,3]))
```

You may need to do some pre-processing before/after reading in the BED file. Below is an example of getting transcription start sites from BED files containing RefSeq transcript locations.

```
# read refseq file
filePathRefseq=system.file("extdata",
                     "refseq.hg19.chr21.bed",
                     package="compGenomRData")


ref.df = read.table(filePathRefseq, header = FALSE,
                     stringsAsFactors=FALSE)
ref.gr=GRanges(seqnames=ref.df[,1],
               ranges=IRanges(start=ref.df[,2],
                              end=ref.df[,3]),
```

```
                         strand=ref.df[,6],name=ref.df[,4])
# get TSS
tss.gr=ref.gr
# end of the + strand genes must be equalized to start pos
end(tss.gr[strand(tss.gr)=="+",])  =start(tss.gr[strand(tss.gr)=="+",])
# startof the - strand genes must be equalized to end pos
start(tss.gr[strand(tss.gr)=="-",])=end(tss.gr[strand(tss.gr)=="-",])
# remove duplicated TSSes ie alternative transcripts
# this keeps the first instance and removes duplicates
tss.gr=tss.gr[!duplicated(tss.gr),]
```

Another way of doing this from a BED file is to use the `readTranscriptfeatures()` function from the `genomation` package. This function takes care of the steps described in the code chunk above.

Reading the genomic features as text files and converting to GRanges is not the only way to create a GRanges object. With the help of the `rtracklayer`[3] package we can directly import BED files.

```
require(rtracklayer)
```

```
# we are reading a BED file, the path to the file
# is stored in filePathRefseq variable
import.bed(filePathRefseq)
```

Next, we will show how to use other methods to automatically obtain the data in the GRanges format from online databases. But you will not be able to use these methods for every data set, so it is good to know how to read data from flat files as well. We will use the `rtracklayer` package to download data from the UCSC Genome Browser. We will download CpG islands as GRanges objects. The `rtracklayer` workflow we show below works like using the UCSC table browser. You need to select which species you are working with, then you need to select which dataset you need to download and lastly you download the UCSC dataset or track as a GRanges object.

```
require(rtracklayer)
session <- browserSession("UCSC",url = 'http://genome-euro.ucsc.edu/cgi-bin/')
genome(session) <- "mm9"
## choose CpG island track on chr12
```

---

[3]http://www.bioconductor.org/packages/release/bioc/html/rtracklayer.html

```
query <- ucscTableQuery(session, track="CpG Islands",table="cpgIslandExt",
        range=GRangesForUCSCGenome("mm9", "chr12"))
## get the GRanges object for the track
track(query)
```

There is also an interface to the Ensembl database called biomaRt[4]. This package will enable you to access and import all of the datasets included in Ensembl. Another similar package is AnnotationHub[5]. This package is an aggregator for different datasets from various sources. Using AnnotationHub one can access data sets from the UCSC browser, Ensembl browser and datasets from genomics consortia such as ENCODE and Roadmap Epigenomics. We provide examples of using Biomart package further into the chapter. In addition, the AnnotationHub package is used in Chapter 9.

#### 6.1.2.1  Frequently used file formats and how to read them into R as a table

There are multiple file formats in genomics but some of them you will see more frequently than others. We already mentioned some of them. Here is a list of files and functions that can read them into R as GRanges objects or something coercible to GRanges objects.

1) **BED**: This format is used and popularized by the UCSC browser, and can hold a variety of information including exon/intron structure of transcripts in a single line. We will be using BED files in this chapter. In its simplest form, the BED file contains the chromosome name, the start position and end position for a genomic feature of interest.
    - genomation::readBed()
    - genomation::readTranscriptFeatures() good for getting intron/exon/promoters from BED12 files
    - rtracklayer::import.bed()
2) **GFF**: GFF format is a tabular text format for genomic features similar to BED. However, it is a more flexible format than BED, which makes it harder to parse at times. Many gene annotation files are in this format.
    - genomation::gffToGranges()
    - rtracklayer::impot.gff()
3) **BAM/SAM**: BAM format is a compressed and indexed tabular file format designed for aligned sequencing reads. SAM is the uncompressed version of the BAM file. We will touch upon BAM files in this chapter. The uncompressed SAM file is similar in spirit to a BED file where you have the basic location of chromosomal location information plus additional

---

[4]https://bioconductor.org/packages/release/bioc/html/biomaRt.html
[5]https://bioconductor.org/packages/release/bioc/html/AnnotationHub.html

columns that are related to the quality of alignment or other relevant information. We will introduce this format in detail later in this chapter.

- •GenomicAlignments::readGAlignments
- •Rsamtools::scanBam returns a data frame with columns from a SAM/BAM file.

4) **BigWig**: This is used to for storing scores associated with genomic intervals. It is an indexed format. Similar to BAM, this makes it easier to query and only necessary portions of the file could be loaded into memory.
    - •rtracklayer::import.bw()

5) **Generic Text files**: This represents any text file with the minimal information of chromosome, start and end coordinates.
    - •genomation::readGeneric()

6) **Tabix/Bcf**: These are tabular file formats indexed and compressed similar to BAM. The following functions return lists rather than tabular data structures. These formats are mostly used to store genomic variation data such as SNPs and indels.
    - •Rsamtools::scanTabix
    - •Rsamtools::scanBcf

### 6.1.3 Finding regions that do/do not overlap with another set of regions

This is one of the most common tasks in genomics. Usually, you have a set of regions that you are interested in and you want to see if they overlap with another set of regions or see how many of them overlap. A good example is transcription factor binding sites determined by ChIP-seq[6] experiments. We will introduce ChIP-seq in more detail in Chapter 9. However, in these types of experiments and the following analysis, one usually ends up with genomic regions that are bound by transcription factors. One of the standard next questions would be to annotate binding sites with genomic annotations such as promoter, exon, intron and/or CpG islands, which are important for gene regulation. Below is a demonstration of how transcription factor binding sites can be annotated using CpG islands. First, we will get the subset of binding sites that overlap with the CpG islands. In this case, binding sites are ChIP-seq peaks.

In the code snippet below, we read the ChIP-seq analysis output files using the genomation::readBroadPeak() function. This function directly outputs a GRanges object. These output files are similar to BED files, where the location of the predicted binding sites are written out in a tabular format with some analysis-related scores

---

[6]http://en.wikipedia.org/wiki/ChIP-sequencing

and/or P-values. After reading the files, we can find the subset of peaks that overlap with the CpG islands using the subsetByoverlaps() function.

```
library(genomation)
filePathPeaks=system.file("extdata",
            "wgEncodeHaibTfbsGm12878Sp1Pcr1xPkRep1.broadPeak.gz",
            package="compGenomRData")

# read the peaks from a bed file
pk1.gr=readBroadPeak(filePathPeaks)

# get the peaks that overlap with CpG islands
subsetByOverlaps(pk1.gr,cpgi.gr)
```

```
## GRanges object with 44 ranges and 5 metadata columns:
##          seqnames            ranges strand |        name     score signalValue
##             <Rle>         <IRanges>  <Rle> | <character> <integer>   <numeric>
##    [1]      chr21   9825360-9826582      * |   peak14562        56      183.11
##    [2]      chr21   9968469-9968984      * |   peak14593       947     3064.92
##    [3]      chr21 15755368-15755956      * |   peak14828        90      291.90
##    [4]      chr21 19191579-19192525      * |   peak14840       290      940.03
##    [5]      chr21 26979619-26980048      * |   peak14854        32      104.67
##    ...        ...               ...    ... .         ...       ...         ...
##   [40]      chr21 46237464-46237809      * |   peak15034        32      106.36
##   [41]      chr21 46707702-46708084      * |   peak15037        67      217.02
##   [42]      chr21 46961552-46961875      * |   peak15039        38      124.31
##   [43]      chr21 47743587-47744125      * |   peak15050       353     1141.58
##   [44]      chr21 47878412-47878891      * |   peak15052       104      338.78
##             pvalue    qvalue
##          <integer> <integer>
##    [1]          -1        -1
##    [2]          -1        -1
##    [3]          -1        -1
##    [4]          -1        -1
##    [5]          -1        -1
##    ...         ...       ...
##   [40]          -1        -1
##   [41]          -1        -1
##   [42]          -1        -1
##   [43]          -1        -1
##   [44]          -1        -1
##   -------
```

```
##    seqinfo: 23 sequences from an unspecified genome; no seqlengths
```

For each CpG island, we can count the number of peaks that overlap with a given
CpG island with GenomicRanges::countOverlaps().

```
counts=countOverlaps(pk1.gr,cpgi.gr)
head(counts)
```

```
## [1] 0 0 0 0 0 0
```

The GenomicRanges::findOverlaps() function can be used to see one-to-one overlaps
between peaks and CpG islands. It returns a matrix showing which peak overlaps
which CpG island.

```
findOverlaps(pk1.gr,cpgi.gr)
```

```
## Hits object with 45 hits and 0 metadata columns:
##        queryHits subjectHits
##        <integer>   <integer>
##   [1]      14562           1
##   [2]      14593           3
##   [3]      14828           8
##   [4]      14840          13
##   [5]      14854          16
##   ...        ...         ...
##  [41]      15034         155
##  [42]      15037         166
##  [43]      15039         176
##  [44]      15050         192
##  [45]      15052         200
##   -------
##   queryLength: 26121 / subjectLength: 205
```

Another interesting thing would be to look at the distances to the nearest CpG
islands for each peak. In addition, just finding the nearest CpG island could also
be interesting. Oftentimes, you will need to find the nearest TSS or gene to your
regions of interest, and the code below is handy for doing that using the nearest()
and distanceToNearest() functions, the resulting plot is shown in Figure 6.2.

```
# find nearest CpGi to each TSS
n.ind=nearest(pk1.gr,cpgi.gr)
# get distance to nearest
```

```
dists=distanceToNearest(pk1.gr,cpgi.gr,select="arbitrary")
dists
```

```
## Hits object with 620 hits and 1 metadata column:
##          queryHits subjectHits |  distance
##          <integer>   <integer> | <integer>
##    [1]       14440           1 |    384188
##    [2]       14441           1 |    382968
##    [3]       14442           1 |    381052
##    [4]       14443           1 |    379311
##    [5]       14444           1 |    376978
##    ...         ...         ... .       ...
##  [616]       15055         205 |     26212
##  [617]       15056         205 |     27402
##  [618]       15057         205 |     30468
##  [619]       15058         205 |     31611
##  [620]       15059         205 |     34090
##    -------
##    queryLength: 26121 / subjectLength: 205
```

```
# histogram of the distances to nearest TSS
dist2plot=mcols(dists)[,1]
hist(log10(dist2plot),xlab="log10(dist to nearest TSS)",
     main="Distances")
```

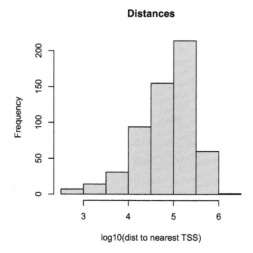

**FIGURE 6.2:** Histogram of distances of CpG islands to the nearest TSSes.

## 6.2  Dealing with mapped high-throughput sequencing reads

The reads from sequencing machines are usually pre-processed and aligned to the genome with the help of specific bioinformatics tools. We have introduced the details of general read processing, quality check and alignment methods in Chapter 7. In this section we will deal with mapped reads. Since each mapped read has a start and end position the genome, mapped reads can be thought of as genomic intervals stored in a file. After mapping, the next task is to quantify the enrichment of those aligned reads in the regions of interest. You may want to count how many reads overlap with your promoter set of interest or you may want to quantify RNA-seq reads overlap with exons. This is similar to operations on genomic intervals which are described previously. If you can read all your alignments into memory and create a GRanges object, you can apply the previously described operations. However, most of the time we can not read all mapped reads into memory, so we have to use specialized tools to query and quantify alignments on a given set of regions. One of the most common alignment formats is SAM/BAM format, most aligners will produce SAM/BAM output or you will be able to convert your specific alignment format to SAM/BAM format. The BAM format is a binary version of the human-readable SAM format. The SAM format has specific columns that contain different kinds of information about the alignment such as mismatches, qualities etc. (see [http://samtools.sourceforge.net/SAM1.pdf] for SAM format specification).

### 6.2.1  Counting mapped reads for a set of regions

The Rsamtools package has functions to query BAM files. The function we will use in the first example is the countBam() function, which takes input of the BAM file and param argument. The param argument takes a ScanBamParam object. The object is instantiated using ScanBamParam() and contains parameters for scanning the BAM file. The example below is a simple example where ScanBamParam() only includes regions of interest, promoters on chr21.

```
promoter.gr=tss.gr
start(promoter.gr)=start(promoter.gr)-1000
end(promoter.gr)  =end(promoter.gr)+1000
promoter.gr=promoter.gr[seqnames(promoter.gr)=="chr21"]

library(Rsamtools)
bamfilePath=system.file("extdata",
          "wgEncodeHaibTfbsGm12878Sp1Pcr1xAlnRep1.chr21.bam",
              package="compGenomRData")
```

```
# get reads for regions of interest from the bam file
param <- ScanBamParam(which=promoter.gr)
counts=countBam(bamfilePath, param=param)
```

Alternatively, aligned reads can be read in using the GenomicAlignments package (which on this occasion relies on the Rsamtools package).

```
library(GenomicAlignments)
alns <- readGAlignments(bamfilePath, param=param)
```

## 6.3 Dealing with continuous scores over the genome

Most high-throughput data can be viewed as a continuous score over the bases of the genome. In case of RNA-seq or ChIP-seq experiments, the data can be represented as read coverage values per genomic base position. In addition, other information (not necessarily from high-throughput experiments) can be represented this way. The GC content and conservation scores per base are prime examples of other data sets that can be represented as scores over the genome. This sort of data can be stored as a generic text file or can have special formats such as Wig (stands for wiggle) from UCSC, or the bigWig format, which is an indexed binary format of the wig files. The bigWig format is great for data that covers a large fraction of the genome with varying scores, because the file is much smaller than regular text files that have the same information and it can be queried more easily since it is indexed.

In R/Bioconductor, continuous data can also be represented in a compressed format, called Rle vector, which stands for run-length encoded vector. This gives superior memory performance over regular vectors because repeating consecutive values are represented as one value in the Rle vector (see Figure 6.3).

Typically, for genome-wide data you will have an RleList object, which is a list of Rle vectors per chromosome. You can obtain such vectors by reading the reads in and calling the coverage() function from the GenomicRanges package. Let's try that on the above data set.

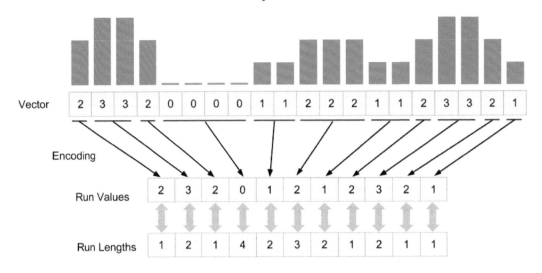

**FIGURE 6.3:** Rle encoding explained.

```
covs=coverage(alns) # get coverage vectors
covs

## RleList of length 24
## $chr1
## integer-Rle of length 249250621 with 1 run
##    Lengths: 249250621
##    Values :         0
##
## $chr2
## integer-Rle of length 243199373 with 1 run
##    Lengths: 243199373
##    Values :         0
##
## $chr3
## integer-Rle of length 198022430 with 1 run
##    Lengths: 198022430
##    Values :         0
##
## $chr4
## integer-Rle of length 191154276 with 1 run
##    Lengths: 191154276
##    Values :         0
##
## $chr5
```

```
## integer-Rle of length 180915260 with 1 run
##   Lengths: 180915260
##   Values :          0
##
## ...
## <19 more elements>
```

Alternatively, you can get the coverage from the BAM file directly. Below, we are getting the coverage directly from the BAM file for our previously defined promoters.

```
covs=coverage(bamfilePath, param=param) # get coverage vectors
```

One of the most common ways of storing score data is, as mentioned, the wig or bigWig format. Most of the ENCODE project data can be downloaded in bigWig format. In addition, conservation scores can also be downloaded in the wig/bigWig format. You can import bigWig files into R using the `import()` function from the `rtracklayer` package. However, it is generally not advisable to read the whole bigWig file in memory as was the case with BAM files. Usually, you will be interested in only a fraction of the genome, such as promoters, exons etc. So it is best that you extract the data for those regions and read those into memory rather than the whole file. Below we read a bigWig file only for the bases on promoters. The operation returns a `GRanges` object with the score column which indicates the scores in the bigWig file per genomic region.

```
library(rtracklayer)

# File from ENCODE ChIP-seq tracks
bwFile=system.file("extdata",
                   "wgEncodeHaibTfbsA549.chr21.bw",
                   package="compGenomRData")
bw.gr=import(bwFile, which=promoter.gr) # get coverage vectors
bw.gr
```

```
## GRanges object with 9205 ranges and 1 metadata column:
##         seqnames            ranges strand |     score
##            <Rle>         <IRanges>  <Rle> | <numeric>
##   [1]      chr21 9825456-9825457      * |         1
##   [2]      chr21 9825458-9825464      * |         2
##   [3]      chr21 9825465-9825466      * |         4
##   [4]      chr21 9825467-9825470      * |         5
##   [5]      chr21         9825471      * |         6
```

```
##      ...         ...            ...       ... .       ...
##    [9201]    chr21 48055809-48055856      * |           2
##    [9202]    chr21 48055857-48055858      * |           1
##    [9203]    chr21 48055872-48055921      * |           1
##    [9204]    chr21 48055944-48055993      * |           1
##    [9205]    chr21 48056069-48056118      * |           1
##    -------
##    seqinfo: 1 sequence from an unspecified genome
```

Following this we can create an RleList object from the GRanges with the coverage()
function.

```
cov.bw=coverage(bw.gr,weight = "score")
```

```
# or get this directly from
cov.bw=import(bwFile, which=promoter.gr,as = "RleList")
```

### 6.3.1  Extracting subsections of Rle and RleList objects

Frequently, we will need to extract subsections of the Rle vectors or RleList objects.
We will need to do this to visualize that subsection or get some statistics out of
those sections. For example, we could be interested in average coverage per base for
the regions we are interested in. We have to extract those regions from the RleList
object and apply summary statistics. Below, we show how to extract subsections
of the RleList object. We are extracting promoter regions from the ChIP-seq read
coverage RleList. Following that, we will plot one of the promoter's coverage values.

```
myViews=Views(cov.bw,as(promoter.gr,"IRangesList")) # get subsets of coverage
# there is a views object for each chromosome
myViews
```

```
## RleViewsList object of length 1:
## $chr21
## Views on a 48129895-length Rle subject
##
## views:
##          start      end width
##    [1] 42218039 42220039  2001 [2 0 0 0 0 0 0 0 0 0 0 0 0 0 0 0 0 0 0 0 0 0 0 0 ...]
##    [2] 17441841 17443841  2001 [0 0 0 0 0 0 0 0 0 0 0 0 0 0 0 0 0 0 0 0 0 0 0 0 ...]
##    [3] 17565698 17567698  2001 [0 0 0 0 0 0 0 0 0 0 0 0 0 0 0 0 0 0 0 0 0 0 0 0 ...]
##    [4] 30395937 30397937  2001 [0 0 0 0 0 0 0 0 0 0 0 0 0 0 0 0 0 0 0 0 0 0 0 0 ...]
##    [5] 27542138 27544138  2001 [1 1 1 1 1 1 1 1 1 1 1 1 1 1 1 2 2 2 2 2 1 1 1 ...]
```

```
##     [6] 27511708 27513708   2001 [0 0 0 0 0 0 0 0 0 0 0 0 0 0 0 0 0 0 0 0 0 0 0 0 ...]
##     [7] 32930290 32932290   2001 [0 0 0 0 0 0 0 0 0 0 0 0 0 0 0 0 0 0 0 0 0 0 0 0 ...]
##     [8] 27542446 27544446   2001 [0 0 0 0 0 0 0 0 0 0 0 0 0 0 0 0 0 0 0 0 0 0 0 0 ...]
##     [9] 28338439 28340439   2001 [0 0 0 0 0 0 0 0 0 0 0 0 0 0 0 0 0 0 0 0 0 0 0 0 ...]
##     ...      ...      ...    ... ...
## [370] 47517032 47519032   2001 [1 1 1 1 1 1 1 1 1 1 1 1 1 1 1 1 1 1 1 1 1 1 1 ...]
## [371] 47648157 47650157   2001 [1 1 1 1 1 1 1 1 1 1 1 1 1 1 1 1 1 1 1 1 1 1 1 ...]
## [372] 47603373 47605373   2001 [0 0 0 0 0 0 0 0 0 0 0 0 0 0 0 0 0 0 0 0 0 0 0 ...]
## [373] 47647738 47649738   2001 [2 2 2 2 2 2 2 2 2 2 2 2 2 2 2 2 2 2 2 2 1 2 ...]
## [374] 47704236 47706236   2001 [0 0 0 0 0 0 0 0 0 0 0 0 0 0 0 0 0 0 0 0 0 0 0 ...]
## [375] 47742785 47744785   2001 [0 0 0 0 0 0 0 0 0 0 0 0 0 0 0 0 0 0 0 0 0 0 0 ...]
## [376] 47881383 47883383   2001 [1 1 1 1 1 1 1 1 1 1 1 1 1 1 1 1 1 1 1 1 1 1 1 ...]
## [377] 48054506 48056506   2001 [0 0 0 0 0 0 0 0 0 0 0 0 0 0 0 0 0 0 0 0 0 0 0 ...]
## [378] 48024035 48026035   2001 [1 1 1 1 1 1 1 1 1 1 1 1 1 1 1 1 1 1 1 1 1 1 1 ...]
```

```
myViews[[1]]
```

```
## Views on a 48129895-length Rle subject
##
## views:
##          start      end width
##    [1] 42218039 42220039   2001 [2 0 0 0 0 0 0 0 0 0 0 0 0 0 0 0 0 0 0 0 0 0 0 0 ...]
##    [2] 17441841 17443841   2001 [0 0 0 0 0 0 0 0 0 0 0 0 0 0 0 0 0 0 0 0 0 0 0 0 ...]
##    [3] 17565698 17567698   2001 [0 0 0 0 0 0 0 0 0 0 0 0 0 0 0 0 0 0 0 0 0 0 0 0 ...]
##    [4] 30395937 30397937   2001 [0 0 0 0 0 0 0 0 0 0 0 0 0 0 0 0 0 0 0 0 0 0 0 0 ...]
##    [5] 27542138 27544138   2001 [1 1 1 1 1 1 1 1 1 1 1 1 2 2 2 2 2 1 1 1 ...]
##    [6] 27511708 27513708   2001 [0 0 0 0 0 0 0 0 0 0 0 0 0 0 0 0 0 0 0 0 0 0 0 0 ...]
##    [7] 32930290 32932290   2001 [0 0 0 0 0 0 0 0 0 0 0 0 0 0 0 0 0 0 0 0 0 0 0 0 ...]
##    [8] 27542446 27544446   2001 [0 0 0 0 0 0 0 0 0 0 0 0 0 0 0 0 0 0 0 0 0 0 0 0 ...]
##    [9] 28338439 28340439   2001 [0 0 0 0 0 0 0 0 0 0 0 0 0 0 0 0 0 0 0 0 0 0 0 0 ...]
##    ...      ...      ...    ... ...
## [370] 47517032 47519032   2001 [1 1 1 1 1 1 1 1 1 1 1 1 1 1 1 1 1 1 1 1 1 1 1 ...]
## [371] 47648157 47650157   2001 [1 1 1 1 1 1 1 1 1 1 1 1 1 1 1 1 1 1 1 1 1 1 1 ...]
## [372] 47603373 47605373   2001 [0 0 0 0 0 0 0 0 0 0 0 0 0 0 0 0 0 0 0 0 0 0 0 ...]
## [373] 47647738 47649738   2001 [2 2 2 2 2 2 2 2 2 2 2 2 2 2 2 2 2 2 2 2 1 2 ...]
## [374] 47704236 47706236   2001 [0 0 0 0 0 0 0 0 0 0 0 0 0 0 0 0 0 0 0 0 0 0 0 ...]
## [375] 47742785 47744785   2001 [0 0 0 0 0 0 0 0 0 0 0 0 0 0 0 0 0 0 0 0 0 0 0 ...]
## [376] 47881383 47883383   2001 [1 1 1 1 1 1 1 1 1 1 1 1 1 1 1 1 1 1 1 1 1 1 1 ...]
## [377] 48054506 48056506   2001 [0 0 0 0 0 0 0 0 0 0 0 0 0 0 0 0 0 0 0 0 0 0 0 ...]
## [378] 48024035 48026035   2001 [1 1 1 1 1 1 1 1 1 1 1 1 1 1 1 1 1 1 1 1 1 1 1 ...]
```

```
# get the coverage vector from the 5th view and plot
plot(myViews[[1]][[5]],type="l")
```

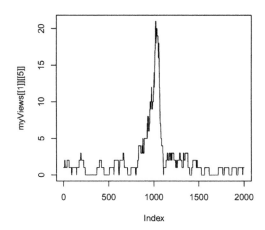

**FIGURE 6.4:** Coverage vector extracted from the RleList via the Views() function is plotted as a line plot.

Next, we are interested in average coverage per base for the promoters using summary functions that work on the views object.

```
# get the mean of the views
head(
  viewMeans(myViews[[1]])
)
```

```
## [1] 0.2258871 0.3498251 1.2243878 0.4997501 2.0904548 0.6996502
```

```
# get the max of the views
head(
  viewMaxs(myViews[[1]])
)
```

```
## [1]  2  4 12  4 21  6
```

## 6.4 Genomic intervals with more information: SummarizedExperiment class

As we have seen, genomic intervals can be mainly contained in a GRanges object. It can also contain additional columns associated with each interval. Here you can save information such as read counts or other scores associated with the interval. However, genomic data often have many layers. With GRanges you can have a table associated with the intervals, but what happens if you have many tables and each table has some metadata associated with it. In addition, rows and columns might have additional annotation that cannot be contained by row or column names. For these cases, the SummarizedExperiment class is ideal. It can hold multi-layered tabular data associated with each genomic interval and the meta-data associated with rows and columns, or associated with each table. For example, genomic intervals associated with the SummarizedExperiment object can be gene locations, and each tabular data structure can be RNA-seq read counts in a time course experiment. Each table could represent different conditions in which experiments are performed. The SummarizedExperiment class is outlined in the figure below (Figure 6.5 ).

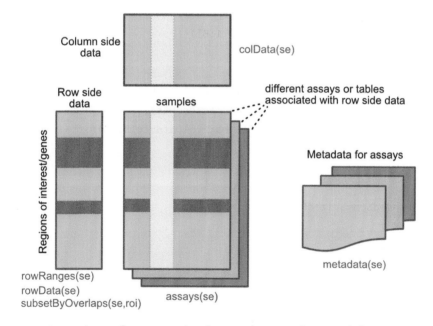

**FIGURE 6.5:** Overview of SummarizedExperiment class and functions. Adapted from the SummarizedExperiment package vignette.

### 6.4.1 Create a SummarizedExperiment object

Here we show how to create a basic SummarizedExperiment object. We will first create a matrix of read counts. This matrix will represent read counts from a series

of RNA-seq experiments from different time points. Following that, we create a
GRanges object to represent the locations of the genes, and a table for column annotations. This will include the names for the columns and any other value we want
to represent. Finally, we will create a SummarizedExperiment object by combining all
those pieces.

```
# simulate an RNA-seq read counts table
nrows <- 200
ncols <- 6
counts <- matrix(runif(nrows * ncols, 1, 1e4), nrows)

# create gene locations
rowRanges <- GRanges(rep(c("chr1", "chr2"), c(50, 150)),
                     IRanges(floor(runif(200, 1e5, 1e6)), width=100),
                     strand=sample(c("+", "-"), 200, TRUE),
                     feature_id=paste0("gene", 1:200))

# create table for the columns
colData <- DataFrame(timepoint=1:6,
                     row.names=LETTERS[1:6])

# create SummarizedExperiment object
se=SummarizedExperiment(assays=list(counts=counts),
                     rowRanges=rowRanges, colData=colData)

se

## class: RangedSummarizedExperiment
## dim: 200 6
## metadata(0):
## assays(1): counts
## rownames: NULL
## rowData names(1): feature_id
## colnames(6): A B ... E F
## colData names(1): timepoint
```

### 6.4.2   Subset and manipulate the SummarizedExperiment object

Now that we have a SummarizedExperiment object, we can subset it and extract/change
parts of it.

#### 6.4.2.1 Extracting parts of the object

`colData()` and `rowData()` extract the column-associated and row-associated tables. `metaData()` extracts the meta-data table if there is any table associated.

```
colData(se) # extract column associated data
```

```
## DataFrame with 6 rows and 1 column
##    timepoint
##    <integer>
## A          1
## B          2
## C          3
## D          4
## E          5
## F          6
```

```
rowData(se) # extrac row associated data
```

```
## DataFrame with 200 rows and 1 column
##         feature_id
##        <character>
## 1            gene1
## 2            gene2
## 3            gene3
## 4            gene4
## 5            gene5
## ...            ...
## 196      gene196
## 197      gene197
## 198      gene198
## 199      gene199
## 200      gene200
```

To extract the main table or tables that contain the values of interest such as read counts, we must use the `assays()` function. This returns a list of `DataFrame` objects associated with the object.

```
assays(se) # extract list of assays
```

```
## List of length 1
## names(1): counts
```

You can use names with $ or [] notation to extract specific tables from the list.

```
assays(se)$counts # get the table named "counts"
```

```
assays(se)[[1]] # get the first table
```

#### 6.4.2.2   Subsetting

Subsetting is easy using [ ] notation. This is similar to the way we subset data frames or matrices.

```
# subset the first five transcripts and first three samples
se[1:5, 1:3]
```

```
## class: RangedSummarizedExperiment
## dim: 5 3
## metadata(0):
## assays(1): counts
## rownames: NULL
## rowData names(1): feature_id
## colnames(3): A B C
## colData names(1): timepoint
```

One can also use the $ operator to subset based on colData() columns. You can extract certain samples or in our case, time points.

```
se[, se$timepoint == 1]
```

In addition, as SummarizedExperiment objects are GRanges objects on steroids, they support all of the findOverlaps() methods and associated functions that work on GRanges objects.

```
# Subset for only rows which are in chr1:100,000-1,100,000
roi <- GRanges(seqnames="chr1", ranges=100000:1100000)
subsetByOverlaps(se, roi)
```

```
## class: RangedSummarizedExperiment
## dim: 50 6
## metadata(0):
## assays(1): counts
## rownames: NULL
## rowData names(1): feature_id
```

```
## colnames(6): A B ... E F
## colData names(1): timepoint
```

## 6.5 Visualizing and summarizing genomic intervals

Data integration and visualization is cornerstone of genomic data analysis. Below, we will show different ways of integrating and visualizing genomic intervals. These methods can be used to visualize large amounts of data in a locus-specific or multi-loci manner.

### 6.5.1 Visualizing intervals on a locus of interest

Oftentimes, we will be interested in a particular genomic locus and try to visualize different genomic datasets over that locus. This is similar to looking at the data over one of the genome browsers. Below we will display genes, GpG islands and read coverage from a ChIP-seq experiment using the Gviz package. For the Gviz package, we first need to set the tracks to display. The tracks can be in various formats. They can be R objects such as IRanges,GRanges and data.frame, or they can be in flat file formats such as bigWig, BED, and BAM. After the tracks are set, we can display them with the plotTracks function, the resulting plot is shown in Figure 6.6.

```r
library(Gviz)
# set tracks to display


# set CpG island track
cpgi.track=AnnotationTrack(cpgi.gr,
                           name = "CpG")


# set gene track
# we will get this from EBI Biomart webservice
gene.track <- BiomartGeneRegionTrack(genome = "hg19",
                                     chromosome = "chr21",
                                     start = 27698681, end = 28083310,
                                     name = "ENSEMBL")


# set track for ChIP-seq coverage
chipseqFile=system.file("extdata",
                        "wgEncodeHaibTfbsA549.chr21.bw",
```

```
                        package="compGenomRData")
cov.track=DataTrack(chipseqFile,type = "l",
                     name="coverage")
```

```
# call the display function plotTracks
track.list=list(cpgi.track,gene.track,cov.track)
plotTracks(track.list,from=27698681,to=28083310,chromsome="chr21")
```

**FIGURE 6.6:** Genomic data tracks visualized using the Gviz functions.

### 6.5.2   Summaries of genomic intervals on multiple loci

Looking at data one region at a time could be inefficient. One can summarize different data sets over thousands of regions of interest and identify patterns. These summaries can include different data types such as motifs, read coverage and other scores associated with genomic intervals. The genomation package can summarize and help identify patterns in the datasets. The datasets can have different kinds of information and multiple file types can be used such as BED, GFF, BAM and bigWig. We will look at H3K4me3 ChIP-seq and DNAse-seq signals from the H1 embryonic stem cell line. H3K4me3 is usually associated with promoters and regions with high DNAse-seq signal are associated with accessible regions, which means mostly regulatory regions. We will summarize those datasets around the transcription start sites (TSS) of genes on chromosome 20 of the human hg19 assembly. We will first read the genes and extract the region around the TSS, 500bp upstream and downstream. We will then create a matrix of ChIP-seq scores for those regions. Each row will represent a region around a specific TSS and columns will be the scores per base. We will then plot average enrichment values around the TSS of genes on chromosome 20.

```r
# get transcription start sites on chr20
library(genomation)
transcriptFile=system.file("extdata",
                    "refseq.hg19.chr20.bed",
                    package="compGenomRData")
feat=readTranscriptFeatures(transcriptFile,
                        remove.unusual = TRUE,
                        up.flank = 500, down.flank = 500)
prom=feat$promoters # get promoters from the features


# get for H3K4me3 values around TSSes
# we use strand.aware=TRUE so - strands will
# be reversed
H3K4me3File=system.file("extdata",
                    "H1.ESC.H3K4me3.chr20.bw",
                    package="compGenomRData")
sm=ScoreMatrix(H3K4me3File,prom,
            type="bigWig",strand.aware = TRUE)


# look for the average enrichment
plotMeta(sm, profile.names = "H3K4me3", xcoords = c(-500,500),
        ylab="H3K4me3 enrichment",dispersion = "se",
        xlab="bases around TSS")
```

The resulting plot is shown in Figure 6.7. The pattern we see is expected, there is a dip just around TSS and the signal is more intense downstream of the TSS.

We can also plot a heatmap where each row is a region around the TSS and color coded by enrichment. This can show us not only the general pattern, as in the meta-region plot, but also how many of the regions produce such a pattern. The heatMatrix() function shown below achieves that. The resulting heatmap plot is shown in Figure 6.8.

```r
heatMatrix(sm,order=TRUE,xcoords = c(-500,500),
        xlab="bases around TSS")
```

Here we saw that about half of the regions do not have any signal. In addition

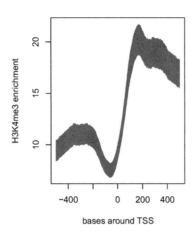

**FIGURE 6.7:** Meta-region plot using genomation.

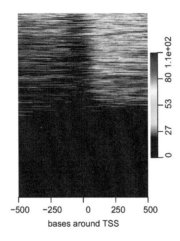

**FIGURE 6.8:** Heatmap of enrichment of H3K4me2 around the TSS.

it seems the multi-modal profile we have observed earlier is more complicated. Certain regions seem to have signal on both sides of the TSS, whereas others have signal mostly on the downstream side.

Normally, there would be more than one experiment or we can integrate datasets from public repositories. In this case, we can see how different signals look in the regions we are interested in. Now, we will also use DNAse-seq data and create a list of matrices with our datasets and plot the average profile of the signals from both datasets. The resulting meta-region plot is shown in Figure 6.9.

```
DNAseFile=system.file("extdata",
                        "H1.ESC.dnase.chr20.bw",
                        package="compGenomRData")

sml=ScoreMatrixList(c(H3K4me3=H3K4me3File,
                        DNAse=DNAseFile),prom,
                        type="bigWig",strand.aware = TRUE)
plotMeta(sml)
```

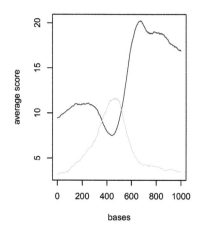

**FIGURE 6.9:** Average profiles of DNAse and H3K4me3 ChIP-seq.

We should now look at the heatmaps side by side and we should also cluster the rows based on their similarity. We will be using multiHeatMatrix since we have multiple ScoreMatrix objects in the list. In this case, we will also use the winsorize argument to limit extreme values, every score above 95th percentile will be equalized the value of the 95th percentile. In addition, heatMatrix and multiHeatMatrix can cluster the rows. Below, we will be using k-means clustering with 3 clusters.

```
set.seed(1029)
multiHeatMatrix(sml,order=TRUE,xcoords = c(-500,500),
                xlab="bases around TSS",winsorize = c(0,95),
                matrix.main = c("H3K4me3","DNAse"),
                column.scale=TRUE,
                clustfun=function(x) kmeans(x, centers=3)$cluster)
```

The resulting heatmaps are shown in Figure 6.10. These plots revealed a different

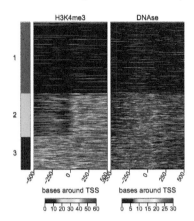

**FIGURE 6.10:** Heatmaps of H3K4me3 and DNAse data.

picture than we have observed before. Almost half of the promoters have no signal for DNAse or H3K4me3; these regions are probably not active and associated genes are not expressed. For regions with the H3K4me3 signal, there are two major patterns: one pattern where both downstream and upstream of the TSS are enriched, and on the other pattern, mostly downstream of the TSS is enriched.

### 6.5.3  Making karyograms and circos plots

Chromosomal karyograms and circos plots are beneficial for displaying data over the whole genome of chromosomes of interest, although the information that can be displayed over these large regions are usually not very clear and only large trends can be discerned by eye, such as loss of methylation in large regions or genome-wide. Below, we show how to use the `ggbio` package for plotting. This package has a slightly different syntax than base graphics. The syntax follows "grammar of graphics" logic, and depends on the `ggplot2` package we introduced in Chapter 2. It is a deconstructed way of thinking about the plot. You add your data and apply mappings and transformations in order to achieve the final output. In `ggbio`, things are relatively easy since a high-level function, the `autoplot` function, will recognize most of the datatypes and guess the most appropriate plot type. You can change its behavior by applying low-level functions. We first get the sizes of chromosomes and make a karyogram template. The empty karyogram is shown in Figure 6.11.

```
library(ggbio)
data(ideoCyto, package = "biovizBase")
p <- autoplot(seqinfo(ideoCyto$hg19), layout = "karyogram")
p
```

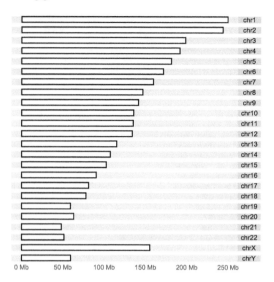

**FIGURE 6.11:** Karyogram example.

Next, we would like to plot CpG islands on this karyogram. We simply do this by adding a layer with the `layout_karyogram()` function. The resulting karyogram is shown in Figure 6.12.

```
# read CpG islands from a generic text file

CpGiFile=filePath=system.file("extdata",
                    "CpGi.hg19.table.txt",
                    package="compGenomRData")
cpgi.gr=genomation::readGeneric(CpGiFile,
        chr = 1, start = 2, end = 3,header=TRUE,
      keep.all.metadata =TRUE,remove.unusual=TRUE )


p + layout_karyogram(cpgi.gr)
```

Next, we would like to plot some data over the chromosomes. This could be the ChIP-seq signal or any other signal over the genome; we will use CpG island scores from the data set we read earlier. We will plot a point proportional to "obsExp" column in the data set. We use the `ylim` argument to squish the chromosomal rectangles and plot on top of those. The `aes` argument defines how the data is mapped to geometry. In this case, the argument indicates that the points will have an x coordinate from CpG island start positions and a y coordinate from the obsExp score of CpG islands. The resulting karyogram is shown in Figure 6.13.

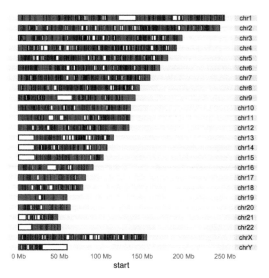

**FIGURE 6.12:** Karyogram of CpG islands over the human genome.

```
p + layout_karyogram(cpgi.gr, aes(x= start, y = obsExp),
                geom="point",
                ylim = c(2,50), color = "red",
                size=0.1,rect.height=1)
```

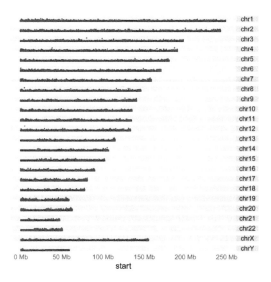

**FIGURE 6.13:** Karyogram of CpG islands and their observed/expected scores over the human genome.

Another way to depict regions or quantitative signals on the chromosomes is circos plots. These are circular plots usually used for showing chromosomal rearrangements, but can also be used for depicting signals. The `ggbio` package can produce all kinds of circos plots. Below, we will show how to use that for our CpG island score example, and the resulting plot is shown in Figure 6.14.

```r
# set the chromsome in a circle
# color set to white to look transparent
p <- ggplot() + layout_circle(ideoCyto$hg19, geom = "ideo", fill = "white",
                              colour="white",cytoband = TRUE,
                              radius = 39, trackWidth = 2)
# plot the scores as points
p <- p + layout_circle(cpgi.gr, geom = "point", grid=TRUE,
                        size = 0.01, aes(y = obsExp),color="red",
                        radius = 42, trackWidth = 10)
# set the chromosome names
p <- p + layout_circle(as(seqinfo(ideoCyto$hg19),"GRanges"),
                        geom = "text", aes(label = seqnames),
                        vjust = 0, radius = 55, trackWidth = 7,
                        size=3)

# display the plot
p
```

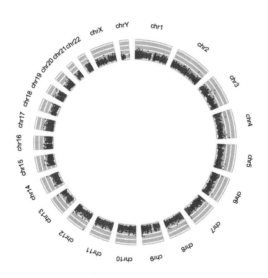

**FIGURE 6.14:** Circos plot for CpG island scores.

## 6.6  Exercises

The data for the exercises is within the `compGenomRData` package.

Run the following to see the data files.

```
dir(system.file("extdata",
          package="compGenomRData"))
```

You will need some of those files to complete the exercises.

### 6.6.1  Operations on genomic intervals with the `GenomicRanges` package

1. Create a `GRanges` object using the information in the table below:[Difficulty: **Beginner**]

   | chr | start | end | strand | score |
   |-----|-------|-----|--------|-------|
   | chr1 | 10000 | 10300 | + | 10 |
   | chr1 | 11100 | 11500 | - | 20 |
   | chr2 | 20000 | 20030 | + | 15 |

2. Use the `start()`, `end()`, `strand()`,`seqnames()` and `width()` functions on the `GRanges` object you created. Figure out what they are doing. Can you get a subset of the `GRanges` object for intervals that are only on the + strand? If you can do that, try getting intervals that are on chr1. *HINT:* `GRanges` objects can be subset using the `[ ]` operator, similar to data frames, but you may need to use `start()`, `end()` and `strand()`,`seqnames()` within the `[]`. [Difficulty: **Beginner/Intermediate**]

3. Import mouse (mm9 assembly) CpG islands and RefSeq transcripts for chr12 from the UCSC browser as `GRanges` objects using `rtracklayer` functions. HINT: Check chapter content and modify the code there as necessary. If that somehow does not work, go to the UCSC browser and download it as a BED file. The track name for Refseq genes is "RefSeq Genes" and the table name is "refGene". [Difficulty: **Beginner/Intermediate**]

4. Following from the exercise above, get the promoters of Refseq transcripts (-1000bp and +1000 bp of the TSS) and calculate what percentage of them overlap with CpG islands. HINT: You have to get the promoter coordinates and use the `findOverlaps()` or `subsetByOverlaps()` from the `GenomicRanges` package. To get promoters, type `?promoters` on the R console and see how to use that function to get promoters or calculate their coordinates as shown in the chapter. [Difficulty: **Beginner/Intermediate**]

5. Plot the distribution of CpG island lengths for CpG islands that overlap with the promoters. [Difficulty: **Beginner/Intermediate**]

6. Get canonical peaks for SP1 (peaks that are in both replicates) on chr21. Peaks for each replicate are located in the `wgEncodeHaibTfbsGm12878Sp1Pcr1xPkRep1.broadPeak.gz` and `wgEncodeHaibTfbsGm12878Sp1Pcr1xPkRep2.broadPeak.gz` files. **HINT**: You need to use `findOverlaps()` or `subsetByOverlaps()` to get the subset of peaks that occur in both replicates (canonical peaks). You can try to read "…broadPeak.gz" files using the `genomation::readBroadPeak()` function; broadPeak is just an extended BED format. In addition, you can try to use the `coverage()` and `slice()` functions to get more precise canonical peak locations. [Difficulty: **Intermediate/Advanced**]

### 6.6.2 Dealing with mapped high-throughput sequencing reads

1. Count the reads overlapping with canonical SP1 peaks using the BAM file for one of the replicates. The following file in the `compGenomRData` package contains the alignments for SP1 ChIP-seq reads: `wgEncodeHaibTfbsGm12878Sp1Pcr1xAlnRep1.chr21.bam`. **HINT**: Use functions from the `GenomicAlignments` package. [Difficulty: **Beginner/Intermediate**]

### 6.6.3 Dealing with contiguous scores over the genome

1. Extract the `Views` object for the promoters on chr20 from the `H1.ESC.H3K4me1.chr20.bw` file available at `CompGenomRData` package. Plot the first "View" as a line plot. **HINT**: See the code in the relevant section in the chapter and adapt the code from there. [Difficulty: **Beginner/Intermediate**]

2. Make a histogram of the maximum signal for the Views in the object you extracted above. You can use any of the view summary functions or use `lapply()` and write your own summary function. [Difficulty: **Beginner/Intermediate**]

3. Get the genomic positions of maximum signal in each view and make a `GRanges` object. **HINT**: See the `?viewRangeMaxs` help page. Try to make a `GRanges` object out of the returned object. [Difficulty: **Intermediate**]

### 6.6.4 Visualizing and summarizing genomic intervals

1. Extract -500,+500 bp regions around the TSSes on chr21; there are refseq files for the hg19 human genome assembly in the `compGenomRData` package. Use SP1 ChIP-seq data in the `compGenomRData` package, ac-

cess the file path via the `system.file()` function, the file name is: `wgEncodeHaibTfbsGm12878Sp1Pcr1xAlnRep1.chr21.bam`. Create an average profile of read coverage around the TSSes. Following that, visualize the read coverage with a heatmap. **HINT:** All of these are possible using the `genomation` package functions. Check `help(ScoreMatrix)` to see how you can use bam files. As an example here is how you can get the file path to refseq annotation on chr21. [Difficulty: **Intermediate/Advanced**]

```
transcriptFilechr21=system.file("extdata",
                       "refseq.hg19.chr21.bed",
                       package="compGenomRData")
```

2.  Extract -500,+500 bp regions around the TSSes on chr20. Use H3K4me3 (`H1.ESC.H3K4me3.chr20.bw`) and H3K27ac (`H1.ESC.H3K27ac.chr20.bw`) ChIP-seq enrichment data in the `compGenomRData` package and create heatmaps and average signal profiles for regions around the TSSes.[Difficulty: **Intermediate/Advanced**]

3.  Download P300 ChIP-seq peaks data from the UCSC browser. The peaks are locations where P300 binds. The P300 binding marks enhancer regions in the genome. (**HINT**: group: "regulation", track: "Txn Factor ChIP", table:"wgEncodeRegTfbsClusteredV3", you need to filter the rows for "EP300" name.) Check enrichment of H3K4me3, H3K27ac and DNase-seq (`H1.ESC.dnase.chr20.bw`) experiments on chr20 on and arounf the P300 binding-sites, use data from `compGenomRData` package. Make multi-heatmaps and metaplots. What is different from the TSS profiles? [Difficulty: **Advanced**]

4.  Cluster the rows of multi-heatmaps for the task above. Are there obvious clusters? **HINT:** Check arguments of the `multiHeatMatrix()` function. [Difficulty: **Advanced**]

5.  Visualize one of the -500,+500 bp regions around the TSS using `Gviz` functions. You should visualize both H3K4me3 and H3K27ac and the gene models. [Difficulty: **Advanced**]

# 7

## Quality Check, Processing and Alignment of High-throughput Sequencing Reads

Advances in sequencing technology are helping researchers sequence the genome deeper than ever. These sequencing experiments typically yield millions of reads. These reads have to be further processed, quality checked and aligned before we can quantify the genomic signal of interest and apply statistics and/or machine learning methods. For example, you may want to count how many reads overlap with your promoter set of interest or you may want to quantify RNA-seq reads that overlap with exons. Post-alignment operations are usually, but not always, similar to operations on genomic intervals. Dealing with mapped reads is described previously in Chapter 6. In addition, we have introduced high-throughput sequencing and its applications in general in Chapter 1. In this chapter we will introduce the fundamentals of read processing and quality check, and we will show how to do those tasks in R. The read quality check and processing is a fundamental step in all high-throughput sequencing analyses. For example, RNA-seq, ChIP-seq and BS-seq analyses shown in Chapters 8, 9 and 10 require these quality check and processing steps prior to further analysis. For a long time, quality check and mapping tasks were outside the R domain. However, nowadays certain packages in R/Bioconductor can accomplish those tasks.

### 7.1  FASTA and FASTQ formats

High-throughput sequencing reads are usually output from sequencing facilities as text files in a format called "FASTQ" or "fastq". This format depends on an earlier format called FASTA. The FASTA format was developed as a text-based format to represent nucleotide or protein sequences (see Figure 7.1 for an example).

The first line in a FASTA file usually starts with a ">" (greater-than) symbol. This first line is called the "description line", and can contain descriptive information about the sequence in the subsequent lines. The description can be the ID or name of the sequence such as gene names. However, very infrequently you may see lines starting with a ";" (semicolon). These lines will be taken as a comment, and can hold additional descriptive information about the sequence in subsequent lines.

```
>NG_008679.1:5001-38170 Homo sapiens paired box 6 (PAX6)
ACCCTCTTTTCTTATCATTGACATTTAAACTCTGGGGCAGGTCCTCGCGTAGAACGCGGCTGTCAGATCT
GCCACTTCCCCTGCCGAGCGGCGGTGAGAAGTGTGGGAACCGGCGCTGCCAGGCTCACCTGCCTCCCCGC
CCTCCGCTCCCAGGTAACCGCCCGGGCTCCGGCCCCGGCCCGGCTCGGGGCCCGCGGGGCCTCTCCGCTG
CCAGCGACTGCTGTCCCCAAATCAAAGCCCGCCCCAAGTGGCCCCGGGGCTTGATTTTTGCTTTTAAAAG
GAGGCATACAAAGATGGAAGCGAGTTACTGAGGGAGGGATAGGAAGGGGGGTGGAGGAGGGACTTGTCTT
TGCCGAGTGTGCTCTTCTGCAAAAGTAGCAAAATGTTCCACTCCTAAGAGTGGACTTCCAGTCCGGCCCT
GAGCTGGGAGTAGGGGGCGGGAGTCTGCTGCTGCTGTCTGCTAAAGCCACTCGCGACCGCGAAAAATGCA
GGAGGTGGGGACGCACTTTGCATCCAGACCTCCTCTGCATCGCAGTTCACGACATCCACGCTTGGGAAAG
TCCGTACCCGCGCCTGGAGCGCTTAAAGACACCCTGCCGCGGGTCGGGCGAGGTGCAGCAGAAGTTTCCC
GCGGTTGCAAAGTGCAGATGGCTGGACCGCAACAAAGTCTAGAGATGGGGTTCGTTTCTCAGAAAGACGC
```

**FIGURE 7.1:** An example fasta file showing the first part of the PAX6 gene.

An extension of the FASTA format is FASTQ format. This format is designed to handle base quality metrics output from sequencing machines. In this format, both the sequence and quality scores are represented as single ASCII characters. The format uses four lines for each sequence, and these four lines are stacked on top of each other in text files output by sequencing workflows. Each of the 4 lines will represent a read. Figure 7.2 shows those four lines with brief explanations for each line.

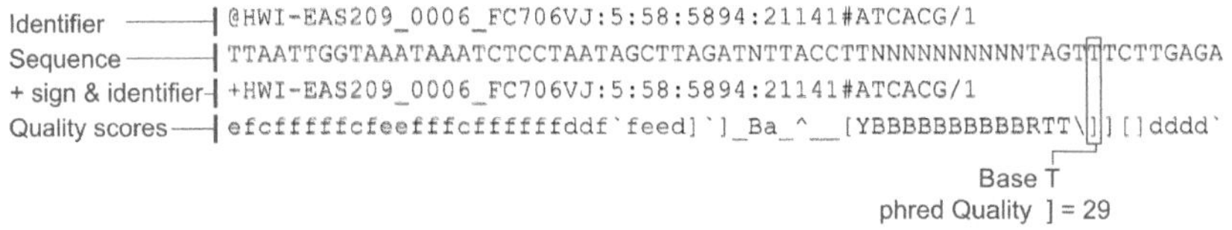

**FIGURE 7.2:** FASTQ format and a brief explanation of each line in the format.

**Line 1** begins with the '@' character and is followed by a sequence identifier and an optional description. This line is utilized by the sequencing technology, and usually contains specific information for the technology. It can contain flow cell IDs, lane numbers, and information on read pairs. **Line 2** is the sequence letters. **Line 3** begins with a '+' character; it marks the end of the sequence and is optionally followed by the same sequence identifier again in line 1. **Line 4** encodes the quality values for the sequence in Line 2, and must contain the same number of symbols as letters in the sequence. Each letter corresponds to a quality score. Although there might be different definitions of the quality scores, a *de facto* standard in the field is to use "Phred quality scores". These scores represent the likelihood of the base being called wrong. Formally, $Q_{\text{phred}} = -10\log_{10} e$, where $e$ is the probability that the base is called wrong. Since the score is in minus log scale, the higher the score, the more unlikely that the base is called wrong.

## 7.2   Quality check on sequencing reads

The sequencing technologies usually produce basecalls with varying quality. In addition, there could be sample-specific issues in your sequencing run, such as adapter contamination. It is standard procedure to check the quality of the reads and identify problems before doing further analysis. Checking the quality and making some decisions for the downstream analysis can influence the outcome of your project.

Below, we will walk you through the quality check steps using the Rqc[1] package. First, we need to feed fastq files to the rqc() function and obtain an object with sequence quality-related results. We are using example fastq files from the ShortRead package.

```
library(Rqc)
folder = system.file(package="ShortRead", "extdata/E-MTAB-1147")

# feeds fastq.qz files in "folder" to quality check function
qcRes=rqc(path = folder, pattern = ".fastq.gz", openBrowser=FALSE)
```

### 7.2.1   Sequence quality per base/cycle

Now that we have the qcRes object, we can plot various sequence quality metrics for our fastq files. We will first plot "sequence quality per base/cycle". This plot, shown in Figure 7.3, depicts the quality scores across all bases at each position in the reads.

```
rqcCycleQualityBoxPlot(qcRes)
```

In our case, the x-axis in the plot is labeled as "cycle". This is because in each sequencing "cycle" a fluorescently labeled nucleotide is added to complement the template sequence, and the sequencing machine identifies which nucleotide is added. Therefore, cycles correspond to bases/nucleotides along the read, and the number of cycles is equivalent to the read length.

Long sequences can have degraded quality towards the ends of the reads. Looking at quality distribution over base positions can help us decide to do trimming towards the end of the reads or not. A good sample will have median quality scores per base

---

[1]https://bioconductor.org/packages/release/bioc/html/Rqc.html

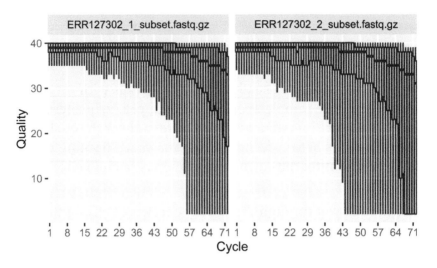

**FIGURE 7.3:** Per base sequence quality boxplot.

above 28. If scores are below 20 towards the ends, you can think about trimming the reads.

### 7.2.2   Sequence content per base/cycle

Per-base sequence content shows nucleotide proportions for each position. In a random sequencing library there should be no nucleotide bias and the lines should be almost parallel with each other. The code below shows how to get this plot. The resulting plot is shown in Figure 7.4.

```
rqcCycleBaseCallsLinePlot(qcRes)
```

However, some types of sequencing libraries can produce a biased sequence composition. For example, in RNA-Seq, it is common to have bias at the beginning of the reads. This happens because of random primers annealing to the start of reads during RNA-Seq library preparation. These primers are not truly random, which leads to a variation at the beginning of the reads. Although RNA-seq experiments will usually have these biases, this will not affect the ability of measuring gene expression.

In addition, some libraries are inherently biased in their sequence composition. For example, in bisulfite sequencing experiments, most of the cytosines will be converted to thymines. This will create a difference in C and T base compositions over the read, however this type of difference is normal for bisulfite sequencing experiments.

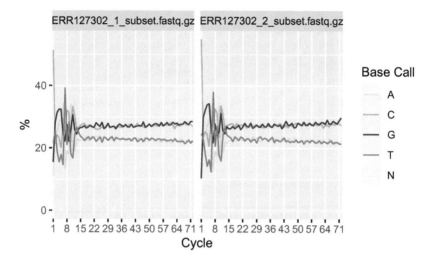

**FIGURE 7.4:** Percentage of nucleotide bases per position across different FASTQ files.

### 7.2.3 Read frequency plot

This plot shows the degree of duplication for every read in the library. We show how to get this plot in the code snippet below and the resulting plot is in Figure 7.5. A high level of duplication, non-unique reads, is likely to indicate an enrichment bias. Technical duplicates arising from PCR artifacts could cause this. PCR is a common step in library preparation which creates many copies of the sequence fragment. In RNA-seq data, the non-unique read proportion can reach more than 20%. However, these duplications may stem from genes simply being expressed at high levels. This means that there will be many copies of transcripts and many copies of the same fragment. Since we cannot be sure these duplicated reads are due to PCR bias or an effect of high transcription, we should not remove duplicated reads in RNA-seq analysis. However, in ChIP-seq experiments duplicated reads are more likely to be due to PCR bias.

```
rqcReadFrequencyPlot(qcRes)
```

### 7.2.4 Other quality metrics and QC tools

Over-represented k-mers along the reads can be an additional check. If there are such sequences it may point to adapter contamination and should be trimmed. Adapters are known sequences that are added to the ends of the reads. This kind of contamination could also be visible at "sequence content per base" plots. In addition, if you know the adapter sequences, you can match it to the end of the reads and trim them. The most popular tool for sequencing quality control is the fastQC tool

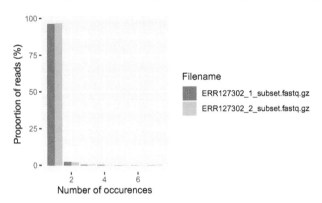

**FIGURE 7.5:** The percent of different duplication levels in FASTQ files. Most of the reads in all libraries have only one copy in this case.

(Andrews, 2010), which is written in Java. It produces the plots that we described above in addition to k-mer overrepresentation and adapter overrepresentation plots. The R package fastqcr[2] can run this Java tool and produce R-based plots and reports. This package simply calls the Java tool and parses its results. Below, we show how to do that.

```
library(fastqcr)

# install the FASTQC java tool
fastqc_install()

# call FASTQC and record the resulting statistics
# in fastqc_results folder
fastqc(fq.dir = folder,qc.dir = "fastqc_results")
```

Now that we have run FastQC on our fastq files, we can read the results to R and construct plots or reports. The `gc_report()` function can create an Rmarkdown-based report from FastQC output.

```
# view the report rendered by R functions
qc_report(qc.path="fastqc_results",
          result.file="reportFile", preview = TRUE)
```

Alternatively, we can read the results with `qc_read()` and make specific plots we are interested in with `qc_plot()`.

---

[2]https://cran.r-project.org/web/packages/fastqcr/index.html

```
# read QC results to R for one fastq file
qc <- qc_read("fastqc_results/ERR127302_1_subset_fastqc.zip")

# make plots, example "Per base sequence quality plot"
qc_plot(qc, "Per base sequence quality")
```

Apart from this, the bioconductor packages Rqc (de Souza et al., 2018) (see `Rqc::rqcReport` function), QuasR (Gaidatzis et al., 2015) (see `QuasR::qQCReport` function), systemPipeR (Backman and Girke, 2016) (see `systemPipeR::seeFastq` function), and ShortRead (Morgan et al., 2009) (see `ShortRead::report` function) can all generate quality reports in a similar fashion to FastQC with some differences in plot content and number.

## 7.3 Filtering and trimming reads

Based on the results of the quality check, you may want to trim or filter the reads. The quality check might have shown the number of reads that have low quality scores. These reads will probably not align very well because of the potential mistakes in base calling, or they may align to wrong places in the genome. Therefore, you may want to remove these reads from your fastq file. Another potential scenario is that parts of your reads need to be trimmed in order to align the reads. In some cases, adapters will be present in either side of the read; in other cases technical errors will lead to decreasing base quality towards the ends of the reads. Both in these cases, the portion of the read should be trimmed so the read can align or better align the genome. We will show how to use the `QuasR` package to trim the reads. Other packages such as `ShortRead` also have capabilities to trim and filter reads. However, the `QuasR::preprocessReads()` function provides a single interface to multiple preprocessing possibilities. With this function, we match adapter sequences and remove them. We can remove low-complexity reads (reads containing repetitive sequences). We can trim the start or ends of the reads by a pre-defined length.

Below we will first set up the file paths to fastq files and filter them based on their length and whether or not they contain the "N" character, which stands for unidentified base. With the same function we will also trim 3 bases from the end of the reads and also trim segments from the start of the reads if they match the "ACCCGGGA" sequence.

```
library(QuasR)
```

```
# obtain a list of fastq file paths
fastqFiles <- system.file(package="ShortRead",
                          "extdata/E-MTAB-1147",
                          c("ERR127302_1_subset.fastq.gz",
                            "ERR127302_2_subset.fastq.gz")
)


# defined processed fastq file names
outfiles <- paste(tempfile(pattern=c("processed_1_",
                           "processed_2_")),".fastq",sep="")


# process fastq files
# remove reads that have more than 1 N, (nBases)
# trim 3 bases from the end of the reads (truncateEndBases)
# Remove ACCCGGGA patern if it occurs at the start (Lpattern)
# remove reads shorter than 40 base-pairs (minLength)
preprocessReads(fastqFiles, outfiles,
                nBases=1,
                truncateEndBases=3,
                Lpattern="ACCCGGGA",
                minLength=40)
```

As we have mentioned, the ShortRead package has low-level functions, which QuasR::preprocessReads() also depends on. We can use these low-level functions to filter reads in ways that are not possible using the QuasR::preprocessReads() function. Below we are going to read in a fastq file and filter the reads where every quality score is below 20.

```
library(ShortRead)


# obtain a list of fastq file paths
fastqFile <- system.file(package="ShortRead",
                         "extdata/E-MTAB-1147",
                         "ERR127302_1_subset.fastq.gz")


# read fastq file
fq = readFastq(fastqFile)


# get quality scores per base as a matrix
qPerBase = as(quality(fq), "matrix")
```

```
# get number of bases per read that have quality score below 20
# we use this
qcount = rowSums( qPerBase <= 20)


# Number of reads where all Phred scores >= 20
fq[qcount == 0]


## class: ShortReadQ
## length: 10699 reads; width: 72 cycles
```

We can finally write out the filtered fastq file with the ShortRead::writeFastq() function.

```
# write out fastq file with only reads where all
# quality scores per base are above 20
writeFastq(fq[qcount == 0],
          paste(fastqFile, "Qfiltered", sep="_"))
```

As fastq files can be quite large, it may not be feasible to read a 30-Gigabyte file into memory. A more memory-efficient way would be to read the file piece by piece. We can do our filtering operations for each piece, write the filtered part out, and read a new piece. Fortunately, this is possible using the ShortRead::FastqStreamer() function. This function enables "streaming" the fastq file in pieces, which are blocks of the fastq file with a pre-defined number of reads. We can access the successive blocks with the yield() function. Each time we call the yield() function after opening the fastq file with FastqStreamer(), a new part of the file will be read to the memory.

```
# set up streaming with block size 1000
# every time we call the yield() function 1000 read portion
# of the file will be read successively.
f <- FastqStreamer(fastqFile,readerBlockSize=1000)

# we set up a while loop to call yield() function to
# go through the file
while(length(fq <- yield(f))) {

    # remove reads where all quality scores are < 20
    # get quality scores per base as a matrix
    qPerBase = as(quality(fq), "matrix")
```

```
# get number of bases per read that have Q score < 20
qcount = rowSums( qPerBase <= 20)


# write fastq file with mode="a", so every new block
# is written out to the same file
writeFastq(fq[qcount == 0],
           paste(fastqFile, "Qfiltered", sep="_"),
           mode="a")
}
```

## 7.4  Mapping/aligning reads to the genome

After the quality check and potential pre-processing, the reads are ready to be mapped or aligned to the reference genome. This process simply finds the most probable origin of each read in the genome. Since there might be errors in sequencing and mutations in the genomes, we may not find exact matches of reads in the genomes. An important feature of the alignment algorithms is to tolerate potential mismatches between reads and the reference genome. In addition, efficient algorithms and data structures are needed for the alignment to be completed in a reasonable amount of time. Alignment methods usually create data structures to store and efficiently search the genome for matching reads. These data structures are called genome indices and creating these indices is the first step for the read alignment. Based on how indices are created, there are two major types of methods. One class of methods relies on "hash tables", to store and search the genomes. Hash tables are simple lookup tables in which all possible k-mers point to locations in the genome. The general idea is that overlapping k-mers constructed from a read go through this lookup table. Each k-mer points to potential locations in the genome. Then, the final location for the read is obtained by optimizing the k-mer chain by their distances in the genome and in the read. This optimization process removes k-mer locations that are distant from other k-mers that map nearby each other.

Another class of algorithms builds genome indices by creating a Burrows-Wheeler transformation of the genome. This in essence creates a compact and searchable data structure for all reads. Although details are out of the scope of this section, these alignment tools provide faster alignment and use less memory. BWA(Li and Durbin, 2009a), Bowtie1/2(Langmead and Salzberg, 2012a) and SOAP(Li et al., 2009) are examples of such algorithms.

The read mapping in R can be done with the gmapR (Barr et al., 2019), QuasR (Gaidatzis et al., 2015), Rsubread (Liao et al., 2013), and systemPipeR (Backman and Girke, 2016) packages. We will demonstrate read mapping with QuasR which uses the Rbowtie package, which wraps the Bowtie aligner. Below, we show how to map reads from a ChIP-seq experiment using QuasR/bowtie.

We will use the qAlign() function which requires two mandatory arguments: 1) a genome file in either fasta format or as a BSgenome package and 2) a sample file which is a text file and contains file paths to fastq files and sample names. In the case below, sample file looks like this:

```
FileName     SampleName
chip_1_1.fq.bz2 Sample1
chip_2_1.fq.bz2 Sample2
```

```
library(QuasR)

# copy example data to current working directory
file.copy(system.file(package="QuasR", "extdata"), ".", recursive=TRUE)

# genome file in fasta format
genomeFile <- "extdata/hg19sub.fa"

# text file containing sample names and fastq file paths
sampleFile <- "extdata/samples_chip_single.txt"

# create alignments
proj <- qAlign(sampleFile, genomeFile)
```

It is good to explain what is going on here as the qAlign() function makes things look simple. This function is designed to be easy. For example, it creates a genome index automatically if it does not exist, and will look for existing indices before it creates one. We provided only two arguments, a text file containing sample names and fastq file paths and a reference genome file. In fact, this function also has many knobs and you can change its behavior by supplying different arguments in order to affect the behavior of Bowtie. For example, you can supply parameters to Bowtie using the alignmentParameter argument. However the qAlign() function is optimized for different types of alignment problems and selects alignment parameters automatically. It is designed to work with alignment and quantification tasks for RNA-seq, ChIP-seq, small-RNA sequencing, Bisulfite sequencing (DNA methylation) and allele-specific analysis. If you want to change the default bowtie parameters, only do it for simple alignment problems such as ChIP-seq and RNA-seq.

**– Want to know more ?**

    \* More on hash tables and Burrows-Wheeler-based aligners

        · A survey of sequence alignment algorithms for next-generation sequencin: (`https://academic.oup.com/bib/article/11/5/473/264166`) H Li, N Homer - Briefings in bioinformatics, 2010

    \* More on QuasR and all the alignment and post-processing capabilities: (`https://bioconductor.org/packages/release/bioc/vignettes/QuasR/inst/doc/QuasR.html`)

## 7.5 Further processing of aligned reads

After alignment, some further processing might be necessary. However, these steps are usually sequencing protocol specific. For example, for methylation obtained via bisulfite sequencing, C->T mismatches should be counted. For gene expression measurements, reads that overlap with transcripts should be counted. These further processing tasks are either done by a specialized alignment-related software or can be done in R in some cases. We will explain these further processing steps when they become relevant in the context of the following chapters.

## 7.6 Exercises

For this set of exercises, we will use the `chip_1_1.fq.bz2` and `chip_2_1.fq.bz2` files from the `QuasR` package. You can reach the folder that contains the files as follows:

```
folder=(system.file(package="QuasR", "extdata"))
dir(folder) # will show the contents of the folder
```

1. Plot the base quality distributions of the ChIP-seq samples `Rqc` package. **HINT**: You need to provide a regular expression pattern for extracting the right files from the folder. `"^chip"` matches the files beginning with "chip". [Difficulty: **Beginner/Intermediate**]

2. Now we will trim the reads based on the quality scores. Let's trim 2-4

bases on the 3' end depending on the quality scores. You can use the `QuasR::preprocessReads()` function for this purpose. [Difficulty: **Beginner/Intermediate**]

3. Align the trimmed and untrimmed reads using `QuasR` and plot alignment statistics, did the trimming improve alignments? [Difficulty: **Intermediate/Advanced**]

# 8

## RNA-seq Analysis

*Chapter Author:* **Bora Uyar**

RNA sequencing (RNA-seq) has proven to be a revolutionary tool since the time it was introduced. The throughput, accuracy, and resolution of data produced with RNA-seq has been instrumental in the study of transcriptomics in the last decade (Wang et al., 2009). There is a variety of applications of transcriptome sequencing and each application may consist of different chains of tools each with many alternatives (Conesa et al., 2016). In this chapter, we are going to demonstrate a common workflow for doing differential expression analysis with downstream applications such as GO term and gene set enrichment analysis. We assume that the sequencing data was generated using one of the NGS sequencing platforms. Where applicable, we will try to provide alternatives to the reader in terms of both the tools to carry out a demonstrated analysis and also the other applications of the same sequencing data depending on the different biological questions.

## 8.1 What is gene expression?

Gene expression is a term used to describe the contribution of a gene to the overall functions and phenotype of a cell through the activity of the molecular products, which are encoded in the specific nucleotide sequence of the gene. RNA is the primary product encoded in a gene, which is transcribed in the nucleus of a cell. A class of RNA molecules, messenger RNAs, are transported from the nucleus to the cytoplasm, where the translation machinery of the cell translates the nucleotide sequence of the mRNA into proteins. The functional protein repertoire in a given cell is the primary factor that dictates the shape, function, and phenotype of a cell. Due to the prime roles of proteins for a cell's fate, most molecular biology literature is focused on protein-coding genes. However, a bigger proportion of a eukaryotic gene repertoire is reserved for non-coding genes, which code for RNA molecules that are not translated into proteins, yet carry out many important cellular functions. All in all, the term gene expression refers to the combined activity of protein-coding or non-coding products of a gene.

In a cell, there are many layers of quality controls and modifications that act upon a gene's product until the end-product attains a particular function. These layers of regulation include epigenetic, transcriptional, post-transcriptional, translational, and post-translational control mechanisms, the latter two applying only to protein-coding genes. A protein or RNA molecule, is only functional if it is produced at the right time, at the right cellular compartment, with the necessary base or amino-acid modifications, with the correct secondary/tertiary structure (or unstructure wherever applicable), among the availability of other metabolites or molecules, which are needed to form complexes to altogether accomplish a certain cellular function. However, traditionally, the number of copies of a gene's products is considered a quantitative measure of a gene's activity. Although this approach does not reflect all of the complexity that defines a functional molecule, quantification of the abundance of transcripts from a gene has proven to be a cost-effective method of understanding genes' functions.

## 8.2   Methods to detect gene expression

Quantification of how much gene expression levels deviate from a baseline gives clues about which genes are actually important for, for instance, disease outcome or cell/tissue identity. The methods of detecting and quantifying gene expression have evolved from low-throughput methods such as the usage of a reporter gene with a fluorescent protein product to find out if a single gene is expressed at all, to high-throughput methods such as massively parallel RNA-sequencing that can profile -at single-nucleotide resolution- the abundance of tens of thousands of distinct transcripts encoded in the largest eukaryotic genomes.

## 8.3   Gene expression analysis using high-throughput sequencing technologies

With the advent of the second-generation (a.k.a next-generation or high-throughput) sequencing technologies, the number of genes that can be profiled for expression levels with a single experiment has increased to the order of tens of thousands of genes. Therefore, the bottleneck in this process has become the data analysis rather than the data generation. Many statistical methods and computational tools are required for getting meaningful results from the data, which comes with a lot of valuable information along with a lot of sources of noise. Fortunately, most of the steps of RNA-seq analysis have become quite mature over the years. Below we will first describe how to reach a read count table from raw fastq reads obtained from an Illumina sequencing run. We will then demonstrate in R how to process the count table, make a case-control differential expression analysis, and do some downstream functional enrichment analysis.

### 8.3.1 Processing raw data

#### 8.3.1.1 Quality check and read processing

The first step in any experiment that involves high-throughput short-read sequencing should be to check the sequencing quality of the reads before starting to do any downstream analysis. The quality of the input sequences holds fundamental importance in the confidence for the biological conclusions drawn from the experiment. We have introduced quality check and processing in Chapter 7, and those tools and workflows also apply in RNA-seq analysis.

#### 8.3.1.2 Improving the quality

The second step in the RNA-seq analysis workflow is to improve the quality of the input reads. This step could be regarded as an optional step when the sequencing quality is very good. However, even with the highest-quality sequencing datasets, this step may still improve the quality of the input sequences. The most common technical artifacts that can be filtered out are the adapter sequences that contaminate the sequenced reads, and the low-quality bases that are usually found at the ends of the sequences. Commonly used tools in the field (trimmomatic (Bolger et al., 2014), trimGalore (Andrews, 2010)) are again not written in R, however there are alternative R libraries for carrying out the same functionality, for instance, QuasR (Gaidatzis et al., 2015) (see `QuasR::preprocessReads` function) and ShortRead (Morgan et al., 2009) (see `ShortRead::filterFastq` function). Some of these approaches are introduced in Chapter 7.

The sequencing quality control and read pre-processing steps can be visited multiple times until achieving a satisfactory level of quality in the sequence data before moving on to the downstream analysis steps.

### 8.3.2 Alignment

Once a decent level of quality in the sequences is reached, the expression level of the genes can be quantified by first mapping the sequences to a reference genome, and secondly matching the aligned reads to the gene annotations, in order to count the number of reads mapping to each gene. If the species under study has a well-annotated transcriptome, the reads can be aligned to the transcript sequences instead of the reference genome. In cases where there is no good quality reference genome or transcriptome, it is possible to de novo assemble the transcriptome from the sequences and then quantify the expression levels of genes/transcripts.

For RNA-seq read alignments, apart from the availability of reference genomes and annotations, probably the most important factor to consider when choosing an alignment tool is whether the alignment method considers the absence of intronic regions in the sequenced reads, while the target genome may contain introns. Therefore, it is important to choose alignment tools that take into account alternative splicing. In the basic setting where a read, which originates from a cDNA sequence corresponding to an exon-exon junction, needs to be split into two parts when aligned against the genome. There are various tools that consider this factor such as STAR (Dobin et al., 2013), Tophat2 (Kim et al., 2013), Hisat2 (Kim et al., 2015), and GSNAP (Wu et al., 2016). Most alignment tools are written in C/C++ languages because of performance concerns. There are also R libraries that can do short read alignments; these are discussed in Chapter 7.

### 8.3.3   Quantification

After the reads are aligned to the target, a SAM/BAM file sorted by coordinates should have been obtained. The BAM file contains all alignment-related information of all the reads that have been attempted to be aligned to the target sequence. This information consists of - most basically - the genomic coordinates (chromosome, start, end, strand) of where a sequence was matched (if at all) in the target, specific insertions/deletions/mismatches that describe the differences between the input and target sequences. These pieces of information are used along with the genomic coordinates of genome annotations such as gene/transcript models in order to count how many reads have been sequenced from a gene/transcript. As simple as it may sound, it is not a trivial task to assign reads to a gene/transcript just by comparing the genomic coordinates of the annotations and the sequences, because of confounding factors such as overlapping gene annotations, overlapping exon annotations from different transcript isoforms of a gene, and overlapping annotations from opposite DNA strands in the absence of a strand-specific sequencing protocol. Therefore, for read counting, it is important to consider:

1.  Strand specificity of the sequencing protocol: Are the reads expected to originate from the forward strand, reverse strand, or unspecific?
2.  Counting mode: - when counting at the gene-level: When there are overlapping annotations, which features should the read be assigned to? Tools usually have a parameter that lets the user select a counting mode. - when counting at the transcript-level: When there are multiple isoforms of a gene, which isoform should the read be assigned to? This consideration is usually an algorithmic consideration that is not modifiable by the end-user.

Some tools can couple alignment to quantification (e.g. STAR), while some assume the alignments are already calculated and require BAM files as input. On the other hand, in the presence of good transcriptome annotations, alignment-free methods

(Salmon (Patro et al., 2017), Kallisto (Bray et al., 2016), Sailfish (Patro et al., 2014)) can also be used to estimate the expression levels of transcripts/genes. There are also reference-free quantification methods that can first de novo assemble the transcriptome and estimate the expression levels based on this assembly. Such a strategy can be useful in discovering novel transcripts or may be required in cases when a good reference does not exist. If a reference transcriptome exists but of low quality, a reference-based transcriptome assembler such as Cufflinks (Trapnell et al., 2010) can be used to improve the transcriptome. In case there is no available transcriptome annotation, a de novo assembler such as Trinity (Haas et al., 2013) or Trans-ABySS (Robertson et al., 2010) can be used to assemble the transcriptome from scratch.

Within R, quantification can be done using: - `Rsubread::featureCounts` - `QuasR::qCount` - `GenomicAlignments::summarizeOverlaps`

### 8.3.4 Within sample normalization of the read counts

The most common application after a gene's expression is quantified (as the number of reads aligned to the gene), is to compare the gene's expression in different conditions, for instance, in a case-control setting (e.g. disease versus normal) or in a time-series (e.g. along different developmental stages). Making such comparisons helps identify the genes that might be responsible for a disease or an impaired developmental trajectory. However, there are multiple caveats that needs to be addressed before making a comparison between the read counts of a gene in different conditions (Maza et al., 2013).

- Library size (i.e. sequencing depth) varies between samples coming from different lanes of the flow cell of the sequencing machine.
- Longer genes will have a higher number of reads.
- Library composition (i.e. relative size of the studied transcriptome) can be different in two different biological conditions.
- GC content biases across different samples may lead to a biased sampling of genes (Risso et al., 2011).
- Read coverage of a transcript can be biased and non-uniformly distributed along the transcript (Mortazavi et al., 2008).

Therefore these factors need to be taken into account before making comparisons.

The most basic normalization approaches address the sequencing depth bias. Such procedures normalize the read counts per gene by dividing each gene's read count by a certain value and multiplying it by $10^6$. These normalized values are usually referred to as CPM (counts per million reads):

- Total Counts Normalization (divide counts by the **sum** of all counts)

- Upper Quartile Normalization (divide counts by the **upper quartile** value of the counts)
- Median Normalization (divide counts by the **median** of all counts)

Popular metrics that improve upon CPM are RPKM/FPKM (reads/fragments per kilobase of million reads) and TPM (transcripts per million). RPKM is obtained by dividing the CPM value by another factor, which is the length of the gene per kilobase. FPKM is the same as RPKM, but is used for paired-end reads. Thus, RPKM/FPKM methods account for, firstly, the **library size**, and secondly, the **gene lengths**.

TPM also controls for both the library size and the gene lengths, however, with the TPM method, the read counts are first normalized by the gene length (per kilobase), and then gene-length normalized values are divided by the sum of the gene-length normalized values and multiplied by 10^6. Thus, the sum of normalized values for TPM will always be equal to 10^6 for each library, while the sum of RPKM/FPKM values do not sum to 10^6. Therefore, it is easier to interpret TPM values than RPKM/FPKM values.

### 8.3.5   Computing different normalization schemes in R

Here we will assume that there is an RNA-seq count table comprising raw counts, meaning the number of reads counted for each gene has not been exposed to any kind of normalization and consists of integers. The rows of the count table correspond to the genes and the columns represent different samples. Here we will use a subset of the RNA-seq count table from a colorectal cancer study. We have filtered the original count table for only protein-coding genes (to improve the speed of calculation) and also selected only five metastasized colorectal cancer samples along with five normal colon samples. There is an additional column width that contains the length of the corresponding gene in the unit of base pairs. The length of the genes are important to compute RPKM and TPM values. The original count tables can be found from the recount2 database (https://jhubiostatistics. shinyapps.io/recount/) using the SRA project code *SRP029880*, and the experimental setup along with other accessory information can be found from the NCBI Trace archive using the SRA project code SRP029880[1].

```
#colorectal cancer
counts_file <- system.file("extdata/rna-seq/SRP029880.raw_counts.tsv",
                           package = "compGenomRData")
coldata_file <- system.file("extdata/rna-seq/SRP029880.colData.tsv",
                            package = "compGenomRData")
```

---

[1]https://trace.ncbi.nlm.nih.gov/Traces/sra/?study=SRP029880

```
counts <- as.matrix(read.table(counts_file, header = T, sep = '\t'))
```

### 8.3.5.1 Computing CPM

Let's do a summary of the counts table. Due to space limitations, the summary for only the first three columns is displayed.

```
summary(counts[,1:3])
```

```
##      CASE_1                CASE_2                CASE_3
## Min.   :        0   Min.   :        0   Min.   :        0
## 1st Qu.:     5155   1st Qu.:     6464   1st Qu.:     3972
## Median :    80023   Median :    85064   Median :    64145
## Mean   :   295932   Mean   :   273099   Mean   :   263045
## 3rd Qu.:   252164   3rd Qu.:   245484   3rd Qu.:   210788
## Max.   :205067466   Max.   :105248041   Max.   :222511278
```

To compute the CPM values for each sample (excluding the `width` column):

```
cpm <- apply(subset(counts, select = c(-width)), 2,
             function(x) x/sum(as.numeric(x)) * 10^6)
```

Check that the sum of each column after normalization equals to 10^6 (except the width column).

```
colSums(cpm)
```

```
## CASE_1 CASE_2 CASE_3 CASE_4 CASE_5 CTRL_1 CTRL_2 CTRL_3 CTRL_4 CTRL_5
##  1e+06  1e+06  1e+06  1e+06  1e+06  1e+06  1e+06  1e+06  1e+06  1e+06
```

### 8.3.5.2 Computing RPKM

```
# create a vector of gene lengths
geneLengths <- as.vector(subset(counts, select = c(width)))

# compute rpkm
rpkm <- apply(X = subset(counts, select = c(-width)),
              MARGIN = 2,
              FUN = function(x) {
```

```
            10^9 * x / geneLengths / sum(as.numeric(x))
        })
```

Check the sample sizes of RPKM. Notice that the sums of samples are all different.

```
colSums(rpkm)
```

```
##     CASE_1    CASE_2    CASE_3    CASE_4    CASE_5    CTRL_1    CTRL_2    CTRL_3
## 158291.0 153324.2 161775.4 173047.4 172761.4 210032.6 301764.2 241418.3
##     CTRL_4    CTRL_5
## 291674.5 252005.7
```

### 8.3.5.3  Computing TPM

```
#find gene length normalized values
rpk <- apply( subset(counts, select = c(-width)), 2,
             function(x) x/(geneLengths/1000))
#normalize by the sample size using rpk values
tpm <- apply(rpk, 2, function(x) x / sum(as.numeric(x)) * 10^6)
```

Check the sample sizes of tpm. Notice that the sums of samples are all equal to 10^6.

```
colSums(tpm)
```

```
## CASE_1 CASE_2 CASE_3 CASE_4 CASE_5 CTRL_1 CTRL_2 CTRL_3 CTRL_4 CTRL_5
##  1e+06  1e+06  1e+06  1e+06  1e+06  1e+06  1e+06  1e+06  1e+06  1e+06
```

None of these metrics (CPM, RPKM/FPKM, TPM) account for the other important confounding factor when comparing expression levels of genes across samples: the **library composition**, which may also be referred to as the **relative size of the compared transcriptomes**. This factor is not dependent on the sequencing technology, it is rather biological. For instance, when comparing transcriptomes of different tissues, there can be sets of genes in one tissue that consume a big chunk of the reads, while in the other tissues they are not expressed at all. This kind of imbalance in the composition of compared transcriptomes can lead to wrong conclusions about which genes are actually differentially expressed. This consideration is addressed in two popular R packages: DESeq2 (Love et al., 2014) and edgeR (Robinson et al., 2010) each with a different algorithm. edgeR uses a normalization procedure called Trimmed Mean of M-values (TMM). DESeq2 implements a normalization procedure using median of Ratios, which is obtained by finding the ratio of the log-transformed count of a gene divided by the average of log-transformed values of the gene in all samples (geometric mean), and then taking the median of these values for all genes. The raw read count of the gene is finally divided by this value (median of ratios) to obtain the normalized counts.

### 8.3.6 Exploratory analysis of the read count table

A typical quality control, in this case interrogating the RNA-seq experiment design, is to measure the similarity of the samples with each other in terms of the quantified expression level profiles across a set of genes. One important observation to make is to see whether the most similar samples to any given sample are the biological replicates of that sample. This can be computed using unsupervised clustering techniques such as hierarchical clustering and visualized as a heatmap with dendrograms. Another most commonly applied technique is a dimensionality reduction technique called Principal Component Analysis (PCA) and visualized as a two-dimensional (or in some cases three-dimensional) scatter plot. In order to find out more about the clustering methods and PCA, please refer to Chapter 4.

#### 8.3.6.1 Clustering

We can combine clustering and visualization of the clustering results by using heatmap functions that are available in a variety of R libraries. The basic R installation comes with the `stats::heatmap` function. However, there are other libraries available in CRAN (e.g. `pheatmap` (Kolde, 2019)) or Bioconductor (e.g. `ComplexHeatmap` (Gu et al., 2016a)) that come with more flexibility and more appealing visualizations.

Here we demonstrate a heatmap using the `pheatmap` package and the previously calculated `tpm` matrix. As these matrices can be quite large, both computing the clustering and rendering the heatmaps can take a lot of resources and time. Therefore, a quick and informative way to compare samples is to select a subset of genes that are, for instance, most variable across samples, and use that subset to do the clustering and visualization.

Let's select the top 100 most variable genes among the samples.

```
#compute the variance of each gene across samples
V <- apply(tpm, 1, var)
#sort the results by variance in decreasing order
#and select the top 100 genes
selectedGenes <- names(V[order(V, decreasing = T)][1:100])
```

Now we can quickly produce a heatmap where samples and genes are clustered (see Figure 8.1 ).

```
library(pheatmap)
pheatmap(tpm[selectedGenes,], scale = 'row', show_rownames = FALSE)
```

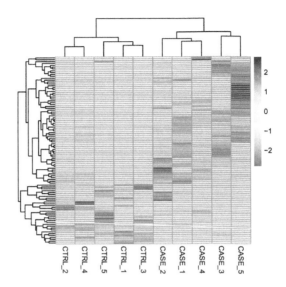

**FIGURE 8.1:** Clustering and visualization of the topmost variable genes as a heatmap.

We can also overlay some annotation tracks to observe the clusters. Here it is important to observe whether the replicates of the same sample cluster most closely with each other, or not. Overlaying the heatmap with such annotation and displaying sample groups with distinct colors helps quickly see if there are samples that don't cluster as expected (see Figure 8.2 ).

```
colData <- read.table(coldata_file, header = T, sep = '\t',
                      stringsAsFactors = TRUE)
pheatmap(tpm[selectedGenes,], scale = 'row',
         show_rownames = FALSE,
         annotation_col = colData)
```

### 8.3.6.2  PCA

Let's make a PCA plot to see the clustering of replicates as a scatter plot in two dimensions (Figure 8.3).

```
library(stats)
library(ggplot2)
#transpose the matrix
```

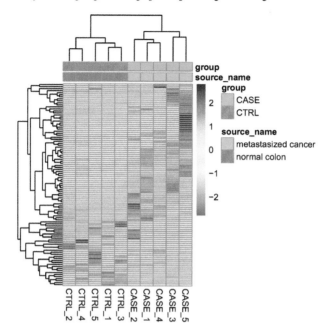

**FIGURE 8.2:** Clustering samples as a heatmap with sample annotations.

```
M <- t(tpm[selectedGenes,])
# transform the counts to log2 scale
M <- log2(M + 1)
#compute PCA
pcaResults <- prcomp(M)

#plot PCA results making use of ggplot2's autoplot function
#ggfortify is needed to let ggplot2 know about PCA data structure.
autoplot(pcaResults, data = colData, colour = 'group')
```

We should observe here whether the samples from the case group (CASE) and samples from the control group (CTRL) can be split into two distinct clusters on the scatter plot of the first two largest principal components.

We can use the summary function to summarize the PCA results to observe the contribution of the principal components in the explained variation.

```
summary(pcaResults)
```

```
## Importance of components:
##                             PC1       PC2       PC3       PC4       PC5      PC6       PC7
## Standard deviation       24.396   2.50514   2.39327   1.93841   1.79193   1.6357   1.46059
```

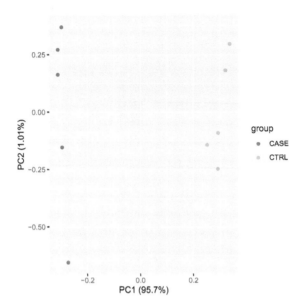

**FIGURE 8.3:** PCA plot of samples using TPM counts.

```
## Proportion of Variance  0.957 0.01009 0.00921 0.00604 0.00516 0.0043 0.00343
## Cumulative Proportion   0.957 0.96706 0.97627 0.98231 0.98747 0.9918 0.99520
##                            PC8     PC9     PC10
## Standard deviation      1.30902 1.12657 4.616e-15
## Proportion of Variance 0.00276 0.00204 0.000e+00
## Cumulative Proportion  0.99796 1.00000 1.000e+00
```

#### 8.3.6.3   Correlation plots

Another complementary approach to see the reproducibility of the experiments is to compute the correlation scores between each pair of samples and draw a correlation plot.

Let's first compute pairwise correlation scores between every pair of samples.

```
library(stats)
correlationMatrix <- cor(tpm)
```

Let's have a look at how the correlation matrix looks (8.1) (showing only two samples each of case and control samples):

We can also draw more visually appealing correlation plots using the `corrplot` package (Figure 8.4). Using the `addrect` argument, we can split clusters into groups and surround them with rectangles. By setting the `addCoef.col` argument to 'white', we can display the correlation coefficients as numbers in white color.

**TABLE 8.1:** Correlation scores between samples

|        | CASE_1    | CASE_2    | CTRL_1    | CTRL_2    |
|--------|-----------|-----------|-----------|-----------|
| CASE_1 | 1.0000000 | 0.9924606 | 0.9594011 | 0.9635760 |
| CASE_2 | 0.9924606 | 1.0000000 | 0.9725646 | 0.9793835 |
| CTRL_1 | 0.9594011 | 0.9725646 | 1.0000000 | 0.9879862 |
| CTRL_2 | 0.9635760 | 0.9793835 | 0.9879862 | 1.0000000 |

```
library(corrplot)
corrplot(correlationMatrix, order = 'hclust',
        addrect = 2, addCoef.col = 'white',
        number.cex = 0.7)
```

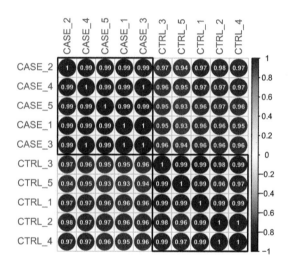

**FIGURE 8.4:** Correlation plot of samples ordered by hierarchical clustering.

Here pairwise correlation levels are visualized as colored circles. Blue indicates positive correlation, while Red indicates negative correlation.

We could also plot this correlation matrix as a heatmap (Figure 8.5). As all the samples have a high pairwise correlation score, using a heatmap instead of a corrplot helps to see the differences between samples more easily. The annotation_col argument helps to display sample annotations and the cutree_cols argument is set to 2 to split the clusters into two groups based on the hierarchical clustering results.

```
library(pheatmap)
# split the clusters into two based on the clustering similarity
pheatmap(correlationMatrix,
         annotation_col = colData,
         cutree_cols = 2)
```

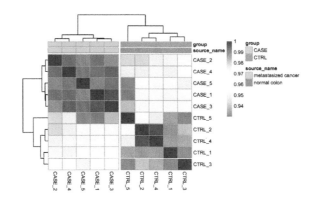

**FIGURE 8.5:** Pairwise correlation of samples displayed as a heatmap.

### 8.3.7  Differential expression analysis

Differential expression analysis allows us to test tens of thousands of hypotheses (one test for each gene) against the null hypothesis that the activity of the gene stays the same in two different conditions. There are multiple limiting factors that influence the power of detecting genes that have real changes between two biological conditions. Among these are the limited number of biological replicates, non-normality of the distribution of the read counts, and higher uncertainty of measurements for lowly expressed genes than highly expressed genes (Love et al., 2014). Tools such as edgeR and DESeq2 address these limitations using sophisticated statistical models in order to maximize the amount of knowledge that can be extracted from such noisy datasets. In essence, these models assume that for each gene, the read counts are generated by a negative binomial distribution. This is a popular distribution that is used for modeling count data. This distribution can be specified with a mean parameter, $m$, and a dispersion parameter, $\alpha$. The dispersion parameter $\alpha$ is directly related to the variance as the variance of this distribution is formulated as: $m + \alpha m^2$. Therefore, estimating these parameters is crucial for differential expression tests. The methods used in edgeR and DESeq2 use dispersion estimates from other genes with similar counts to precisely estimate the per-gene dispersion values. With accurate dispersion parameter estimates, one can estimate the variance more precisely, which in turn improves the result of the differential expression test. Although statistical models are different, the process here is similar to the moderated t-test and qualifies as an empirical Bayes method which we introduced in Chapter 3. There, we calculated gene-wise variability and shrunk each gene-wise variability towards the median variability of all genes. In the case of RNA-seq the dispersion coefficient $\alpha$ is shrunk towards the value of dispersion from other genes with similar read counts.

Now let us take a closer look at the DESeq2 workflow and how it calculates differential expression:

1. The read counts are normalized by computing size factors, which addresses the differences not only in the library sizes, but also the library compositions.
2. For each gene, a dispersion estimate is calculated. The dispersion value computed by DESeq2 is equal to the squared coefficient of variation (variation divided by the mean).
3. A line is fit across the dispersion estimates of all genes computed in step 2 versus the mean normalized counts of the genes.
4. Dispersion values of each gene are shrunk towards the fitted line in step 3.
5. A Generalized Linear Model is fitted which considers additional confounding variables related to the experimental design such as sequencing batches, treatment, temperature, patient's age, sequencing technology, etc., and uses negative binomial distribution for fitting count data.
6. For a given contrast (e.g. treatment type: drug-A versus untreated), a test for differential expression is carried out against the null hypothesis that the log fold change of the normalized counts of the gene in the given pair of groups is exactly zero.
7. It adjusts p-values for multiple-testing.

In order to carry out a differential expression analysis using DESeq2, three kinds of inputs are necessary:

1. The **read count table**: This table must be raw read counts as integers that are not processed in any form by a normalization technique. The rows represent features (e.g. genes, transcripts, genomic intervals) and columns represent samples.
2. A **colData** table: This table describes the experimental design.
3. A **design formula**: This formula is needed to describe the variable of interest in the analysis (e.g. treatment status) along with (optionally) other covariates (e.g. batch, temperature, sequencing technology).

Let's define these inputs:

```
#remove the 'width' column
countData <- as.matrix(subset(counts, select = c(-width)))
#define the experimental setup
colData <- read.table(coldata_file, header = T, sep = '\t',
                        stringsAsFactors = TRUE)
#define the design formula
designFormula <- "~ group"
```

Now, we are ready to run DESeq2.

```
library(DESeq2)
library(stats)
#create a DESeq dataset object from the count matrix and the colData
dds <- DESeqDataSetFromMatrix(countData = countData,
                              colData = colData,
                              design = as.formula(designFormula))
#print dds object to see the contents
print(dds)
```

```
## class: DESeqDataSet
## dim: 19719 10
## metadata(1): version
## assays(1): counts
## rownames(19719): TSPAN6 TNMD ... MYOCOS HSFX3
## rowData names(0):
## colnames(10): CASE_1 CASE_2 ... CTRL_4 CTRL_5
## colData names(2): source_name group
```

The DESeqDataSet object contains all the information about the experimental setup, the read counts, and the design formulas. Certain functions can be used to access this information separately: rownames(dds) shows which features are used in the study (e.g. genes), colnames(dds) displays the studied samples, counts(dds) displays the count table, and colData(dds) displays the experimental setup.

Remove genes that have almost no information in any of the given samples.

```
#For each gene, we count the total number of reads for that gene in all samples
#and remove those that don't have at least 1 read.
dds <- dds[ rowSums(DESeq2::counts(dds)) > 1, ]
```

Now, we can use the DESeq() function of DESeq2, which is a wrapper function that implements estimation of size factors to normalize the counts, estimation of dispersion values, and computing a GLM model based on the experimental design formula. This function returns a DESeqDataSet object, which is an updated version of the dds variable that we pass to the function as input.

```
dds <- DESeq(dds)
```

Now, we can compare and contrast the samples based on different variables of interest. In this case, we currently have only one variable, which is the group variable that determines if a sample belongs to the CASE group or the CTRL group.

```
#compute the contrast for the 'group' variable where 'CTRL'
#samples are used as the control group.
DEresults = results(dds, contrast = c("group", 'CASE', 'CTRL'))
#sort results by increasing p-value
DEresults <- DEresults[order(DEresults$pvalue),]
```

Thus we have obtained a table containing the differential expression status of case samples compared to the control samples.

It is important to note that the sequence of the elements provided in the contrast argument determines which group of samples are to be used as the control. This impacts the way the results are interpreted, for instance, if a gene is found up-regulated (has a positive log2 fold change), the up-regulation status is only relative to the factor that is provided as control. In this case, we used samples from the "CTRL" group as control and contrasted the samples from the "CASE" group with respect to the "CTRL" samples. Thus genes with a positive log2 fold change are called up-regulated in the case samples with respect to the control, while genes with a negative log2 fold change are down-regulated in the case samples. Whether the deregulation is significant or not, warrants assessment of the adjusted p-values.

Let's have a look at the contents of the DEresults table.

```
#shows a summary of the results
print(DEresults)
```

```
## log2 fold change (MLE): group CASE vs CTRL
## Wald test p-value: group CASE vs CTRL
## DataFrame with 19097 rows and 6 columns
##              baseMean log2FoldChange      lfcSE       stat       pvalue
##             <numeric>      <numeric>  <numeric>  <numeric>    <numeric>
## CYP2E1        4829889        9.36024   0.215223    43.4909  0.00000e+00
```

```
## FCGBP        10349993        -7.57579  0.186433   -40.6355  0.00000e+00
## ASGR2          426422         8.01830  0.216207    37.0863 4.67898e-301
## GCKR           100183         7.82841  0.233376    33.5442 1.09479e-246
## APOA5          438054        10.20248  0.312503    32.6477 8.64906e-234
## ...               ...             ...       ...        ...          ...
## CCDC195        20.4981       -0.215607   2.89255 -0.0745386          NA
## SPEM3          23.6370      -22.154765   3.02785 -7.3170030          NA
## AC022167.5     21.8451       -2.056240   2.89545 -0.7101618          NA
## BX276092.9     29.9636        0.407326   2.89048  0.1409199          NA
## ETDC           22.5675       -1.795274   2.89421 -0.6202983          NA
##                    padj
##               <numeric>
## CYP2E1      0.00000e+00
## FCGBP       0.00000e+00
## ASGR2      2.87741e-297
## GCKR       5.04945e-243
## APOA5      3.19133e-230
## ...                 ...
## CCDC195              NA
## SPEM3                NA
## AC022167.5           NA
## BX276092.9           NA
## ETDC                 NA
```

The first three lines in this output show the contrast and the statistical test that were used to compute these results, along with the dimensions of the resulting table (number of columns and rows). Below these lines is the actual table with 6 columns: baseMean represents the average normalized expression of the gene across all considered samples. log2FoldChange represents the base-2 logarithm of the fold change of the normalized expression of the gene in the given contrast. lfcSE represents the standard error of log2 fold change estimate, and stat is the statistic calculated in the contrast which is translated into a pvalue and adjusted for multiple testing in the padj column. To find out about the importance of adjusting for multiple testing, see Chapter 3.

### 8.3.7.1  Diagnostic plots

At this point, before proceeding to do any downstream analysis and jumping to conclusions about the biological insights that are reachable with the experimental data at hand, it is important to do some more diagnostic tests to improve our confidence about the quality of the data and the experimental setup.

### 8.3.7.1.1 MA plot

An MA plot is useful to observe if the data normalization worked well (Figure 8.6). The MA plot is a scatter plot where the x-axis denotes the average of normalized counts across samples and the y-axis denotes the log fold change in the given contrast. Most points are expected to be on the horizontal 0 line (most genes are not expected to be differentially expressed).

```
library(DESeq2)
DESeq2::plotMA(object = dds, ylim = c(-5, 5))
```

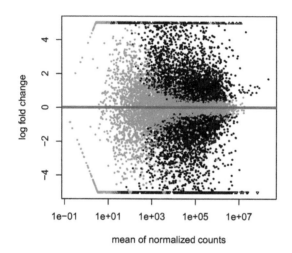

**FIGURE 8.6:** MA plot of differential expression results.

### 8.3.7.1.2 P-value distribution

It is also important to observe the distribution of raw p-values (Figure 8.7). We expect to see a peak around low p-values and a uniform distribution at P-values above 0.1. Otherwise, adjustment for multiple testing does not work and the results are not meaningful.

```
library(ggplot2)
ggplot(data = as.data.frame(DEresults), aes(x = pvalue)) +
  geom_histogram(bins = 100)
```

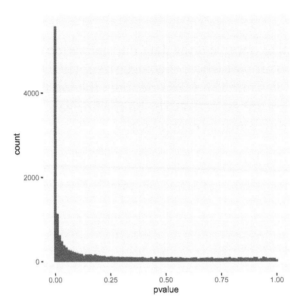

**FIGURE 8.7:** P-value distribution genes before adjusting for multiple testing.

*8.3.7.1.3   PCA plot*

A final diagnosis is to check the biological reproducibility of the sample replicates in a PCA plot or a heatmap. To plot the PCA results, we need to extract the normalized counts from the DESeqDataSet object. It is possible to color the points in the scatter plot by the variable of interest, which helps to see if the replicates cluster well (Figure 8.8).

```
library(DESeq2)
# extract normalized counts from the DESeqDataSet object
countsNormalized <- DESeq2::counts(dds, normalized = TRUE)

# select top 500 most variable genes
selectedGenes <- names(sort(apply(countsNormalized, 1, var),
                        decreasing = TRUE)[1:500])

plotPCA(countsNormalized[selectedGenes,],
        col = as.numeric(colData$group), adj = 0.5,
        xlim = c(-0.5, 0.5), ylim = c(-0.5, 0.6))
```

Alternatively, the normalized counts can be transformed using the DESeq2::rlog function and DESeq2::plotPCA() can be readily used to plot the PCA results (Figure 8.9).

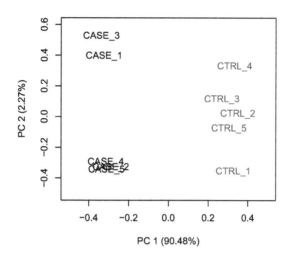

**FIGURE 8.8:** Principle component analysis plot based on top 500 most variable genes.

```
rld <- rlog(dds)
DESeq2::plotPCA(rld, ntop = 500, intgroup = 'group') +
  ylim(-50, 50) + theme_bw()
```

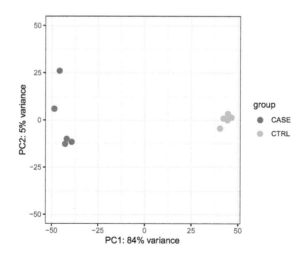

**FIGURE 8.9:** PCA plot of top 500 most variable genes.

8.3.7.1.4    *Relative Log Expression (RLE) plot*

A similar plot to the MA plot is the RLE (Relative Log Expression) plot that is useful in finding out if the data at hand needs normalization (Gandolfo and Speed, 2018). Sometimes, even the datasets normalized using the explained methods above may need further normalization due to unforeseen sources of variation that might stem from the library preparation, the person who carries out the experiment, the date of sequencing, the temperature changes in the laboratory at the time of library preparation, and so on and so forth. The RLE plot is a quick diagnostic that can be applied on the raw or normalized count matrices to see if further processing is required.

Let's do RLE plots on the raw counts and normalized counts using the EDASeq package (Risso et al., 2011) (see Figure 8.10).

```
library(EDASeq)
par(mfrow = c(1, 2))
plotRLE(countData, outline=FALSE, ylim=c(-4, 4),
        col=as.numeric(colData$group),
        main = 'Raw Counts')
plotRLE(DESeq2::counts(dds, normalized = TRUE),
        outline=FALSE, ylim=c(-4, 4),
        col = as.numeric(colData$group),
        main = 'Normalized Counts')
```

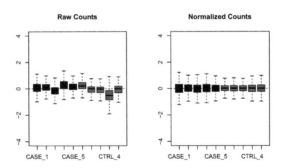

**FIGURE 8.10:** Relative log expression plots based on raw and normalized count matrices.

Here the RLE plot is comprised of boxplots, where each box-plot represents the distribution of the relative log expression of the genes expressed in the corresponding sample. Each gene's expression is divided by the median expression value of that gene across all samples. Then this is transformed to log scale, which gives the relative log expression value for a single gene. The RLE values for all the genes from a sample are visualized as a boxplot.

Ideally the boxplots are centered around the horizontal zero line and are as tightly distributed as possible (Risso et al., 2014). From the plots that we have made for the raw and normalized count data, we can observe how the normalized dataset has improved upon the raw count data for all the samples. However, in some cases, it is important to visualize RLE plots in combination with other diagnostic plots such as PCA plots, heatmaps, and correlation plots to see if there is more unwanted variation in the data, which can be further accounted for using packages such as RUVSeq (Risso et al., 2014) and sva (Leek et al., 2012). We will cover details about the RUVSeq package to account for unwanted sources of noise in RNA-seq datasets in later sections.

### 8.3.8 Functional enrichment analysis

#### 8.3.8.1 GO term analysis

In a typical differential expression analysis, thousands of genes are found differentially expressed between two groups of samples. While prior knowledge of the functions of individual genes can give some clues about what kind of cellular processes have been affected, e.g. by a drug treatment, manually going through the whole list of thousands of genes would be very cumbersome and not be very informative in the end. Therefore a commonly used tool to address this problem is to do enrichment analyses of functional terms that appear associated to the given set of differentially expressed genes more often than expected by chance. The functional terms are usually associated to multiple genes. Thus, genes can be grouped into sets by shared functional terms. However, it is important to have an agreed upon controlled vocabulary on the list of terms used to describe the functions of genes. Otherwise, it would be impossible to exchange scientific results globally. That's why initiatives such as the Gene Ontology Consortium have collated a list of Gene Ontology (GO) terms for each gene. GO term analysis is probably the most common analysis applied after a differential expression analysis. GO term analysis helps quickly find out systematic changes that can describe differences between groups of samples.

In R, one of the simplest ways to do functional enrichment analysis for a set of genes is via the gProfileR package.

Let's select the genes that are significantly differentially expressed between the case and control samples. Let's extract genes that have an adjusted p-value below 0.1 and that show a 2-fold change (either negative or positive) in the case compared to control. We will then feed this gene set into the gProfileR function. The top 10 detected GO terms are displayed in Table 8.2.

**TABLE 8.2:** Top GO terms sorted by p-value.

|    | p.value | term.size | precision | domain | term.name |
|----|---------|-----------|-----------|--------|-----------|
| 52 | 0 | 2740 | 0.223 | CC | plasma membrane part |
| 3  | 0 | 1609 | 0.136 | BP | ion transport |
| 32 | 0 | 3656 | 0.258 | BP | regulation of biological quality |
| 26 | 0 | 385  | 0.042 | BP | extracellular structure organization |
| 17 | 0 | 7414 | 0.452 | BP | multicellular organismal process |
| 68 | 0 | 1069 | 0.090 | MF | transmembrane transporter activity |
| 43 | 0 | 1073 | 0.090 | BP | organic acid metabolic process |
| 6  | 0 | 975  | 0.083 | BP | response to drug |
| 9  | 0 | 1351 | 0.107 | BP | biological adhesion |
| 49 | 0 | 4760 | 0.302 | BP | system development |

```r
library(DESeq2)
library(gProfileR)
library(knitr)
# extract differential expression results
DEresults <- results(dds, contrast = c('group', 'CASE', 'CTRL'))

#remove genes with NA values
DE <- DEresults[!is.na(DEresults$padj),]
#select genes with adjusted p-values below 0.1
DE <- DE[DE$padj < 0.1,]
#select genes with absolute log2 fold change above 1 (two-fold change)
DE <- DE[abs(DE$log2FoldChange) > 1,]

#get the list of genes of interest
genesOfInterest <- rownames(DE)

#calculate enriched GO terms
goResults <- gprofiler(query = genesOfInterest,
                organism = 'hsapiens',
                src_filter = 'GO',
                hier_filtering = 'moderate')
```

#### 8.3.8.2   Gene set enrichment analysis

A gene set is a collection of genes with some common property. This shared property among a set of genes could be a GO term, a common biological pathway, a shared interaction partner, or any biologically relevant commonality that is meaningful in the context of the pursued experiment. Gene set enrichment analysis (GSEA) is a valuable exploratory analysis tool that can associate systematic changes to a high-level function rather than individual genes. Analysis of coordinated changes of expression levels of gene sets can provide complementary benefits on top of per-gene-based differential expression analyses. For instance, consider a gene set belonging to a biological pathway where each member of the pathway displays a slight deregulation in a disease sample compared to a normal sample. In such a case, individual genes might not be picked up by the per-gene-based differential expression analysis. Thus, the GO/Pathway enrichment on the differentially expressed list of genes would not show an enrichment of this pathway. However, the additive effect of slight changes of the genes could amount to a large effect at the level of the gene set, thus the pathway could be detected as a significant pathway that could explain the mechanistic problems in the disease sample.

We use the bioconductor package gage (Luo et al., 2009) to demonstrate how to do GSEA using normalized expression data of the samples as input. Here we are using only two gene sets: one from the top GO term discovered from the previous GO analysis, one that we compile by randomly selecting a list of genes. However, annotated gene sets can be used from databases such as MSIGDB (Subramanian et al., 2005), which compile gene sets from a variety of resources such as KEGG (Kanehisa et al., 2016) and REACTOME (Fabregat et al., 2018a).

```r
#Let's define the first gene set as the list of genes from one of the
#significant GO terms found in the GO analysis. order go results by pvalue
goResults <- goResults[order(goResults$p.value),]
#restrict the terms that have at most 100 genes overlapping with the query
go <- goResults[goResults$overlap.size < 100,]
# use the top term from this table to create a gene set
geneSet1 <- unlist(strsplit(go[1,]$intersection, ','))


#Define another gene set by just randomly selecting 25 genes from the counts
#table get normalized counts from DESeq2 results
normalizedCounts <- DESeq2::counts(dds, normalized = TRUE)
geneSet2 <- sample(rownames(normalizedCounts), 25)


geneSets <- list('top_GO_term' = geneSet1,
                 'random_set' = geneSet2)
```

Using the defined gene sets, we'd like to do a group comparison between the case samples with respect to the control samples.

**TABLE 8.3:** Up-regulation statistics

|              | p.geomean | stat.mean | p.val  | q.val  | set.size | exp1   |
|--------------|-----------|-----------|--------|--------|----------|--------|
| top_GO_term  | 0.0000    | 7.1994    | 0.0000 | 0.0000 | 32       | 0.0000 |
| random_set   | 0.4761    | 0.0604    | 0.4761 | 0.4761 | 25       | 0.4761 |

**TABLE 8.4:** Down-regulation statistics

|              | p.geomean | stat.mean | p.val  | q.val | set.size | exp1   |
|--------------|-----------|-----------|--------|-------|----------|--------|
| random_set   | 0.5239    | 0.0604    | 0.5239 | 1     | 25       | 0.5239 |
| top_GO_term  | 1.0000    | 7.1994    | 1.0000 | 1     | 32       | 1.0000 |

```
library(gage)
#use the normalized counts to carry out a GSEA.
gseaResults <- gage(exprs = log2(normalizedCounts+1),
        ref = match(rownames(colData[colData$group == 'CTRL',]),
                  colnames(normalizedCounts)),
        samp = match(rownames(colData[colData$group == 'CASE',]),
                  colnames(normalizedCounts)),
        gsets = geneSets, compare = 'as.group')
```

We can observe if there is a significant up-regulation or down-regulation of the gene set in the case group compared to the controls by accessing gseaResults$greater as in Table 8.3 or gseaResults$less as in Table 8.4.

We can see that the random gene set shows no significant up- or down-regulation (Tables 8.3 and (8.4), while the gene set we defined using the top GO term shows a significant up-regulation (adjusted p-value < 0.0007) (8.3). It is worthwhile to visualize these systematic changes in a heatmap as in Figure 8.11.

```
library(pheatmap)
# get the expression data for the gene set of interest
M <- normalizedCounts[rownames(normalizedCounts) %in% geneSet1, ]
# log transform the counts for visualization scaling by row helps visualizing
# relative change of expression of a gene in multiple conditions
pheatmap(log2(M+1),
        annotation_col = colData,
        show_rownames = TRUE,
        fontsize_row = 8,
        scale = 'row',
```

```
    cutree_cols = 2,
    cutree_rows = 2)
```

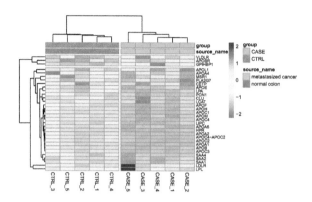

**FIGURE 8.11:** Heatmap of expression value from the genes with the top GO term.

We can see that almost all genes from this gene set display an increased level of expression in the case samples compared to the controls.

### 8.3.9    Accounting for additional sources of variation

When doing a differential expression analysis in a case-control setting, the variable of interest, i.e. the variable that explains the separation of the case samples from the control, is usually the treatment, genotypic differences, a certain phenotype, etc. However, in reality, depending on how the experiment and the sequencing were designed, there may be additional factors that might contribute to the variation between the compared samples. Sometimes, such variables are known, for instance, the date of the sequencing for each sample (batch information), or the temperature under which samples were kept. Such variables are not necessarily biological but rather technical, however, they still impact the measurements obtained from an RNA-seq experiment. Such variables can introduce systematic shifts in the obtained measurements. Here, we will demonstrate: firstly how to account for such variables using DESeq2, when the possible sources of variation are actually known; secondly, how to account for such variables when all we have is just a count table but we observe that the variable of interest only explains a small proportion of the differences between case and control samples.

#### 8.3.9.1    Accounting for covariates using DESeq2

For demonstration purposes, we will use a subset of the count table obtained for a heart disease study, where there are RNA-seq samples from subjects with normal and failing hearts. We again use a subset of the samples, focusing on 6 case and 6 control samples and we only consider protein-coding genes (for speed concerns).

Let's import count and colData for this experiment.

```r
counts_file <- system.file('extdata/rna-seq/SRP021193.raw_counts.tsv',
                           package = 'compGenomRData')
colData_file <- system.file('extdata/rna-seq/SRP021193.colData.tsv',
                            package = 'compGenomRData')

counts <- read.table(counts_file)
colData <- read.table(colData_file, header = T, sep = '\t',
                      stringsAsFactors = TRUE)
```

Let's take a look at how the samples cluster by calculating the TPM counts as displayed as a heatmap in Figure 8.12.

```r
library(pheatmap)
#find gene length normalized values
geneLengths <- counts$width
rpk <- apply( subset(counts, select = c(-width)), 2,
             function(x) x/(geneLengths/1000))
#normalize by the sample size using rpk values
tpm <- apply(rpk, 2, function(x) x / sum(as.numeric(x)) * 10^6)

selectedGenes <- names(sort(apply(tpm, 1, var),
                            decreasing = T)[1:100])
pheatmap(tpm[selectedGenes,],
         scale = 'row',
         annotation_col = colData,
         show_rownames = FALSE)
```

Here we can see from the clusters that the dominating variable is the 'Library Selection' variable rather than the 'diagnosis' variable, which determines the state of the organ from which the sample was taken. Case and control samples are all mixed in both two major clusters. However, ideally, we'd like to see a separation of the case and control samples regardless of the additional covariates. When testing for differential gene expression between conditions, such confounding variables can be accounted for using DESeq2. Below is a demonstration of how we instruct DESeq2 to account for the 'library selection' variable:

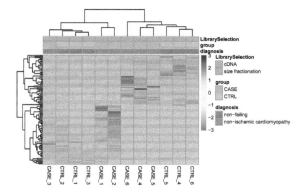

**FIGURE 8.12:** Visualizing batch effects in an experiment.

```
library(DESeq2)
# remove the 'width' column from the counts matrix
countData <- as.matrix(subset(counts, select = c(-width)))
# set up a DESeqDataSet object
dds <- DESeqDataSetFromMatrix(countData = countData,
                             colData = colData,
                             design = ~ LibrarySelection + group)
```

When constructing the design formula, it is very important to pay attention to the sequence of variables. We leave the variable of interest to the last and we can add as many covariates as we want to the beginning of the design formula. Please refer to the DESeq2 vignette if you'd like to learn more about how to construct design formulas.

Now, we can run the differential expression analysis as has been demonstrated previously.

```
# run DESeq
dds <- DESeq(dds)
# extract results
DEresults <- results(dds, contrast = c('group', 'CASE', 'CTRL'))
```

### 8.3.9.2 Accounting for estimated covariates using RUVSeq

In cases when the sources of potential variation are not known, it is worthwhile to use tools such as RUVSeq or sva that can estimate potential sources of variation and clean up the counts table from those sources of variation. Later on, the estimated covariates can be integrated into DESeq2's design formula.

Let's see how to utilize the RUVseq package to first diagnose the problem and then solve it. Here, for demonstration purposes, we'll use a count table from a lung carcinoma study in which a transcription factor (Ets homologous factor - EHF) is overexpressed and compared to the control samples with baseline EHF expression. Again, we only consider protein coding genes and use only five case and five control samples. The original data can be found on the recount2 database with the accession 'SRP049988'.

```r
counts_file <- system.file('extdata/rna-seq/SRP049988.raw_counts.tsv',
                           package = 'compGenomRData')
colData_file <- system.file('extdata/rna-seq/SRP049988.colData.tsv',
                            package = 'compGenomRData')

counts <- read.table(counts_file)
colData <- read.table(colData_file, header = T,
                      sep = '\t', stringsAsFactors = TRUE)
# simplify condition descriptions
colData$source_name <- ifelse(colData$group == 'CASE',
                              'EHF_overexpression', 'Empty_Vector')
```

Let's start by making heatmaps of the samples using TPM counts (see Figure 8.13).

```r
#find gene length normalized values
geneLengths <- counts$width
rpk <- apply( subset(counts, select = c(-width)), 2,
             function(x) x/(geneLengths/1000))
#normalize by the sample size using rpk values
tpm <- apply(rpk, 2, function(x) x / sum(as.numeric(x)) * 10^6)
selectedGenes <- names(sort(apply(tpm, 1, var),
                            decreasing = T)[1:100])
pheatmap(tpm[selectedGenes,],
         scale = 'row',
         annotation_col = colData,
         cutree_cols = 2,
         show_rownames = FALSE)
```

We can see that the overall clusters look fine, except that one of the case samples (CASE_5) clusters more closely with the control samples than the other case samples. This mis-clustering could be a result of some batch effect, or any other technical preparation steps. However, the colData object doesn't contain any variables that we can use to pinpoint the exact cause of this. So, let's use RUVSeq to estimate potential covariates to see if the clustering results can be improved.

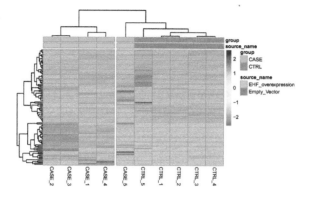

**FIGURE 8.13:** Diagnostic plot to observe.

First, we set up the experiment:

```
library(EDASeq)
# remove 'width' column from counts
countData <- as.matrix(subset(counts, select = c(-width)))
# create a seqExpressionSet object using EDASeq package
set <- newSeqExpressionSet(counts = countData,
                           phenoData = colData)
```

Next, let's make a diagnostic RLE plot on the raw count table.

```
# make an RLE plot and a PCA plot on raw count data and color samples by group
par(mfrow = c(1,2))
plotRLE(set, outline=FALSE, ylim=c(-4, 4), col=as.numeric(colData$group))
plotPCA(set, col = as.numeric(colData$group), adj = 0.5,
        ylim = c(-0.7, 0.5), xlim = c(-0.5, 0.5))
```

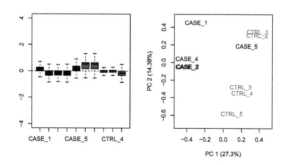

**FIGURE 8.14:** Diagnostic RLE and PCA plots based on raw count table.

```
## make RLE and PCA plots on TPM matrix
par(mfrow = c(1,2))
plotRLE(tpm, outline=FALSE, ylim=c(-4, 4), col=as.numeric(colData$group))
plotPCA(tpm, col=as.numeric(colData$group), adj = 0.5,
        ylim = c(-0.3, 1), xlim = c(-0.5, 0.5))
```

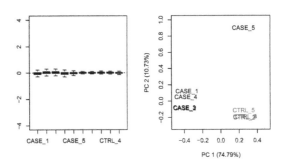

**FIGURE 8.15:** Diagnostic RLE and PCA plots based on TPM normalized count table.

Both RLE and PCA plots look better on normalized data (Figure 8.15) compared to raw data (Figure 8.14), but still suggest the necessity of further improvement, because the CASE_5 sample still clusters with the control samples. We haven't yet accounted for the source of unwanted variation.

### 8.3.9.3  Removing unwanted variation from the data

RUVSeq has three main functions for removing unwanted variation: RUVg(), RUVs(), and RUVr(). Here, we will demonstrate how to use RUVg and RUVs. RUVr will be left as an exercise for the reader.

#### 8.3.9.3.1  Using RUVg

One way of removing unwanted variation depends on using a set of reference genes that are not expected to change by the sources of technical variation. One strategy along this line is to use spike-in genes, which are artificially introduced into the sequencing run (Jiang et al., 2011). However, there are many sequencing datasets that don't have this spike-in data available. In such cases, an empirical set of genes can be collected from the expression data by doing a differential expression analysis and discovering genes that are unchanged in the given conditions. These unchanged genes are used to clean up the data from systematic shifts in expression due to the unwanted sources of variation. Another strategy could be to use a set of house-keeping genes as negative controls, and use them as a reference to correct the systematic biases in the data. Let's use a list of ~500 house-keeping genes compiled here: https://www.tau.ac.il/~elieis/HKG/HK_genes.txt.

```
library(RUVSeq)

#source for house-keeping genes collection:
#https://m.tau.ac.il/~elieis/HKG/HK_genes.txt
HK_genes <- read.table(file = system.file("extdata/rna-seq/HK_genes.txt",
                                            package = 'compGenomRData'),
                        header = FALSE)
# let's take an intersection of the house-keeping genes with the genes available
# in the count table
house_keeping_genes <- intersect(rownames(set), HK_genes$V1)
```

We will now run RUVg() with the different number of factors of unwanted variation. We will plot the PCA after removing the unwanted variation. We should be able to see which k values, number of factors, produce better separation between sample groups.

```
# now, we use these genes as the empirical set of genes as input to RUVg.
# we try different values of k and see how the PCA plots look

par(mfrow = c(2, 2))
for(k in 1:4) {
  set_g <- RUVg(x = set, cIdx = house_keeping_genes, k = k)
  plotPCA(set_g, col=as.numeric(colData$group), cex = 0.9, adj = 0.5,
          main = paste0('with RUVg, k = ',k),
          ylim = c(-1, 1), xlim = c(-1, 1), )
}
```

Based on the separation of case and control samples in the PCA plots in Figure 8.16, we choose k = 1 and re-run the RUVg() function with the house-keeping genes to do more diagnostic plots.

```
# choose k = 1

set_g <- RUVg(x = set, cIdx = house_keeping_genes, k = 1)
```

Now let's do diagnostics: compare the count matrices with or without RUVg processing, comparing RLE plots (Figure 8.17) and PCA plots (Figure 8.18) to see the effect of RUVg on the normalization and separation of case and control samples.

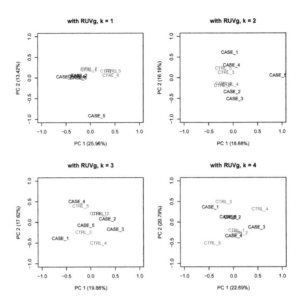

**FIGURE 8.16:** PCA plots on RUVg normalized data with varying number of covariates (k).

```
# RLE plots

par(mfrow = c(1,2))
plotRLE(set, outline=FALSE, ylim=c(-4, 4),
        col=as.numeric(colData$group), main = 'without RUVg')
plotRLE(set_g, outline=FALSE, ylim=c(-4, 4),
        col=as.numeric(colData$group), main = 'with RUVg')
```

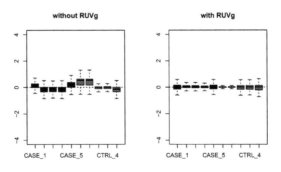

**FIGURE 8.17:** RLE plots to observe the effect of RUVg.

```
# PCA plots
```

```
par(mfrow = c(1,2))
plotPCA(set, col=as.numeric(colData$group), adj = 0.5,
        main = 'without RUVg',
        ylim = c(-1, 0.5), xlim = c(-0.5, 0.5))
plotPCA(set_g, col=as.numeric(colData$group), adj = 0.5,
        main = 'with RUVg',
        ylim = c(-1, 0.5), xlim = c(-0.5, 0.5))
```

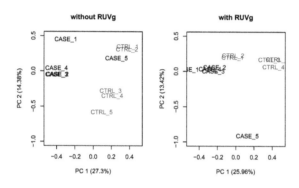

**FIGURE 8.18:** PCA plots to observe the effect of RUVg.

We can observe that using RUVg() with house-keeping genes as reference has improved the clusters, however not yielded ideal separation. Probably the effect that is causing the 'CASE_5' to cluster with the control samples still hasn't been completely eliminated.

### 8.3.9.3.2 Using RUVs

There is another strategy of RUVSeq that works better in the presence of replicates in the absence of a confounded experimental design, which is the RUVs() function. Let's see how that performs with this data. This time we don't use the house-keeping genes. We rather use all genes as input to RUVs(). This function estimates the correction factor by assuming that replicates should have constant biological variation, rather, the variation in the replicates are the unwanted variation.

```
# make a table of sample groups from colData
differences <- makeGroups(colData$group)
## looking for two different sources of unwanted variation (k = 2)
## use information from all genes in the expression object
par(mfrow = c(2, 2))
```

```
for(k in 1:4) {
  set_s <- RUVs(set, unique(rownames(set)),
                k=k, differences) #all genes
  plotPCA(set_s, col=as.numeric(colData$group),
          cex = 0.9, adj = 0.5,
        main = paste0('with RUVs, k = ',k),
        ylim = c(-1, 1), xlim = c(-0.6, 0.6))
}
```

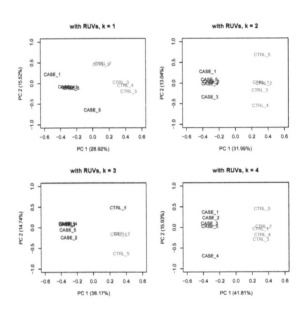

**FIGURE 8.19:** PCA plots on RUVs normalized data with varying number of covariates (k).

Based on the separation of case and control samples in the PCA plots in Figure 8.19, we can see that the samples are better separated even at k = 2 when using `RUVs()`. Here, we re-run the `RUVs()` function using k = 2, in order to do more diagnostic plots. We try to pick a value of k that is good enough to distinguish the samples by condition of interest. While setting the value of k to higher values could improve the percentage of explained variation by the first principle component to up to 61%, we try to avoid setting the value unnecessarily high to avoid removing factors that might also correlate with important biological differences between conditions.

```
# choose k = 2
set_s <- RUVs(set, unique(rownames(set)), k=2, differences) #
```

Now let's do diagnostics again: compare the count matrices with or without RUVs processing, comparing RLE plots (Figure 8.20) and PCA plots (Figure 8.21) to see the effect of RUVg on the normalization and separation of case and control samples.

```
## compare the initial and processed objects
## RLE plots

par(mfrow = c(1,2))
plotRLE(set, outline=FALSE, ylim=c(-4, 4),
        col=as.numeric(colData$group),
        main = 'without RUVs')
plotRLE(set_s, outline=FALSE, ylim=c(-4, 4),
        col=as.numeric(colData$group),
        main = 'with RUVs')
```

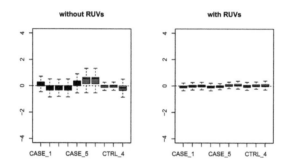

**FIGURE 8.20:** RLE plots to observe the effect of RUVs.

```
## PCA plots

par(mfrow = c(1,2))
plotPCA(set, col=as.numeric(colData$group),
        main = 'without RUVs', adj = 0.5,
        ylim = c(-0.75, 0.75), xlim = c(-0.75, 0.75))
plotPCA(set_s, col=as.numeric(colData$group),
        main = 'with RUVs', adj = 0.5,
        ylim = c(-0.75, 0.75), xlim = c(-0.75, 0.75))
```

Let's compare PCA results from RUVs and RUVg with the initial raw counts matrix. We will simply run the plotPCA() function on different normalization schemes. The resulting plots are in Figure 8.22:

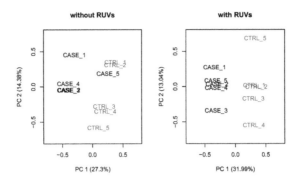

**FIGURE 8.21:** PCA plots to observe the effect of RUVs.

```
par(mfrow = c(1,3))
plotPCA(countData, col=as.numeric(colData$group),
        main = 'without RUV - raw counts', adj = 0.5,
        ylim = c(-0.75, 0.75), xlim = c(-0.75, 0.75))
plotPCA(set_g, col=as.numeric(colData$group),
        main = 'with RUVg', adj = 0.5,
        ylim = c(-0.75, 0.75), xlim = c(-0.75, 0.75))
plotPCA(set_s, col=as.numeric(colData$group),
        main = 'with RUVs', adj = 0.5,
        ylim = c(-0.75, 0.75), xlim = c(-0.75, 0.75))
```

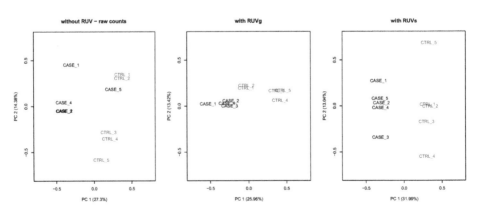

**FIGURE 8.22:** PCA plots to observe the before/after effect of RUV functions.

It looks like RUVs() has performed better than RUVg() in this case. So, let's use count data that is processed by RUVs() to re-do the initial heatmap. The resulting heatmap is in Figure 8.23.

```
library(EDASeq)
library(pheatmap)
# extract normalized counts that are cleared from unwanted variation using RUVs
normCountData <- normCounts(set_s)
selectedGenes <- names(sort(apply(normCountData, 1, var),
                       decreasing = TRUE))[1:500]
pheatmap(normCountData[selectedGenes,],
        annotation_col = colData,
        show_rownames = FALSE,
        cutree_cols = 2,
        scale = 'row')
```

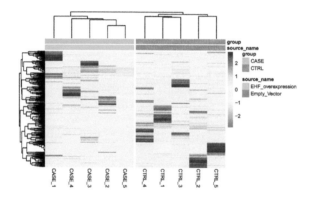

**FIGURE 8.23:** Clustering samples using the top 500 most variable genes normalized using RUVs (k = 2).

As can be observed the replicates from different groups cluster much better with each other after processing with RUVs(). It is important to note that RUVs uses information from replicates to shift the expression data and it would not work in a confounding design where the replicates of case samples and replicates of the control samples are sequenced in different batches.

#### 8.3.9.4 Re-run DESeq2 with the computed covariates

Having computed the sources of variation using RUVs(), we can actually integrate these variables with DESeq2 to re-do the differential expression analysis.

```
library(DESeq2)
#set up DESeqDataSet object
dds <- DESeqDataSetFromMatrix(countData = countData,
                       colData = colData,
                       design = ~ group)
```

```
# filter for low count genes
dds <- dds[rowSums(DESeq2::counts(dds)) > 10]

# insert the covariates W1 and W2 computed using RUVs into DESeqDataSet object
colData(dds) <- cbind(colData(dds),
                  pData(set_s)[rownames(colData(dds)),
                          grep('W_[0-9]',
                              colnames(pData(set_s)))])

# update the design formula for the DESeq analysis (save the variable of
# interest to the last!)
design(dds) <- ~ W_1 + W_2 + group
# repeat the analysis
dds <- DESeq(dds)
# extract deseq results
res <- results(dds, contrast = c('group', 'CASE', 'CTRL'))
res <- res[order(res$padj),]
```

## 8.4    Other applications of RNA-seq

RNA-seq generates valuable data that contains information not only at the gene level but also at the level of exons and transcripts. Moreover, the kind of information that we can extract from RNA-seq is not limited to expression quantification. It is possible to detect alternative splicing events such as novel isoforms (Trapnell et al., 2010), and differential usage of exons (Anders et al., 2012). It is also possible to observe sequence variants (substitutions, insertions, deletions, RNA-editing) that may change the translated protein product (McKenna et al., 2010). In the context of cancer genomes, gene-fusion events can be detected with RNA-seq (McPherson et al., 2011). Finally, for the purposes of gene prediction or improving existing gene predictions, RNA-seq is a valuable method (Stanke and Morgenstern, 2005). In order to learn more about how to implement these, it is recommended that you go through the tutorials of the cited tools.

## 8.5 Exercises

### 8.5.1 Exploring the count tables

Here, import an example count table and do some exploration of the expression data.

```
counts_file <- system.file("extdata/rna-seq/SRP029880.raw_counts.tsv",
                           package = "compGenomRData")
coldata_file <- system.file("extdata/rna-seq/SRP029880.colData.tsv",
                            package = "compGenomRData")
```

1. Normalize the counts using the TPM approach. [Difficulty: **Beginner**]
2. Plot a heatmap of the top 500 most variable genes. Compare with the heatmap obtained using the 100 most variable genes. [Difficulty: **Beginner**]
3. Re-do the heatmaps setting the `scale` argument to `none`, and `column`. Compare the results with `scale = 'row'`. [Difficulty: **Beginner**]
4. Draw a correlation plot for the samples depicting the sample differences as 'ellipses', drawing only the upper end of the matrix, and order samples by hierarchical clustering results based on `average` linkage clustering method. [Difficulty: **Beginner**]
5. How else could the count matrix be subsetted to obtain quick and accurate clusters? Try selecting the top 100 genes that have the highest total expression in all samples and re-draw the cluster heatmaps and PCA plots. [Difficulty: **Intermediate**]
6. Add an additional column to the annotation data.frame object to annotate the samples and use the updated annotation data.frame to plot the heatmaps. (Hint: Assign different batch values to CASE and CTRL samples). Make a PCA plot and color samples by the added variable (e.g. batch). [Difficulty: Intermediate]
7. Try making the heatmaps using all the genes in the count table, rather than sub-selecting. [Difficulty: **Advanced**]
8. Use the `Rtsne` package[2] to draw a t-SNE plot of the expression values. Color the points by sample group. Compare the results with the PCA plots. [Difficulty: **Advanced**]

---

[2]https://cran.r-project.org/web/packages/Rtsne/Rtsne.pdf

### 8.5.2 Differential expression analysis

Firstly, carry out a differential expression analysis starting from raw counts. Use the following datasets:

```
counts_file <- system.file("extdata/rna-seq/SRP029880.raw_counts.tsv",
                           package = "compGenomRData")
coldata_file <- system.file("extdata/rna-seq/SRP029880.colData.tsv",
                            package = "compGenomRData")
```

- Import the read counts and colData tables.
- Set up a DESeqDataSet object.
- Filter out genes with low counts.
- Run DESeq2 contrasting the CASE sample with CONTROL samples.

Now, you are ready to do the following exercises:

1. Make a volcano plot using the differential expression analysis results. (Hint: x-axis denotes the log2FoldChange and the y-axis represents the -log10(pvalue)). [Difficulty: **Beginner**]
2. Use DESeq2::plotDispEsts to make a dispersion plot and find out the meaning of this plot. (Hint: Type ?DESeq2::plotDispEsts) [Difficulty: **Beginner**]
3. Explore lfcThreshold argument of the DESeq2::results function. What is its default value? What does it mean to change the default value to, for instance, 1? [Difficulty: **Intermediate**]
4. What is independent filtering? What happens if we don't use it? Google independent filtering statquest and watch the online video about independent filtering. [Difficulty: **Intermediate**]
5. Re-do the differential expression analysis using the edgeR package. Find out how much DESeq2 and edgeR agree on the list of differentially expressed genes. [Difficulty: **Advanced**]
6. Use the compcodeR package to run the differential expression analysis using at least three different tools and compare and contrast the results following the compcodeR vignette. [Difficulty: **Advanced**]

### 8.5.3 Functional enrichment analysis

1. Re-run gProfileR, this time using pathway annotations such as KEGG, REACTOME, and protein complex databases such as CORUM, in addition to the GO terms. Sort the resulting tables by columns precision and/or recall. How do the top GO terms change when sorted for precision, recall, or p.value? [Difficulty: **Beginner**]
2. Repeat the gene set enrichment analysis by trying different options for

the `compare` argument of the `GAGE::gage` function. How do the results differ? [Difficulty: **Beginner**]
3. Make a scatter plot of GO term sizes and obtained p-values by setting the `gProfiler::gprofiler` argument `significant = FALSE`. Is there a correlation of term sizes and p-values? (Hint: Take -log10 of p-values). If so, how can this bias be mitigated? [Difficulty: **Intermediate**]
4. Do a gene-set enrichment analysis using gene sets from top 10 GO terms. [Difficulty: **Intermediate**]
5. What are the other available R packages that can carry out gene set enrichment analysis for RNA-seq datasets? [Difficulty: **Intermediate**]
6. Use the topGO package (`https://bioconductor.org/packages/release/bioc/html/topGO.html`) to re-do the GO term analysis. Compare and contrast the results with what has been obtained using the `gProfileR` package. Which tool is faster, `gProfileR` or topGO? Why? [Difficulty: **Advanced**]
7. Given a gene set annotated for human, how can it be utilized to work on *C. elegans* data? (Hint: See `biomaRt::getLDS`). [Difficulty: **Advanced**]
8. Import curated pathway gene sets with Entrez identifiers from the MSIGDB database[3] and re-do the GSEA for all curated gene sets. [Difficulty: **Advanced**]

### 8.5.4 Removing unwanted variation from the expression data

For the exercises below, use the datasets at:

```
counts_file <- system.file('extdata/rna-seq/SRP049988.raw_counts.tsv',
                           package = 'compGenomRData')
colData_file <- system.file('extdata/rna-seq/SRP049988.colData.tsv',
                            package = 'compGenomRData')
```

1. Run RUVSeq using multiple values of `k` from 1 to 10 and compare and contrast the PCA plots obtained from the normalized counts of each RUVSeq run. [Difficulty: **Beginner**]
2. Re-run RUVSeq using the `RUVr()` function. Compare PCA plots from `RUVs`, `RUVg` and `RUVr` using the same `k` values and find out which one performs the best. [Difficulty: **Intermediate**]
3. Do the necessary diagnostic plots using the differential expression results from the EHF count table. [Difficulty: **Intermediate**]
4. Use the `sva` package to discover sources of unwanted variation and re-do the differential expression analysis using variables from the output of `sva` and compare the results with `DESeq2` results using `RUVSeq` corrected normalization counts. [Difficulty: **Advanced**]

---

[3]`http://software.broadinstitute.org/gsea/msigdb/collections.jsp`

# 9

## ChIP-seq analysis

*Chapter Author*: **Vedran Franke**

Protein-DNA interactions are responsible for a large part of the gene expression regulation. Proteins such as transcription factors as well as histones are directly related to how much and in which contexts the genes are expressed. Some of these concepts are already introduced in Chapter 1 if readers need a more in-depth introduction. In this chapter, we will introduce how to process and analyze ChIP-seq data in order to identify genome-wide protein binding sites and to discover underlying sequence context via transcription factor binding-site motifs.

### 9.1 Regulatory protein-DNA interactions

One of the most fascinating biological phenomena is the fact that a myriad of different cell types, in a multicellular organism, are encoded by one single genome. How exactly this is achieved is still a major unanswered question in biology. Cell types differ based on a multitude of features: their size, shape, mobility, surface receptors, metabolic content. However, the main predominant feature, which influences all of the above, is which genes are expressed in each cell type. Therefore, if we can understand what controls which genes will be expressed, and where they will be expressed, we can start forming a picture of how a single genomic template, can give rise to a complex organism.

As explained in Chapter 1, gene expression is controlled by a special class of genes called transcription factors - genes which control other genes. Transcription factor genes encode proteins which can bind to the DNA, and control whether a certain part of DNA will be transcribed (expressed), or stay silent (repressed). They program the expression patterns in each cell. Transcription factors contain DNA binding domains, which are specifically folded protein sequences which recognize specific DNA motifs (a short nucleotide sequence). Such sequence binding imparts transcription factors with specificity, transcription factors do not bind everywhere on the DNA, rather they are localized to short stretches which contain the corresponding DNA motif.

DNA in the nucleus is wrapped around a protein complex called the histone complex.

Histones form a chain of beads along the DNA. By changing their position, histones can make certain parts of the DNA more or less accessible to transcription factors. Histone complexes can be chemically modified with different post-translational modifications (see Chapter 1). Such modifications change histone mobility, and their interactions with different proteins, thereby creating an additional regulatory layer on top of the DNA sequence.

In order to understand the target genes of a certain transcription factor, and how they control the gene expression, we need to know where on the DNA the transcription factor is located.

## 9.2    Measuring protein-DNA interactions with ChIP-seq

ChIP-seq stands for chromatin immunoprecipitation followed by sequencing, and is an experimental method for finding locations on DNA which are bound by proteins. It has been extensively used to study in-vivo binding preferences of transcription factors, and genomic distribution of modified histones.

In the remainder of this chapter, you will learn how to assess quality control of ChIP-seq data sets, perform peak calling to find bound regions, and assess the quality of the peak calling.

Once you have obtained peaks, you will learn how to perform sequence analysis to construct motif models, and compare signals between experiments. Biological experiments often contain a multitude of consecutive steps. Each step can profoundly influence the quality of the data, and the subsequent analysis. The computational biologist has to have an in-depth knowledge of the experimental design, and the underlying experimental steps, in order to choose the proper tools and the type of analysis, which will give proper and correct results (Kharchenko et al., 2008; Kidder et al., 2011; Landt et al., 2012; Chen et al., 2012b; Felsani et al., 2015). In this chapter we will go through the main experimental steps in the ChIP-seq analysis and address the most common experimental pitfalls.

The main principle of the method is to use a specific antibody to enrich DNA fragments which are bound by the protein of interest.

The DNA fragments are then sequenced, mapped onto the corresponding reference genome, and computationally analyzed to distinguish regions which were really bound by the protein, from the background regions.

The experimental methodology is depicted in Figure 9.1, and consists of the following steps:

1. Cross linking of cells with formaldehyde to bind the proteins to the DNA. This process covalently links the proteins to the DNA.

2. Fragmentation of DNA using sonication or enzymatic digestion, shearing of DNA into small fragments (ranging from 50 - 500 bp).

3. Immunoprecipitation using a specific antibody. An immunoprecipitation step which enriches fragments bound by the protein.

4. Cross-link reversal. Frees the DNA fragments for further processing.

5. Size selection of DNA fragments. Only fragments of certain length are used in the library preparation and sequencing.

6. Fragment amplification using PCR. The amount of DNA is a limiting step for the protocol. Therefore the fragments need to be amplified using PCR.

7. DNA fragment sequencing

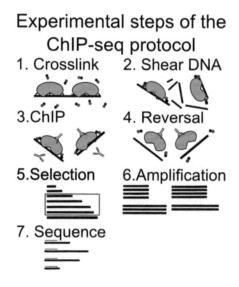

**FIGURE 9.1:** Main experimental steps in the ChIP-seq protocol.

After sequencing, the role of the computational biologist is to assess the quality of the experiment, find the location of the protein of interest, and finally, to integrate with existing data sets.

Each step of the experimental protocol can affect the quality of the data set, and the subsequent analysis steps. It is, therefore, of crucial importance to perform quality control for every sequenced experiment.

## 9.3 Factors that affect ChIP-seq experiment and analysis quality

### 9.3.1 Antibody specificity

Antibody specificity is a term which refers to how strongly an antibody binds to its preferred target, with respect to everything else present in the cell. It is the paramount measure influencing the successful execution of a ChIP experiment. Antibodies can bind multiple proteins with the same affinity. This is called antibody cross-reactivity. If an antibody cross-reacts with multiple proteins, the results of a ChIP experiment will be ambiguous. Instead of finding where our protein binds to the DNA, we will get a superposition of binding of multiple proteins. Such data are impossible to analyze correctly, and will produce false conclusions. There are many experimental procedures for validating antibody specificities, and an antibody should pass multiple tests in order to be considered valid. The exact recommendations are listed by the ENCODE consortium (Landt et al., 2012).

Every time we are analyzing a new ChIP-seq experiment, we have to take our time to convince ourselves that all of the appropriate experimental controls were performed to validate the antibody specificity (Wardle and Tan, 2015).

### 9.3.2 Sequencing depth

Variation in sequencing depth is the first systematic technical bias we encounter in ChIP-seq experiments. Namely, different samples will contain different number of sequenced reads. Different sequencing depth influences our ability to detect enriched regions, and complicates comparisons between samples (Jung et al., 2014). The statistical procedure of removing the influence of sequencing depth on the quantification is called depth scaling; we calculate a scaling factor which is used to multiply the signal strength before the comparison. There are multiple methods for normalization, and each method comes with its assumptions. **Scale normalization** is done by dividing the read counts (in certain genomic locations) by the total amount of sequenced reads. This method presumes that the ChIP efficiency worked equally well in all studied conditions. Because the ChIP efficiency differs in different antibodies, it is often unsuitable for comparisons of ChIP-seq experiments done on different proteins. **Robust normalization** tries to locate genomic regions which do not change between different biological conditions (regions where the protein is constantly bound), and then uses the sum of the reads in those regions as the scaling factor. This method presumes that we can reliably identify regions which do not change (Shao et al., 2012). **Background normalization** presumes that the genome can be split into two categories: background regions and true signal regions. It then uses the number of reads in the background regions to define the scaling factor (Liang and Keleş, 2012). **External normalization** uses external reference for normalization; we add known amounts of chromatin from a distant species, or artificial spike-ins which are then used as a scaling reference. This is used when we think there are global changes in the biding profiles between two biological conditions – very large changes in the signal profile (Bonhoure et al., 2014).

The choice of normalization method depends on the type of analysis (Angelini et al., 2015); if we want to quantitatively compare the abundance of different histone marks in different cell types, we will need the different normalization procedure than if we want to compare TF binding in the same setting.

### 9.3.3 PCR duplication

The amounts of the DNA obtained after the ChIP experiment are quite often lower than the minimal amount which can be sequenced. Polymerase chain reaction (PCR) is a procedure used for amplification of DNA fragments. It is used to increase the amount of DNA in our sample prior to sequencing. PCR is a stochastic procedure, meaning that the results of each PCR reaction cannot be predicted. Due to its stochastic nature, PCR can be a significant source of variability in the ChIP-seq experiments (Aird et al., 2011; Benjamini and Speed, 2012; Teng and Irizarry, 2016). A quality control is necessary to check whether all of our samples have the same sequence properties, i.e. the same enrichment of dinucleotides, such as CpG. If the samples differ in their sequence properties, that means we have to account for them during the analysis (Teng and Irizarry, 2017).

### 9.3.4 Biological replicates

Biological replicates are independently executed ChIP-seq experiments from different samples, corresponding to the same biological conditions. They are indispensable for estimating ChIP quality, and give us an estimate of the variability in the experiment which we can expect due to unknown biological variables. Without biological replicates, it is statistically impossible to compare ChIP-seq samples from different biological conditions, because we do not know whether the observed changes are a result of the inherent biological variability (the source of which we do not understand), or they result from the change in the biological condition (different tissue or transcription factor used in the experiment). If we encounter an experimental setup which does not include biological replicates, we should be extremely skeptical about all conclusions derived from such analysis.

### 9.3.5 Control experiments

There are three types of control experiments which can be performed to control for known and unknown experimental biases:

1.  **Input control**: Sequencing of genomic DNA without the immunoprecipitation step.

2.  **IgG control**: Using a polyclonal mixture of non-specific IgG antibodies instead of a specific antibody.

3.  **Knockout control**: Performing the ChIP experiment in a biological system which does not contains our protein of interest (i.e. in a cell line where the transcription factor was knocked out) (Krebs et al., 2014).

Each type of control experiment controls for a certain set of experimental biases.

**Input control** is the most frequent type of control performed. It shows the differential susceptibility of genomic regions to the ChIP-procedure. Due to the hierarchical structure of chromatin, different genomic regions have different propensities for cross-linking, sonication, and immunoprecipitation. This causes an uneven probability of observing DNA fragments originating from different genomic regions. Because different cell types (cell lines, and cancer cell lines), have different chromatin structure, ChIP samples will show a cell-type-specific bias in observed enrichment profiles. An important note to consider is that the input control is basically a reduced whole genome sequencing experiment, while the ChIP enriches for only a subset of genomic regions. If both ChIP and Input samples are sequenced to the same depth (same number of reads), the background distribution in the input sample will be under sampled. It is recommended to sequence the input sample deeper than the ChIP sample (Chen et al., 2012a).

**IgG control** uses a soup of nonspecific antibodies to control for background binding. In principle, the antibodies should be isolated from the same batch of serum which was used to create the specific antibody (used for ChIP). It should, in theory, give a background profile of non-specific binding. The proper control, is however, seldom available. Additionally, because the antibodies are unspecific, the amount of precipitated DNA will be low, and the samples will require additional rounds of PCR amplification.

**KO control** is a ChIP experiment performed in the biological system where the native protein is not present. Such an experiment profiles the non-specific binding of the antibody to other proteins, and directly to the DNA. The primary, and only, concern is that the perturbation caused by the knock-out (or knock-down), changes the cell so much, that the ChIP profile is not comparable to the original cell. This is the most accurate type of control experiment, however, it is frequently technically challenging to perform if the cells are not viable after the knock-out, or if the knock-out is impossible to perform.

### 9.3.6 Using tagged proteins

If an antibody of sufficient quality is not available, it is possible to resort to constructs where the protein of interest gets engineered with a ChIP-able tag. The proper control for such experiments is to perform the ChIP in the cell line containing the engineered protein, and without the protein. It must be noted that the tagging procedure can change the binding preferences of the protein, and therefore the experimental conclusions.

## 9.4 Pre-processing ChIP data

The focus of ChIP preprocessing is to check the quality of the sequencing experiment, remove sequencing artifacts, and find the genomic location of sequenced fragments using read mapping. The quality control consists of read quality control and adapter trimming. These methods are described in depth in Chapter 7.

### 9.4.1 Mapping of ChIP-seq data

Mapping is a procedure of trying to locate the exact genomic location which created each genomic fragment, each sequenced read. Several tools are available for mapping ChIP-seq data sets: Bowtie, Bowtie2, BWA (Langmead et al., 2009; Langmead and Salzberg, 2012b; Li and Durbin, 2009b), and all of them have comparable sensitivity and specificity (Ruffalo et al., 2011). Read length is the variable with the biggest effect on the mapping procedure. The longer the sequenced reads, the more uniquely can the read be assigned to a position on the genome. Reads which are assigned ambiguously to multiple locations in the genome are called multi-mapping reads. Such fragments are most often produced by repetitive genomic regions, such as retrotransposons, pseudogenes or paralogous genes (Li and Freudenberg, 2014). It is important to, a priori, decide whether such duplicated regions are of interest for the current experimental setup (i.e. whether we want to study transcription factor binding in olfactory receptors). If they are, then the multi-mapping reads should be included in the analysis. If they are not, they should be omitted. This is done during the mapping step, by limiting the number of locations to which a read can map. The methodology of working with multi-mapping reads differs according to the use case, and will not be considered in this chapter. For more information, please see the references (Chung et al., 2011).

Current Illumina sequencing procedures enable sequencing of DNA fragments from just one, or both ends. Sequencing from both ends is called **paired-end** sequencing and greatly enhances the sample **mappability**, the percentage of genome which can be uniquely mapped. Additionally, it provides an out-of-the-box estimate of the average DNA fragment length, a parameter which is important for quality control and peak calling. Although it would always be preferable to do paired-end sequencing it substantially increases the sequencing costs, which can be prohibitive.

Different reads, which map to the same genomic location (same chromosome, position, and strand), are called **duplicated reads**. Such reads are an indication that the same DNA fragment was present multiple times during the library preparation. This can happen due to high enrichment with highly specific antibodies, or such fragments can be artificially produced during PCR amplification. Because we do not know the exact origin of the duplicated fragments, they are most often collapsed during the peak calling procedure, i.e. when multiple reads map to the same chromosome, position, and strand, only one read is used. If the transcription factor binds to a small number of regions in the genome, such data reduction might be too stringent, and we can increase the sensitivity by allowing up to **N** different reads, per position (i.e. if more than **N** reads map to the same location, only **N** reads are kept for the downstream analysis).

Some peak calling algorithms have automated statistical methods for determining the number of reads, per position, which will be used in the analysis (Zhang et al., 2008).

An important consideration to take into account is the genome which was used in the experiment. Cell lines, cancer samples, and personal genomes usually contain structural genomic alterations which are not present in the reference genome (duplications, insertions, and deletions). Such regions can cause false negatives, and false positives in the ChIP-seq experiment. If a region was present multiple times in the experimental system, and only a single time in the reference genome, it will be relatively enriched in the final sequencing library. Such fragments will pile up on a single location during the mapping step, and create an artificial peak, which can be falsely characterized as a binding event. Such regions are called **blacklisted** regions and should be removed from the downstream analysis. The UCSC browser database[1] contains tables with such regions for the most commonly used model organism species.

This chapter presumes that the user is already familiar with the following technical and conceptual knowledge in computational data processing. From Chapters 7 and 6, you should be familiar with the concept of multi-mapping reads, and the following file formats BED, GTF, WIG, bigWig, BAM. You should also be familiar with PCR, what are PCR duplicates, positive and negative DNA strands, and technical and biological replicates.

---

[1]http://genome.ucsc.edu

## 9.5 ChIP quality control

While the goal of the read quality assessment is to check whether the sequencing produced a high enough number of high-quality reads the goal of ChIP quality control is to ascertain whether the chromatin immunoprecipitation enrichment was successful. This is a crucial step in the ChIP-seq analysis because it can help us identify low-quality ChIP samples, and give information about which experimental steps went wrong.

There are four steps in ChIP quality control:

1. Sample correlation clustering: Clustering of the pair-wise correlations between genome-wide signal profiles.

2. Data visualization in a genomic browser.

3. Average fragment length determination: Determining whether the ChIP was enriched for fragments of a certain length.

4. Visualization of GC bias. Here we will plot the ChIP enrichment versus the average GC content in the corresponding genomic bin.

### 9.5.1 The data

Here we will familiarize ourselves with the datasets that will be used in the chapter. Experimental data was downloaded from the public ENCODE (ENCODE Project Consortium, 2012) database of ChIP-seq experiments. The experiments were performed on a lymphoblastoid cell line, GM12878, and mapped to the GRCh38 (hg38) version of the human genome, using the standard ENCODE ChIP-seq pipeline. In this chapter, due to compute time considerations, we have taken a subset of the data which corresponds to the human chromosome 21 (chr21).

The data sets are located in the compGenomRData package. The location of the data sets can be accessed using the system.file() command, in the following way:

```
data_path = system.file('extdata/chip-seq',package='compGenomRData')
```

The available datasets can be listed using the list.files() function:

```
chip_files = list.files(data_path, full.names=TRUE)
```

The dataset consists of the following ChIP experiments:

1. **Transcription factors**: CTCF, SMC3, ZNF143, PolII (RNA polymerase 2)

2. **Histone modifications**: H3K4me3, H3K36me3, H3k27ac, H3k27me3

3. Various input samples

### 9.5.2 Sample clustering

Clustering is an ordering procedure which groups samples by similarity; the more similar samples are grouped closer to one another. The details of clustering methodologies are described in Chapter 4. Clustering of ChIP signal profiles is used for two purposes: The first one is to ascertain whether there is concordance between biological replicates; biological replicates should show greater similarity than ChIP of different proteins. The second function is to see whether our experiments conform to known prior knowledge. For example, we would expect to see greater similarity between proteins which belong to the same protein complex.

To quantify the ChIP signal we will firstly construct 1-kilobase-wide tilling windows over the genome, and subsequently count the number of reads in each window, for each experiment. We will then normalize the counts, to account for a different total number of reads in each experiment, and finally calculate the correlation between all pairs of samples. Although this procedure represents a crude way of data quantification, it provides sufficient information to ascertain the data quality.

Using the GenomeInfoDb we will first fetch the chromosome lengths corresponding to the hg38 version of the human genome, and filter the length for human chromosome 21.

```
# load the chromosome info package
library(GenomeInfoDb)

# fetch the chromosome lengths for the human genome
hg_chrs = getChromInfoFromUCSC('hg38')

# find the length of chromosome 21
hg_chrs = subset(hg_chrs, grepl('chr21$',chrom))
```

The tileGenome() function from the GenomicRanges package constructs equally sized windows over the genome of interest. The function takes two arguments:

1. A vector of chromosome lengths

2. Window size

Firstly, we convert the chromosome lengths *data.frame* into a *named vector*.

```
# downloaded hg_chrs is a data.frame object,
# we need to convert the data.frame into a named vector
seqlengths = with(hg_chrs, setNames(size, chrom))
```

Then we construct the windows.

```
# load the genomic ranges package
library(GenomicRanges)

# tileGenome function returns a list of GRanges of a given width,
# spanning the whole chromosome
tilling_window = tileGenome(seqlengths, tilewidth=1000)

# unlist converts the list to one GRanges object
tilling_window = unlist(tilling_window)
```

```
## GRanges object with 46710 ranges and 0 metadata columns:
##              seqnames            ranges strand
##                 <Rle>         <IRanges>  <Rle>
##      [1]        chr21         1-1000        *
##      [2]        chr21      1001-2000        *
##      [3]        chr21      2001-3000        *
##      [4]        chr21      3001-4000        *
##      [5]        chr21      4001-5000        *
##      ...          ...            ...      ...
##  [46706]        chr21 46704985-46705984   *
##  [46707]        chr21 46705985-46706984   *
##  [46708]        chr21 46706985-46707984   *
##  [46709]        chr21 46707985-46708984   *
##  [46710]        chr21 46708985-46709983   *
##  -------
##  seqinfo: 1 sequence from an unspecified genome
```

We will use the `summarizeOverlaps()` function from the `GenomicAlignments` package to count the number of reads in each genomic window. The function will do the counting automatically for all our experiments. The `summarizeOverlaps()` function returns a `SummarizedExperiment` object. The object contains the counts, genomic ranges which were used for the quantification, and the sample descriptions.

```
# load GenomicAlignments
library(GenomicAlignments)

# fetch bam files from the data folder
bam_files = list.files(
    path        = data_path,
    full.names = TRUE,
    pattern     = 'bam$'
)

# use summarizeOverlaps to count the reads
so = summarizeOverlaps(tilling_window, bam_files)

# extract the counts from the SummarizedExperiment
counts = assays(so)[[1]]
```

Different ChIP experiments were sequenced to different depths; each experiment contains a different number of reads. To remove the effect of the experimental depth on the quantification, the samples need to be normalized. The standard normalization procedure, for ChIP data, is to divide the counts in each tilling window by the total number of sequenced reads, and multiply it by a constant factor (to avoid extremely small numbers). This normalization procedure is called the **cpm** - counts per million.

$$CPM = counts * (10^6/total\ number\ of\ reads)$$

```
# calculate the cpm from the counts matrix
# the following command works because
# R calculates everything by columns
cpm = t(t(counts)*(1000000/colSums(counts)))
```

We remove all tiles which do not have overlapping reads. Tiles with o counts do not provide any additional discriminatory power, rather, they introduce artificial similarity between the samples (i.e. samples with only a handful of bound regions will have a lot of tiles with 0 counts, while they do not have to have any overlapping enriched tiles).

```
# remove all tiles which do not contain reads
cpm = cpm[rowSums(cpm) > 0,]
```

We use the `sub()` function to shorten the column names of the cpm matrix.

```
# change the formatting of the column names
# remove the .chr21.bam suffix
colnames(cpm) = sub('.chr21.bam','',   colnames(cpm))


# remove the GM12878_hg38 prefix
colnames(cpm) = sub('GM12878_hg38_','',colnames(cpm))
```

Finally, we calculate the pairwise Pearson correlation coefficient using the `cor()` function. The function takes as input a region-by-sample count matrix, and returns a sample X sample matrix, where each field contains the correlation coefficient between two samples.

```
# calculates the pearson correlation coefficient between the samples
correlation_matrix = cor(cpm, method='pearson')
```

The `Heatmap()` function from the `ComplexHeatmap` (Gu et al., 2016b) package is used to visualize the correlation coefficient. The function automatically performs hierarchical clustering - it groups the samples which have the highest pairwise correlation. The diagonal represents the correlation of each sample with itself.

```
# load ComplexHeatmap
library(ComplexHeatmap)


# load the circlize package, and define
# the color palette which will be used in the heatmap
library(circlize)
heatmap_col = circlize::colorRamp2(
    breaks = c(-1,0,1),
    colors = c('blue','white','red')
)


# plot the heatmap using the Heatmap function
Heatmap(
    matrix = correlation_matrix,
    col    = heatmap_col
)
```

In Figure 9.2 we can see a perfect example of why quality control is important. **CTCF** is a zinc finger protein which co-localizes with the Cohesin complex. **SMC3** is a sub unit of the Cohesin complex, and we would therefore expect to see that the **SMC3**

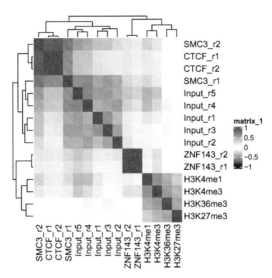

**FIGURE 9.2:** Heatmap showing ChIP-seq sample similarity using the Pearson correlation coefficient.

signal profile has high correlation with the **CTCF** signal profile. This is true for the second biological replicate of **SMC3**, while the first replicate (SMC3_r1) clusters with the input samples. This indicates that the sample likely has low enrichment. We can see that the ChIP and Input samples form separate clusters. This implies that the ChIP samples have an enrichment of fragments. Additionally, we see that the biological replicates of other experiments cluster together.

### 9.5.3   Visualization in the genome browser

One of the first steps in any ChIP-seq analysis should be looking at the data. By looking at the data we get an intuition about the quality of the experiment, and start seeing preliminary correlations between the samples, which we can use to guide our analysis. This can be achieved either by plotting signal profiles around regions of interest, or by loading data into a genome browser (such as IGV, or UCSC genome browsers).

Genome browsers are standalone applications which represent the genome as a one-dimensional (1D) coordinate system. The browsers enable simultaneous visualization and comparison of multiple types of annotations and experimental data.

Genome browsers can visualize most of the commonly used genomic data formats: BAM, BED, wig, and bigWig. The easiest way to access our data would be to load the .bam files into the browser. This will show us the sequence and position of every mapped read. If we want to view multiple samples in parallel, loading every mapped read can be restrictive. It takes up a lot of computational resources, and the amount of information makes the visual comparison hard to do. We would like to convert our data so that we get a compressed visualization, which would show us the main properties of our samples, namely, the quality and the location of the enrichment. This is achieved by summarizing the read enrichment into a signal profile - the whole experiment is converted into a numeric vector - a coverage vector. The vector contains information on how many reads overlap each position in the genome.

We will proceed as follows: Firstly, we will import a **.bam** file into **R**. Then we will calculate the signal profile (construct the coverage vector), and finally, we export the vector as a **.bigWig** file.

First we select one of the ChIP samples.

```
# list the bam files in the directory
# the '$' sign tells the pattern recognizer to omit bam.bai files
bam_files = list.files(
    path       = data_path,
    full.names = TRUE,
    pattern    = 'bam$'
)


# select the first bam file
chip_file = bam_files[1]
```

We will use the readGAlignments() function from the GenomicAlignments package to load the reads into **R**, and then the GRanges() function to convert them into a GRanges object.

```
# load the genomic alignments package
library(GenomicAlignments)

# read the ChIP reads into R
reads = readGAlignments(chip_file)

# the reads need to be converted to a granges object
reads = granges(reads)
```

Because DNA fragments are being sequenced from their ends (both the 3' and 5' end), the read enrichment does not correspond to the exact location of the bound protein. Rather, reads end to form clusters of enrichment upstream and downstream of the true binding location. To correct for this, we use a small hack. Before we create the signal profiles, we will extend the reads towards their **3'** end. The reads are extended to form fragments of 200 base pairs. This is an empiric measure, which corresponds to the average fragment size of the Illumina sample preparation kit. The exact average fragment size will differ from 200 base pairs, but if the deviation is not large (i.e. more than 200 base pairs), it will not affect the visual properties of our samples.

Read extension is done using the resize() function. The function takes two arguments:

1.  width: resulting fragment width
2.  fix: which position of the fragment should not be changed (if fix is set to start, the reads will be extended towards the **3'** end. If fix is set to end, they will be extended towards the **5'** end)

```
# extends the reads towards the 3' end
reads = resize(reads, width=200, fix='start')


# keeps only chromosome 21
reads = keepSeqlevels(reads, 'chr21', pruning.mode='coarse')
```

Conversion of reads into coverage vectors is done with the coverage() function. The function takes only one argument (width), which corresponds to chromosome sizes. For this purpose we can use the, previously created, seqlengths variable. The coverage() function converts the reads into a compressed Rle object. We have introduced these workflows in Chapter 6.

```
# convert the reads into a signal profile
cov = coverage(reads, width = seqlengths)


## RleList of length 1
## $chr21
## integer-Rle of length 46709983 with 199419 runs
##    Lengths: 5038228     200   63546      20 ...     200    1203     200   27856
##    Values :       0       1       0       1 ...       1       0       1       0
```

The name of the output file is created by changing the file suffix from **.bam** to **.bigWig**.

```
# change the file extension from .bam to .bigWig
output_file = sub('.bam','.bigWig', chip_file)
```

Now we can use the `export.bw()` function from the rtracklayer package to write the bigWig file.

```
# load the rtracklayer package
library(rtracklayer)

# export the bigWig output file
export.bw(cov, 'output_file')
```

### 9.5.3.1 Vizualization of track data using Gviz

We can create genome browserlike visualizations using the `Gviz` package, which was introduced in Chapter 6. The `Gviz` is a tool which enables exhaustive customized visualization of genomics experiments. The basic usage principle is to define tracks, where each track can represent genomic annotation, or a signal profile; subsequently we define the order of the tracks and plot them. Here we will define two tracks, a genome axis, which will show the position along the human chromosome 21; and a signal track from our CTCF experiment.

```
library(Gviz)
# define the genome axis track
axis    = GenomeAxisTrack(
    range = GRanges('chr21', IRanges(1, width=seqlengths))
)

# convert the signal into genomic ranges and define the signal track
gcov    = as(cov, 'GRanges')
dtrack = DataTrack(gcov, name = "CTCF", type='l')

# define the track ordering
track_list = list(axis,dtrack)
```

Tracks are plotted with the `plotTracks()` function. The `sizes` argument needs to be the same size as the track_list, and defines the relative size of each track. Figure 9.3 shows the output of the `plotTracks()` function.

```
# plot the list of browser tracks
# sizes argument defines the relative sizes of tracks
# background title defines the color for the track labels
plotTracks(
    trackList          = track_list,
```

```
    sizes             = c(.1,1),
    background.title = "black"
)
```

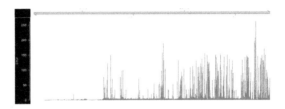

**FIGURE 9.3:** ChIP-seq signal visualized as a browser track using Gviz.

### 9.5.4   Plus and minus strand cross-correlation

Cross-correlation between plus and minus strands is a method which quantifies whether the DNA library was enriched for fragments of a certain length.

Similarity between the plus and minus strands defined as the correlation of the signal profiles for the reads that map to the + and the - strands. The distribution of reads is shown in Figure 9.4.

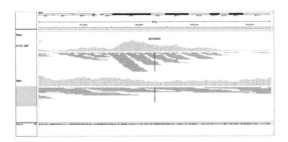

**FIGURE 9.4:** Browser screenshot of aligned reads for one ChIP, and control sample. ChIP samples have an asymetric distribution of reads; reads mapping to the + strand are located on the left side of the peak, while the reads mapping to the - strand are found on the right side of the peak.

Due to the sequencing properties, reads which correspond to the **5'** fragment ends will map to the opposite strand from the reads coming from the **3'** ends. Most often (depending on the sequencing protocol) the reads from the **5'** fragment ends map to the **+** strand, while the reads from the **3'** ends map to the **-** strand.

We calculate the cross-correlation by shifting the signal on the **+** strand, by a pre-defined amount (i.e. shift by 1 - 400 nucleotides), and calculating, for each shift, the correlation between the **+**, and the **-** strands. Subsequently we plot the correlation versus shift, and locate the maximum value. The maximum value should correspond to the average DNA fragment length which was present in the library. This value tells us whether the ChIP enriched for fragments of certain length (i.e. whether the ChIP was successful).

Due to the size of genomic data, it might be computationally prohibitive to calculate the Pearson correlation between whole genome (or even whole chromosome) signal profiles. To get around this problem, we will resort to a trick; we will disregard the dynamic range of the signal profiles, and only keep the information of which genomic bases contained the ends of the fragments. This is done by calculating the coverage vector of the read starting position (separately for each strand), and converting the coverage vector into a Boolean vector. The Boolean vector contains the information of which genomic positions contained the DNA fragment ends.

Similarity between two Boolean vectors can be promptly computed using the Jaccard index. The Jaccard index is defined as an intersection between two Boolean vectors, divided by their union as shown in Figure 9.5.

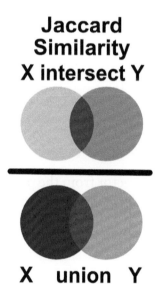

**FIGURE 9.5:** Jaccard similarity is defined as the ratio of the intersection and union of two sets.

Firstly, we load the reads for one of the CTCF ChIP experiments. Then we create signal profiles, separately for reads on the + and - strands. Unlike before, we do not extend the reads to the average expected fragment length (200 base pairs); we keep only the starting position of each read.

```
# load the reads
reads = readGAlignments(chip_file)
reads = granges(reads)

# keep only the starting position of each read
reads = resize(reads, width=1, fix='start')

reads = keepSeqlevels(reads, 'chr21', pruning.mode='coarse')
```

Now we can calculate the coverage vector of the read starting position. The coverage vector is then automatically converted into a Boolean vector by asking which genomic positions have *coverage* > 0.

```
# calculate the coverage profile for plus and minus strand
reads = split(reads, strand(reads))

# coverage(x, width = seqlengths)[[1]] > 0
# calculates the coverage and converts
# the coverage vector into a boolean
cov   = lapply(reads, function(x){
    coverage(x, width = seqlengths)[[1]] > 0
})
cov   = lapply(cov, as.vector)
```

We will now shift the coverage vector from the plus strand by 1 to 400 base pairs, and for each pair shift we will calculate the Jaccard index between the vectors on the plus and minus strand.

```
# defines the shift range
wsize = 1:400

# defines the jaccard similarity
jaccard = function(x,y)sum((x & y)) / sum((x | y))

# shifts the + vector by 1 - 400 nucleotides and
# calculates the correlation coefficient
cc = shiftApply(
    SHIFT = wsize,
    X     = cov[['+']],
    Y     = cov[['-']],
```

```
    FUN    = jaccard
)
```

```
# converts the results into a data frame
cc = data.frame(fragment_size = wsize, cross_correlation = cc)
```

We can finally plot the shift in base pairs versus the correlation coefficient:

```
library(ggplot2)
ggplot(data = cc, aes(fragment_size, cross_correlation)) +
    geom_point() +
    geom_vline(xintercept = which.max(cc$cross_correlation),
                size=2, color='red', linetype=2) +
    theme_bw() +
    theme(
        axis.text = element_text(size=10, face='bold'),
        axis.title = element_text(size=14,face="bold"),
        plot.title = element_text(hjust = 0.5)) +
    xlab('Shift in base pairs') +
    ylab('Jaccard similarity')
```

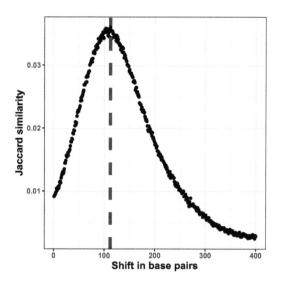

**FIGURE 9.6:** The figure shows the correlation coefficient between the ChIP-seq signal on + and − strands. The peak of the distribution designates the fragment size.

Figure 9.6 shows the shift in base pairs, which corresponds to the maximum value of the correlation coefficient gives us an approximation to the expected average DNA fragment length. Because this value is not 0, or monotonically decreasing, we can conclude that there was substantial enrichment of certain fragments in the ChIP samples.

### 9.5.5 GC bias quantification

The PCR amplification procedure can cause a significant bias in the ChIP experiments. The bias can be influenced by the DNA fragment size distribution, sequence composition, hexamer distribution of PCR primers, and the number of cycles used for the amplification. One way to determine whether some of the samples have significantly different sequence composition is to look at whether regions with differing GC composition were equally enriched in all experiments.

We will do the following: Firstly we will calculate the GC content of each of the tilling windows, and then we will compare the GC content with the corresponding cpm (count per million reads) value, for each tile.

```
# fetches the chromosome lengths and constructs the tiles
library(GenomeInfoDb)
library(GenomicRanges)


hg_chrs        = getChromInfoFromUCSC('hg38')
hg_chrs        = subset(hg_chrs, grepl('chr21$',chrom))
seqlengths     = with(hg_chrs, setNames(size, chrom))


# tileGenome produces a list per chromosome
# unlist combines the elements of the list
# into one GRanges object
tilling_window = unlist(tileGenome(
    seqlengths = seqlengths,
    tilewidth  = 1000
))
```

We will extract the sequence information from the BSgenome.Hsapiens.UCSC.hg38 package. BSgenome are generic Bioconductor containers for genomic sequences. Sequences are extracted from the BSgenome container using the getSeq() function. The getSeq() function takes as input the genome object, and the ranges with the regions of interest; in our case, the tilling windows. The function returns a DNAString object.

```
# loads the human genome sequence
library(BSgenome.Hsapiens.UCSC.hg38)
```

```
# extracts the sequence from the human genome
seq = getSeq(BSgenome.Hsapiens.UCSC.hg38, tilling_window)
```

To calculate the GC content, we will use the `oligonucleotideFrequency()` function on the DNAString object. By setting the width parameter to 2 we will calculate the **dinucleotide** frequency. Each row in the resulting table will contain the number of all possible dinucleotides observed in each tilling window. Because we have tilling windows of the same length, we do not necessarily need to normalize the counts by the window length. If all of the windows have different lengths (i.e. when at the ChIP-seq peaks), then normalization is a prerequisite.

```
# calculates the frequency of all possible dimers
# in our sequence set
nuc = oligonucleotideFrequency(seq, width = 2)
```

```
# converts the matrix into a data.frame
nuc = as.data.frame(nuc)
```

```
# calculates the percentages, and rounds the number
nuc = round(nuc/1000,3)
```

Now we can combine the GC frequency with the cpm values. We will convert the cpm values to the log10 scale. To avoid taking the $log(0)$, we add a pseudo count of 1 to cpm.

```
# counts the number of reads per tilling window
# for each experiment
so = summarizeOverlaps(tilling_window, bam_files)
```

```
# converts the raw counts to cpm values
counts  = assays(so)[[1]]
cpm     = t(t(counts)*(1000000/colSums(counts)))
```

```
# because the cpm scale has a large dynamic range
# we transform it using the log function
cpm_log = log10(cpm+1)
```

Combine the cpm values with the GC content,

```
gc = cbind(data.frame(cpm_log), GC = nuc['GC'])
```

and plot the results.

```
ggplot(
    data = gc,
    aes(
        x = GC,
        y = GM12878_hg38_CTCF_r1.chr21.bam
    )) +
  geom_point(size=2, alpha=.3) +
  theme_bw() +
  theme(
    axis.text  = element_text(size=10, face='bold'),
    axis.title = element_text(size=14,face="bold"),
    plot.title = element_text(hjust = 0.5)) +
  xlab('GC content in one kilobase windows') +
  ylab('log10( cpm + 1 )') +
  ggtitle('CTCF Replicate 1')
```

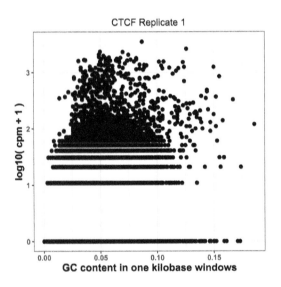

**FIGURE 9.7:** GC content abundance in a ChIP-seq experiment.

Figure 9.7 visualizes the CPM versus GC content, and gives us two important pieces of information. Firstly, it shows whether there was a specific amplification of regions with extremely high or extremely low GC content. This would be a strong indication that either the PCR or the size selection procedure were not successfully executed. The second piece of information comes by comparison of plots corresponding to multiple experiments. If different ChIP-samples have highly diverging enrichment of different ChIP regions, then some of the samples were affected by unknown batch effects. Such effects need to be taken into account in downstream analysis.

Firstly, we will reorder the columns of the data.frame using the pivot_longer() function from the tidyr package.

```
# load the tidyr package
library(tidyr)

# pivot_longer converts a fat data.frame into a tall data.frame,
# which is the format used by the ggplot package
gcd = pivot_longer(
    data      = gc,
    cols      = -GC,
    names_to  = 'experiment',
    values_to = 'cpm'
)

# we select the ChIP files corresponding to the ctcf experiment
gcd = subset(gcd, grepl('CTCF', experiment))

# remove the chr21 suffix
gcd$experiment = sub('chr21.','',gcd$experiment)
```

We can now visualize the relationship using a scatter plot. Figure 9.8 compares the GC content dependency on the CPM between the first and the second CTCF replicate. In this case, the replicate looks similar.

```
ggplot(data = gcd, aes(GC, log10(cpm+1))) +
  geom_point(size=2, alpha=.05) +
  theme_bw() +
  facet_wrap(~experiment, nrow=1)+
  theme(
    axis.text = element_text(size=10, face='bold'),
    axis.title = element_text(size=14,face="bold"),
    plot.title = element_text(hjust = 0.5)) +
  xlab('GC content in one kilobase windows') +
```

```
ylab('log10( cpm + 1 )') +
ggtitle('CTCF Replicates 1 and 2')
```

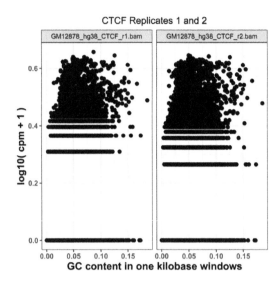

**FIGURE 9.8:** Comparison of GC content and signal abundance between two CTCF biological replicates.

### 9.5.6  Sequence read genomic distribution

The fourth way to look at the ChIP quality control is to visualize the genomic distribution of reads in different functional genomic regions. If the ChIP samples have the same distribution of reads as the Input samples, this implies a lack of specific enrichment. Additionally, if we have prior knowledge of where our proteins should be located, we can use the visualization to judge how well the genomic distributions conform to our priors. For example, the trimethylation of histone H3 on lysine 36 - **H3K36me3** is associated with elongating polymerase and productive transcription. If we performed a successful ChIP experiment with an anti-**H3K36me3** antibody, we would expect most of the reads to fall within gene bodies (introns and exons).

#### 9.5.6.1  Hierarchical annotation of genomic features

Overlapping genomic features (a transcription start site of one gene might be in an intron of another gene) will cause an ambiguity during the read annotation. If a read overlaps more than one functional category, we are not certain which category it should be assigned to. To solve the problem of multiple assignments, we need to construct a set of annotation rules. A heuristic solution is to organize the genomic annotation into a hierarchy which will imply prioritization. We can then look, for each read, which functional categories it overlaps, and if it is within multiple categories, we assign the read to the topmost category. As an example, let's say that we have 4 genomic categories: 1) TSS (transcription start sites), 2) exon, 3) intron, and 4) intergenic with the following hierarchy: **TSS -> exon -> intron -> intergenic**. This means that if a read overlaps a TSS and an intron, it will be annotated as TSS. This approach is shown in Figure 9.9.

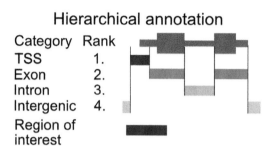

FIGURE 9.9: Principle of hierarchical annotation. The region of interest is annotated as the topmost ranked category that it overlaps. In this case, our region overlaps a TSS, an exon, and an intergenic region. Because the TSS has the topmost rank, it is annotated as a TSS.

Now we will construct the set of functional genomic regions, and annotate the reads.

#### 9.5.6.2   Finding annotations

There are multiple sources of genomic annotation. **UCSC**, **Genbank**, and **Ensembl** databases represent stable resources, from which the annotation can be easily obtained.

AnnotationHub is a Bioconductor-based online resource which contains a large number of experiments from various sources. We will use the AnnotationHub to download the location of genes corresponding to the **hg38** genome. The hub is accessed in the following way:

```
# load the AnnotationHub package
library(AnnotationHub)

# connect to the hub object
hub = AnnotationHub()
```

The hub variable contains the programming interface towards the online database. We can use the query() function to find out the ID of the "ENSEMBL" gene annotation.

```
# query the hub for the human annotation
AnnotationHub::query(
    x       = hub,
    pattern = c('ENSEMBL','Homo','GRCh38','chr','gtf')
)
```

```
## AnnotationHub with 32 records
## # snapshotDate(): 2020-04-27
## # $dataprovider: Ensembl
## # $species: Homo sapiens
## # $rdataclass: GRanges
## # additional mcols(): taxonomyid, genome, description,
## #   coordinate_1_based, maintainer, rdatadateadded, preparerclass, tags,
## #   rdatapath, sourceurl, sourcetype
## # retrieve records with, e.g., 'object[["AH50842"]]'
##
##              title
##   AH50842 | Homo_sapiens.GRCh38.84.chr.gtf
##   AH50843 | Homo_sapiens.GRCh38.84.chr_patch_hapl_scaff.gtf
##   AH51012 | Homo_sapiens.GRCh38.85.chr.gtf
##   AH51013 | Homo_sapiens.GRCh38.85.chr_patch_hapl_scaff.gtf
##   AH51953 | Homo_sapiens.GRCh38.86.chr.gtf
##   ...         ...
##   AH75392 | Homo_sapiens.GRCh38.98.chr_patch_hapl_scaff.gtf
##   AH79159 | Homo_sapiens.GRCh38.99.chr.gtf
##   AH79160 | Homo_sapiens.GRCh38.99.chr_patch_hapl_scaff.gtf
##   AH80075 | Homo_sapiens.GRCh38.100.chr.gtf
##   AH80076 | Homo_sapiens.GRCh38.100.chr_patch_hapl_scaff.gtf
```

We are interested in the version **GRCh38.92**, which is available under **AH61126**. To download the data from the hub, we use the `[[` operator on the hub API. We will download the annotation in the **GTF** format, into a GRanges object.

```
# retrieve the human gene annotation
gtf = hub[['AH61126']]
```

```
## GRanges object with 6 ranges and 3 metadata columns:
##      seqnames      ranges strand |   source        type      score
##         <Rle>   <IRanges>  <Rle> | <factor>    <factor>  <numeric>
##   [1]        1 11869-14409      + |   havana        gene         NA
##   [2]        1 11869-14409      + |   havana  transcript         NA
##   [3]        1 11869-12227      + |   havana        exon         NA
```

```
##    [4]         1 12613-12721      + |    havana exon             NA
##    [5]         1 13221-14409      + |    havana exon             NA
##    [6]         1 12010-13670      + |    havana transcript       NA
##    -------
##    seqinfo: 25 sequences (1 circular) from GRCh38 genome
```

By default the ENSEMBL project labels chromosomes using numeric identifiers
(i.e. 1,2,3 ... X), without the **chr** prefix. We need to therefore append the prefix to
the chromosome names (seqlevels). `pruning.mode = 'coarse'` designates that the
chromosome names will be replaced in the gtf object.

```
# extract ensemel chromosome names
ensembl_seqlevels = seqlevels(gtf)


# paste the chr prefix to the chromosome names
ucsc_seqlevels    = paste0('chr', ensembl_seqlevels)


# replace ensembl with ucsc chromosome names
seqlevels(gtf, pruning.mode='coarse') = ucsc_seqlevels
```

And finally we subset only regions which correspond to chromosome 21.

```
# keep only chromosome 21
gtf = gtf[seqnames(gtf) == 'chr21']
```

### 9.5.6.3  Constructing genomic annotation

Once we have downloaded the annotation we can define the functional hierarchy.
We will use the previously mentioned ordering: **TSS -> exon -> intron -> intergenic**,
with **TSS** having the highest priority and the intergenic regions having the lowest
priority.

```
# construct a GRangesList with human annotation
annotation_list = GRangesList(

    # promoters function extends the gtf around the TSS
    # by an upstream and downstream amounts
    tss    = promoters(
        x = subset(gtf, type=='gene'),
        upstream   = 1000,
        downstream = 1000),
    exon   = subset(gtf, type=='exon'),
```

```
    intron = subset(gtf, type=='gene')
)
```

#### 9.5.6.4  Annotating reads

To annotate the reads we will define a function that takes as input a **.bam** file, and an annotation list, and returns the frequency of reads in each genomic category. We will then loop over all of the **.bam** files to annotate each experiment.

The `annotateReads()` function works in the following way:

1.  Load the **.bam** file.
2.  Find overlaps between the reads and the annotation categories.
3.  Arrange the annotated reads based on the hierarchy, and remove duplicated assignments.
4.  Count the number of reads in each category.

The crucial step to understand here is using the `arrange()` and `filter()` functions to keep only one annotated category per read.

```
annotateReads = function(bam_file, annotation_list){

    library(dplyr)
    message(basename(bam_file))

    # load the reads into R
    bam    = readGAlignments(bam_file)

    # find overlaps between reads and annotation
    result = as.data.frame(
        findOverlaps(bam, annotation_list)
    )

    # appends to the annotation index the corresponding
    # annotation name
    annotation_name   = names(annotation_list)[result$subjectHits]
    result$annotation = annotation_name

    # order the overlaps based on the hierarchy
    result = result[order(result$subjectHits),]
```

```r
    # select only one category per read
    result = subset(result, !duplicated(queryHits))

    # count the number of reads in each category
    # group the result data frame by the corresponding category
    result = group_by(.data=result, annotation)

    # count the number of reads in each category
    result = summarise(.data = result, counts = length(annotation))

    # classify all reads which are outside of
    # the annotation as intergenic
    result = rbind(
        result,
        data.frame(
            annotation = 'intergenic',
            counts     = length(bam) - sum(result$counts)
        )
    )

    # calculate the frequency
    result$frequency  = with(result, round(counts/sum(counts),2))

    # append the experiment name
    result$experiment = basename(bam_file)

    return(result)
}
```

We execute the annotation function on all files.

```r
# list all bam files in the folder
bam_files   = list.files(data_path, full.names=TRUE, pattern='bam$')

# calculate the read distribution for every file
annot_reads_list = lapply(bam_files, function(x){
    annotateReads(
        bam_file        = x,
        annotation_list = annotation_list
```

```
    )
})
```

First, we combine the results in one data frame, and reformat the experiment
names.

```
# collapse the per-file read distributions into one data.frame
annot_reads_df = dplyr::bind_rows(annot_reads_list)

# format the experiment names
experiment_name = annot_reads_df$experiment
experiment_name = sub('.chr21.bam','', experiment_name)
experiment_name = sub('GM12878_hg38_','',experiment_name)
annot_reads_df$experiment = experiment_name
```

And plot the results.

```
ggplot(data = annot_reads_df,
       aes(
            x    = experiment,
            y    = frequency,
            fill = annotation
        )) +
    geom_bar(stat='identity') +
    theme_bw() +
    scale_fill_brewer(palette='Set2') +
    theme(
        axis.text = element_text(size=10, face='bold'),
        axis.title = element_text(size=14,face="bold"),
        plot.title = element_text(hjust = 0.5),
        axis.text.x = element_text(angle = 90, hjust = 1)) +
    xlab('Sample') +
    ylab('Percentage of reads') +
    ggtitle('Percentage of reads in annotation')
```

Figure 9.10 shows a slight increase of **H3K36me3** on the exons and introns, and
**H3K4me3** on the **TSS**. Interestingly, both replicates of the **ZNF143** transcription
factor show increased read abundance around the TSS.

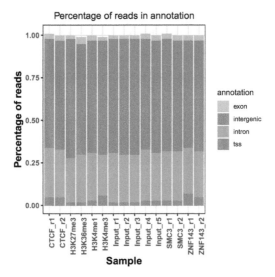

**FIGURE 9.10:** Read distribution in genomice functional annotation categories.

## 9.6 Peak calling

After we are convinced that the data is of sufficient quality, we can proceed with the downstream analysis. One of the first steps in the ChIP-seq analysis is peak calling. Peak calling is a statistical procedure, which uses coverage properties of ChIP and Input samples to find regions which are enriched due to protein binding. The procedure requires mapped reads, and outputs a set of regions, which represent the putative binding locations. Each region is usually associated with a significance score which is an indicator of enrichment.

For peak calling we will use the normR Bioconductor package. normR uses a binomial mixture model, and performs simultaneous normalization and peak finding. Due to the nature of the model, it is quite flexible and can be used for different types of ChIP experiments.

One of the caveats of normR is that it does not inherently support multiple biological replicates, for the same biological sample. Therefore, the peak calling procedure needs to be done on each replicate separately, and the peaks need to be combined in post-processing.

### 9.6.1 Types of ChIP-seq experiments

Based on the binding properties of ChIP-ped proteins, ChIP-seq signal profiles can be divided into three classes:

1. **Sharp** (point signal): A signal profile which is localized to specific short genomic regions (up to a couple of hundred base pairs) It is usually obtained from transcription factors, or highly localized posttranslational histone modifications (H3K4me3, which is found on gene promoters).

2. **Broad** (wide signal): The signal covers broad genomic domains spanning up to several kilobases. Usually produced by disperse histone modifications (H3K36me3, located on gene bodies, or H3K23me3, which is deposited by the Polycomb complex in large genomic regions).

3. **Mixed**: The signal consists of a mixture of sharp and broad regions. It is produced by proteins which have dynamic behavior. Most often these are ChIP experiments of RNA Polymerase 2.

Different types of ChIP experiments usually require specialized analysis tools. Some peak callers are developed to specifically detect narrow peaks (Zhang et al., 2008; Xu et al., 2010; Shao et al., 2012), while others detect enrichment in diffuse broad regions (Zang et al., 2009; Micsinai et al., 2012; Beck et al., 2012; Song and Smith, 2011; Xing et al., 2012), or mixed (Polymerase 2) signals (Han et al., 2012). Recent developments in peak calling methods (such as normR) can however accommodate multiple types of ChIP experiments (Rashid et al., 2011). The choice of the algorithm will largely depend on the type of the wanted results, and the peculiarities of the experimental design and execution (Laajala et al., 2009; Wilbanks and Facciotti, 2010).

If you are not certain what kind of signal profile to expect from a ChIP-seq experiment, the best solution is to visualize the data. We will now use the data from **H3K4me3** (Sharp), **H3K36me3** (Broad), and **POL2** (Mixed) ChIP experiments to show the differences in the signal profiles. We will use the bigWig files to visualize the signal profiles around a highly expressed human gene from chromosome 21. This will give us an indication of how the profiles for different types of ChIP experiments differ. First we select the files of interest:

```
# set names for chip-seq bigWig files
chip_files = list(
    H3K4me3  = 'GM12878_hg38_H3K4me3.chr21.bw',

    H3K36me3 = 'GM12878_hg38_H3K36me3.chr21.bw',

    POL2     = 'GM12878_hg38_POLR2A.chr21.bw'
)
# get full paths to the files
chip_files = lapply(chip_files, function(x){
```

```
    file.path(data_path, x)
})
```

Next we import the coverage profiles into **R**:

```
# load rtracklayer
library(rtracklayer)

# import the ChIP bigWig files
chip_profiles = lapply(chip_files, rtracklayer::import.bw)
```

We fetch the reference annotation for human chromosome 21.

```
library(AnnotationHub)
hub = AnnotationHub()
gtf = hub[['AH61126']]

# select only chromosome 21
seqlevels(gtf, pruning.mode='coarse') = '21'

# extract chromosome names
ensembl_seqlevels = seqlevels(gtf)

# paste the chr prefix to the chromosome names
ucsc_seqlevels    = paste0('chr', ensembl_seqlevels)

# replace ensembl with ucsc chromosome names
seqlevels(gtf, pruning.mode='coarse') = ucsc_seqlevels
```

To enable Gviz to work with genomic annotation we will convert the GRanges object into a transcript database using the following function:

```
# load the GenomicFeatures object
library(GenomicFeatures)

# convert the gtf annotation into a data.base
txdb         = makeTxDbFromGRanges(gtf)
```

And convert the transcript database into a Gviz track.

```
# define the gene track object
gene_track  = GeneRegionTrack(txdb, chr='chr21', genome='hg38')
```

Once we have downloaded the annotation, and imported the signal profiles into **R** we are ready to visualize the data. We will again use the Gviz library. We firstly define the coordinate system. The ideogram track which will show the position of our current viewpoint on the chromosome, and a genome axis track, which will show the exact coordinates.

```
# load Gviz package
library(Gviz)
# fetches the chromosome length information
hg_chrs = getChromInfoFromUCSC('hg38')
hg_chrs = subset(hg_chrs, (grepl('chr21$',chrom)))

# convert data.frame to named vector
seqlengths = with(hg_chrs, setNames(size, chrom))

# constructs the ideogram track
chr_track  = IdeogramTrack(
    chromosome = 'chr21',
    genome     = 'hg38'
)


# constructs the coordinate system
axis = GenomeAxisTrack(
    range = GRanges('chr21', IRanges(1, width=seqlengths))
)
```

We use a loop to convert the signal profiles into a DataTrack object.

```
# use a lapply on the imported bw files to create the track objects
# we loop over experiment names, and select the corresponding object
# within the function
data_tracks = lapply(names(chip_profiles), function(exp_name){

    # chip_profiles[[exp_name]] - selects the
    # proper experiment using the exp_name
    DataTrack(
        range = chip_profiles[[exp_name]],
```

```
        name  = exp_name,

        # type of the track
        type  = 'h',

        # line width parameter
        lwd   = 5
    )
})
```

We are finally ready to create the genome screenshot. We will focus on an extended region around the URB1 gene.

```
# select the start coordinate for the URB1 gene
start = min(start(subset(gtf, gene_name == 'URB1')))

# select the end coordinate for the URB1 gene
end   = max(end(subset(gtf, gene_name == 'URB1')))

# plot the signal profiles around the URB1 gene
plotTracks(
    trackList = c(chr_track, axis, gene_track, data_tracks),

    # relative track sizes
    sizes     = c(1,1,1,1,1,1),

    # background color
    background.title     = "black",

    # controls visualization of gene sets
    collapseTranscripts  = "longest",
    transcriptAnnotation = "symbol",

    # coordinates to visualize
    from = start - 5000,
    to   = end   + 5000
)
```

Figure 9.11 shows the signal profile around the URB1 gene. H3K4me3 signal profile contains a strong narrow peak on the transcription start site. H3K36me3 shows

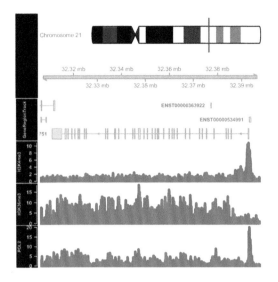

**FIGURE 9.11:** ChIP-seq signal around the URB1 gene.

strong enrichment in the gene body, while the POL2 ChIP shows a mixed profile, with a strong peak at the TSS and an enrichment over the gene body.

### 9.6.2 Peak calling: Sharp peaks

We will now use the normR (Helmuth et al., 2016) package for peak calling in sharp and broad peak experiments.

Select the input files. Since normR does not support the usage of biological replicates, we will showcase the peak calling on one of the CTCF samples.

```
# full path to the ChIP data file
chip_file    = file.path(data_path, 'GM12878_hg38_CTCF_r1.chr21.bam')

# full path to the Control data file
control_file = file.path(data_path, 'GM12878_hg38_Input_r5.chr21.bam')
```

To understand the dynamic range of enrichment, we will create a scatter plot showing the strength of signal in the CTCF and Input.

Let us first count the reads in 1-kb windows, and normalize them to counts per million sequenced reads.

```
# as previously done, we calculate the cpm for each experiment
library(GenomicRanges)
library(GenomicAlignments)
```

```
# select the chromosome
hg_chrs = getChromInfoFromUCSC('hg38')
hg_chrs = subset(hg_chrs, grepl('chr21$',chrom))

seqlengths = with(hg_chrs, setNames(size, chrom))

# define the windows
tilling_window = unlist(tileGenome(seqlengths, tilewidth=1000))

# count the reads
counts          = summarizeOverlaps(
    features = tilling_window,
    reads    = c(chip_file, control_file)
)

# normalize read counts
counts          = assays(counts)[[1]]
cpm = t(t(counts)*(1000000/colSums(counts)))
```

## We can now plot the ChIP versus Input signal:

```
library(ggplot2)
# convert the matrix into a data.frame for ggplot
cpm = data.frame(cpm)
ggplot(
    data = cpm,
    aes(
        x = GM12878_hg38_Input_r5.chr21.bam,
        y = GM12878_hg38_CTCF_r1.chr21.bam)
    ) +
    geom_point() +
    geom_abline(slope = 1) +
    theme_bw() +
    theme_bw() +
    scale_fill_brewer(palette='Set2') +
    theme(
        axis.text   = element_text(size=10, face='bold'),
        axis.title  = element_text(size=14,face="bold"),
        plot.title  = element_text(hjust = 0.5),
        axis.text.x = element_text(angle = 90, hjust = 1)) +
```

```
xlab('Input CPM') +
ylab('CTCF CPM') +
ggtitle('ChIP versus Input')
```

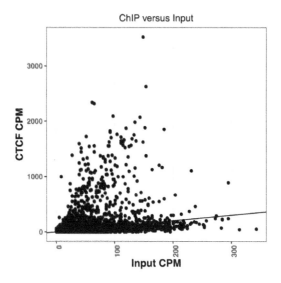

**FIGURE 9.12:** Comparison of CPM values between ChIP and Input experiments. Good ChIP experiments should always show enrichment.

Regions above the diagonal, in Figure 9.12, show higher enrichment in the ChIP samples, while the regions below the diagonal show higher enrichment in the Input samples.

Let us now perform for peak calling. normR usage is deceivingly simple; we need to provide the location ChIP and Control read files, and the genome version to the enrichR() function. The function will automatically create tilling windows (250bp by default), count the number of reads in each window, and fit a mixture of binomial distributions.

```
library(normr)
# peak calling using chip and control
ctcf_fit = enrichR(

        # ChIP file
        treatment = chip_file,

        # control file
        control  = control_file,
```

```
                      # genome version
                      genome   = "hg38",

                      # print intermediary steps during the analysis
                      verbose  = FALSE)
```

With the summary function we can take a look at the results:

**summary**(ctcf_fit)

```
## NormRFit-class object
##
## Type:                   'enrichR'
## Number of Regions:      12353090
## Number of Components:   2
## Theta* (naive bg):       0.137
## Background component B: 1
##
## +++ Results of fit +++
## Mixture Proportions:
## Background        Class 1
##     97.72%          2.28%
## Theta:
## Background        Class 1
##      0.103          0.695
##
## Bayesian Information Criterion:   539882
##
## +++ Results of binomial test +++
## T-Filter threshold: 4
## Number of Regions filtered out: 12267164
## Significantly different from background B based on q-values:
## TOTAL:
##              ***        **        *         .              n.s.
## Bins          0        627       120       195        87   84897
## %         0.000      0.711     0.847     1.068     1.166   96.209
## ---
## Signif. codes:  0 '***' 0.001 '**' 0.01 '*' 0.05 '.' 0.1 ' ' 1 'n.s.'
```

The summary function shows that most of the regions of chromosome 21 correspond to the background: $97.72$. In total we have $1029 = (627 + 120 + 195 + 87)$ significantly enriched regions.

We will now extract the regions into a GRanges object. The getRanges() function extracts the regions from the model. Using the getQvalue(), and getEnrichment() function we assign to our regions the statistical significance and calculated enrichment. In order to identify only highly significant regions, we keep only ranges where the false discovery rate (q value) is below $0.01$.

```
# extracts the ranges
ctcf_peaks = getRanges(ctcf_fit)


# annotates the ranges with the supporting p value
ctcf_peaks$qvalue    = getQvalues(ctcf_fit)


# annotates the ranges with the calculated enrichment
ctcf_peaks$enrichment = getEnrichment(ctcf_fit)


# selects the ranges which correspond to the enriched class
ctcf_peaks = subset(ctcf_peaks, !is.na(component))


# filter by a stringent q value threshold
ctcf_peaks = subset(ctcf_peaks, qvalue < 0.01)


# order the peaks based on the q value
ctcf_peaks = ctcf_peaks[order(ctcf_peaks$qvalue)]


## GRanges object with 724 ranges and 3 metadata columns:
##           seqnames            ranges strand | component       qvalue enrichment
##              <Rle>         <IRanges>  <Rle> | <integer>    <numeric>  <numeric>
##     [1]     chr21 43939251-43939500      * |         1 4.69885e-140    1.37891
##     [2]     chr21 43646751-43647000      * |         1 2.52008e-137    1.42361
##     [3]     chr21 43810751-43811000      * |         1 1.86406e-121    1.30519
##     [4]     chr21 43939001-43939250      * |         1 2.10824e-121    1.19820
##     [5]     chr21 37712251-37712500      * |         1 6.35715e-118    1.70989
##     ...       ...               ...    ... .       ...          ...        ...
##   [720]     chr21 38172001-38172250      * |         1   0.00867374   0.951189
##   [721]     chr21 38806001-38806250      * |         1   0.00867374   0.951189
##   [722]     chr21 42009501-42009750      * |         1   0.00867374   0.656253
##   [723]     chr21 46153001-46153250      * |         1   0.00867374   0.951189
##   [724]     chr21 46294751-46295000      * |         1   0.00867374   0.722822
##   -------
```

```
##    seqinfo: 24 sequences from an unspecified genome
```

After stringent q value filtering we are left with 724 peaks. For the ease of downstream analysis, we will limit the sequence levels to chromosome 21.

```
seqlevels(ctcf_peaks, pruning.mode='coarse') = 'chr21'
```

Let's export the peaks into a .txt file which we can use the downstream in the analysis.

```
# write the peaks loacations into a txt table
write.table(ctcf_peaks, file.path(data_path, 'CTCF_peaks.txt'),
            row.names=F, col.names=T, quote=F, sep='\t')
```

We can now repeat the CTCF versus Input plot, and label significantly marked peaks. Using the count overlaps we mark which of our 1-kb regions contained significant peaks.

```
# find enriched tilling windows
enriched_regions = countOverlaps(tilling_window, ctcf_peaks) > 0
```

```
library(ggplot2)
cpm$enriched_regions = enriched_regions

ggplot(
    data = cpm,
    aes(
        x = GM12878_hg38_Input_r5.chr21.bam,
        y = GM12878_hg38_CTCF_r1.chr21.bam,
        color = enriched_regions
    )) +
  geom_point() +
  geom_abline(slope = 1) +
  theme_bw() +
  scale_fill_brewer(palette='Set2') +
  theme(
    axis.text   = element_text(size=10, face='bold'),
    axis.title  = element_text(size=14,face="bold"),
    plot.title  = element_text(hjust = 0.5),
    axis.text.x = element_text(angle = 90, hjust = 1)) +
  xlab('Input CPM') +
```

```
ylab('CTCF CPM') +
ggtitle('ChIP versus Input') +
scale_color_manual(values=c('gray','red'))
```

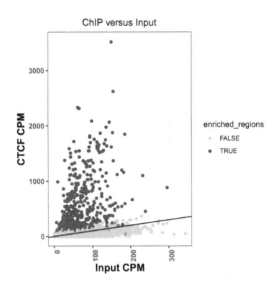

**FIGURE 9.13:** Comparison of signal between ChIP and input samples. Red labeled dots correspond to called peaks.

Figure 9.13 shows that normR identified all of the regions above the diagonal as statistically significant. It has, however, labeled a significant number of regions below the diagonal. Because of the sophisticated statistical model, normR has greater sensitivity, and these peaks might really be enriched regions, it is worth investigating the nature of these regions. This is left as an exercise to the reader.

We can now create a genome browser screenshot around a peak region. This will show us what kind of signal properties have contributed to the peak calling. We would expect to see a strong, bell-shaped, enrichment in the ChIP sample, and uniform noise in the Input sample.

Let us now visualize the signal around the most enriched peak. The following function takes as input a **.bam** file, and loads the bam into R. It extends the reads to a size of 200 bp, and creates the coverage vector.

```
# calculate the coverage for one bam file
calculateCoverage = function(
  bam_file,
  extend = 200
){
```

```
# load reads into R
reads = readGAlignments(bam_file)

# convert reads into a GRanges object
reads = granges(reads)

# resize the reads to 200bp
reads = resize(reads, width=extend, fix='start')

# get the coverage vector
cov   = coverage(reads)

# normalize the coverage vector to the sequencing depth
cov = round(cov * (1000000/length(reads)),2)

# convert the coverage go a GRanges object
cov   = as(cov, 'GRanges')

# keep only chromosome 21
seqlevels(cov, pruning.mode='coarse') = 'chr21'
return(cov)
}
```

Let's apply the function to the ChIP and input samples.

```
# calculate coverage for the ChIP file
ctcf_cov = calculateCoverage(chip_file)

# calculate coverage for the control file
cont_cov = calculateCoverage(control_file)
```

Using Gviz, we will construct the layered tracks. First, we layout the genome coordinates:

```
# load Gviz and get the chromosome coordinates
library(Gviz)
chr_track  = IdeogramTrack('chr21', 'hg38')
axis       = GenomeAxisTrack(
    range = GRanges('chr21', IRanges(1, width=seqlengths))
)
```

Then, the peak locations:

```
# peaks track
peaks_track = AnnotationTrack(ctcf_peaks, name = "CTCF Peaks")
```

And finally, the signal files:

```
chip_track  = DataTrack(
    range = ctcf_cov,
    name  = "CTCF",
    type  = 'h',
    lwd   = 3
)

cont_track  = DataTrack(
    range = cont_cov,
    name  = "Input",
    type  = 'h',
    lwd=3
)


plotTracks(
    trackList = list(chr_track, axis, peaks_track, chip_track, cont_track),
    sizes     = c(.2,.5,.5,1,1),
    background.title = "black",
    from = start(ctcf_peaks)[1] - 1000,
    to   = end(ctcf_peaks)[1]   + 1000
)
```

In Figure 9.14, the ChIP sample looks as expected. Although the Input sample shows an enrichment, it is important to compare the scales on both samples. The normalized ChIP signal goes up to 2500, while the maximum value in the input sample is only 60.

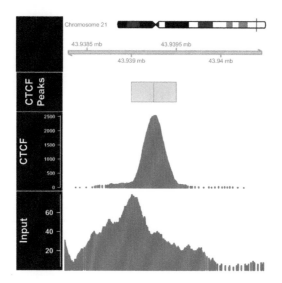

**FIGURE 9.14:** ChIP and Input signal profile around the peak centers.

### 9.6.3 Peak calling: Broad regions

We will now use normR to call peaks for the H3K36me3 histone modification, which is associated with gene bodies of expressed genes. We define the ChIP and Input files:

```
# fetch the ChIP-file for H3K36me3
chip_file    = file.path(data_path, 'GM12878_hg38_H3K36me3.chr21.bam')

# fetch the corresponding input file
control_file = file.path(data_path, 'GM12878_hg38_Input_r5.chr21.bam')
```

Because H3K36 regions span broad domains, it is necessary to increase the tilling window size which will be used for counting. Using the countConfiguration() function, we will set the tilling window size to 5000 base pairs.

```
library(normr)
# define the window width for the counting
countConfiguration = countConfigSingleEnd(binsize = 5000)


# find broad peaks using enrichR
h3k36_fit = enrichR(

            # ChIP file
```

```
        treatment   = chip_file,

        # control file
        control     = control_file,

        # genome version
        genome      = "hg38",
        verbose     = FALSE,

        # window size for counting
        countConfig = countConfiguration)



summary(h3k36_fit)

## NormRFit-class object
##
## Type:                  'enrichR'
## Number of Regions:     617665
## Number of Components:  2
## Theta* (naive bg):     0.197
## Background component B: 1
##
## +++ Results of fit +++
## Mixture Proportions:
## Background      Class 1
##      85.4%        14.6%
## Theta:
## Background      Class 1
##      0.138        0.442
##
## Bayesian Information Criterion:  741525
##
## +++ Results of binomial test +++
## T-Filter threshold: 5
## Number of Regions filtered out: 610736
## Significantly different from background B based on q-values:
## TOTAL:
##              ***      **      *       .              n.s.
## Bins           0     1005    314     381     237     4992
## %           0.00     9.18   12.04   15.52   17.68   45.58
```

```
## ---
## Signif. codes:  0 '***' 0.001 '**' 0.01 '*' 0.05 '.' 0.1 ' ' 1 'n.s.'
```

The summary function shows that we get 1937 enriched regions. We will extract enriched regions, and plot them in the same way we did for the CTCF.

```
# get the locations of broad peaks
h3k36_peaks              = getRanges(h3k36_fit)

# extract the qvalue and enrichment
h3k36_peaks$qvalue       = getQvalues(h3k36_fit)
h3k36_peaks$enrichment = getEnrichment(h3k36_fit)

# select proper peaks
h3k36_peaks = subset(h3k36_peaks, !is.na(component))
h3k36_peaks = subset(h3k36_peaks, qvalue < 0.01)
h3k36_peaks = h3k36_peaks[order(h3k36_peaks$qvalue)]

# collapse nearby enriched regions
h3k36_peaks = reduce(h3k36_peaks)


# construct the data tracks for the H3K36me3 and Input files
h3k36_cov = calculateCoverage(chip_file)
data_tracks = list(
    h3k36 = DataTrack(h3k36_cov,  name = 'h3k36_cov',  type='h', lwd=3),
    input = DataTrack(cont_cov,   name = 'Input',      type='h', lwd=3)
)


# define the window for the visualization
start = min(start(h3k36_peaks[2])) - 25000
end   = max(end(h3k36_peaks[2])) + 25000

# create the peak track
peak_track = AnnotationTrack(reduce(h3k36_peaks), name='H3K36me3')

# plots the enriched region
plotTracks(
    trackList = c(chr_track, axis, gene_track, peak_track, data_tracks),
    sizes     = c(.5,.5,.5,.1,1,1),
    background.title    = "black",
```

```
    collapseTranscripts   = "longest",
    transcriptAnnotation = "symbol",
    from = start,
    to   = end
)
```

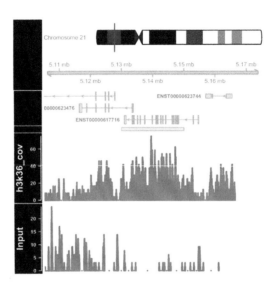

**FIGURE 9.15:** Visualization of H3K36me3 ChIP signal on a called broad peak.

Figure 9.15 shows a highly enriched H3K36me3 region covering the gene body, as expected.

### 9.6.4   Peak quality control

Peak calling is not a mathematically defined procedure; it is impossible to unambiguously define what a "peak" is. Therefore all of the peak calling procedures use heuristics, and statistical models which have been shown to work well in specific use cases. After peak calling, it is always necessary to check whether the defined peaks really are located in enriched regions, and in addition, use prior knowledge to ascertain whether the peaks correspond to known biology.

Peak calling can falsely identify enriched regions if the input sample is not sequenced to the proper depth. Because the input samples correspond to **de facto** whole genome sequencing, and the ChIP procedure enriches for a subset of the genome, it can often happen that many regions in the genome are not sufficiently covered by the Input sample. Such variability in the signal profile of Input samples can cause a region to be defined as a peak, enriched in the ChIP sample, while in reality it is depleted in the Input, due to under-sampling. For example, the figure in the previous chapter, showing an enriched region H3K36me3 over a gene body, shows a large depletion in the Input sample over the same region. Such depletion should be a concern and merit further investigation.

The quality of enrichment can be checked by calculating the percentage of reads within peaks for both ChIP and Input samples. ChIP samples should have a high percentage of reads in peaks, while for the input samples, the percentage of reads should correspond to the percentage of genome covered by peaks.

For transcription factor ChIP experiments, an important control is to determine whether the peak regions contain sequences which are known to be bound by the corresponding transcription factor - whether they contain known transcription factor binding motifs. Transcription factor binding motifs are sequence models which model the propensity of binding DNA sequences. Such sequence models can be downloaded from public databases and compared to see whether there is a positional enrichment around our peaks.

We will now calculate the percentage of reads within peaks for the H3K36me3 experiment. Subsequently, we will download the known CTCF sequence model, and compare it to our peak regions.

### 9.6.4.1   Percentage of reads in peaks

To calculate the reads in peaks, we will firstly extract the number of reads in each tilling window from the normR produced fit object. This is done using the getCounts() function. We will then use the q-value to define which tilling windows correspond to peaks, and count the number of reads within and outside peaks.

```
# extract, per tilling window, counts from the fit object
h3k36_counts = data.frame(getCounts(h3k36_fit))


# change the column names of the data.frame
colnames(h3k36_counts) = c('Input','H3K36me3')


# extract the q-value corresponding to each bin
h3k36_counts$qvalue = getQvalues(h3k36_fit)


# define which regions are peaks using a q value cutoff
h3k36_counts$enriched[is.na(h3k36_counts$qvalue)]  = 'Not Peak'
h3k36_counts$enriched[h3k36_counts$qvalue > 0.05]  = 'Not Peak'
h3k36_counts$enriched[h3k36_counts$qvalue <= 0.05] = 'Peak'


# remove the q value column
h3k36_counts$qvalue = NULL
```

```r
# reshape the data.frame into a long format
h3k36_counts_df = tidyr::pivot_longer(
    data       = h3k36_counts,
    cols       = -enriched,
    names_to   = 'experiment',
    values_to  = 'counts'
)


# sum the number of reads in the Peak and Not Peak regions
h3k36_counts_df = group_by(.data = h3k36_counts_df, experiment, enriched)
h3k36_counts_df = summarize(.data = h3k36_counts_df, num_of_reads = sum(counts))


# calculate the percentage of reads.
h3k36_counts_df          = group_by(.data = h3k36_counts_df, experiment)
h3k36_counts_df          = mutate(.data = h3k36_counts_df, total=sum(num_of_reads))
h3k36_counts_df$percentage = with(h3k36_counts_df, round(num_of_reads/total,2))
```

```
## # A tibble: 4 x 5
## # Groups:    experiment [2]
##    experiment enriched num_of_reads  total percentage
##    <chr>      <chr>           <int>  <int>      <dbl>
## 1 H3K36me3    Not Peak        67623 158616       0.43
## 2 H3K36me3    Peak            90993 158616       0.570
## 3 Input       Not Peak       492369 648196       0.76
## 4 Input       Peak           155827 648196       0.24
```

We can now plot the percentage of reads in peaks:

```r
ggplot(
    data = h3k36_counts_df,
    aes(
        x = experiment,
        y = percentage,
        fill = enriched
    )) +
    geom_bar(stat='identity', position='dodge') +
    theme_bw() +
    theme(
        axis.text = element_text(size=10, face='bold'),
        axis.title = element_text(size=12,face="bold"),
        plot.title = element_text(hjust = 0.5)) +
```

```
        xlab('Experiment') +
        ylab('Percetage of reads in region') +
        ggtitle('Percentage of reads in peaks for H3K36me3') +
        scale_fill_manual(values=c('gray','red'))
```

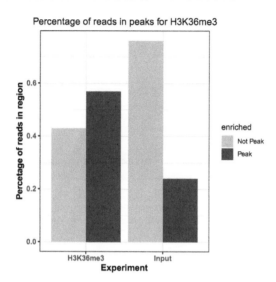

**FIGURE 9.16:** Percentage of ChIP reads in called peaks. Higher percentage indicates higher ChIP quality.

Figure 9.16 shows that the ChIP sample is clearly enriched in the peak regions. The percentage of reads in peaks will depend on the quality of the antibody (strength of enrichment), and the size of peaks which are bound by the protein of interest. If the total size of peaks is small, relative to the genome size, we can expect that the percentage of reads in peaks will be small.

#### 9.6.4.2 DNA motifs on peaks

Well-studied transcription factors have publicly available transcription factor binding motifs. If such a model is available for our transcription factor of interest, we can use it to check the quality of our ChIP data. Two common measures are used for this purpose:

1. Percentage of peaks containing the motif of interest.
2. Positional distribution of the motif - the distribution of motif locations should be centered on the peak centers.

##### 9.6.4.2.1 Representing motifs as matrices

In order to calculate the percentage of CTCF peaks which contain a known CTCF motif. We need to find the CTCF motif and have the computational tools to search for that motif. The DNA binding motifs can be extracted from the MotifDB Bioconductor database. The MotifDB is an agglomeration of multiple motif databases.

```r
# load the MotifDB package
library(MotifDb)

# fetch the CTCF motif from the data base
motifs = query(query(MotifDb, 'Hsapiens'), 'CTCF')

# show all available ctcf motifs
motifs
```

```
## MotifDb object of length 12
## | Created from downloaded public sources: 2013-Aug-30
## | 12 position frequency matrices from 8 sources:
## |        HOCOMOCOv10:   2
## |  HOCOMOCOv11-core-A:   2
## |        JASPAR_2014:   1
## |        JASPAR_CORE:   1
## |       SwissRegulon:   2
## |         jaspar2016:   1
## |         jaspar2018:   2
## |          jolma2013:   1
## | 1 organism/s
## |           Hsapiens:   12
## Hsapiens-SwissRegulon-CTCFL.SwissRegulon
## Hsapiens-SwissRegulon-CTCF.SwissRegulon
## Hsapiens-HOCOMOCOv10-CTCFL_HUMAN.H10MO.A
## Hsapiens-HOCOMOCOv10-CTCF_HUMAN.H10MO.A
## Hsapiens-HOCOMOCOv11-core-A-CTCFL_HUMAN.H11MO.0.A
## ...
## Hsapiens-JASPAR_2014-CTCF-MA0139.1
## Hsapiens-jaspar2016-CTCF-MA0139.1
## Hsapiens-jaspar2018-CTCF-MA0139.1
## Hsapiens-jaspar2018-CTCFL-MA1102.1
## Hsapiens-jolma2013-CTCF
```

We will extract the CTCF from the MotifDB (Khan et al., 2018) database.

```r
# based on the MotifDB version, the location of the CTCF motif
# might change, if you do not get the expected results please try
```

**TABLE 9.1:** Position Frequency Matrix (PFM) for the CTCF motif

| | 1 | 2 | 3 | 4 | 5 | 6 | 7 | 8 | 9 | 10 | 11 | 12 | 13 | 14 | 15 | 16 | 17 | 18 | 19 | 20 |
|---|---|---|---|---|---|---|---|---|---|---|---|---|---|---|---|---|---|---|---|---|
| A | 0.17 | 0.23 | 0.29 | 0.10 | 0.33 | 0.06 | 0.05 | 0.04 | 0.02 | 0 | 0.25 | 0.00 | 0 | 0.05 | 0.25 | 0.00 | 0.17 | 0 | 0.02 | 0.19 |
| C | 0.42 | 0.28 | 0.30 | 0.32 | 0.11 | 0.33 | 0.56 | 0.00 | 0.96 | 1 | 0.67 | 0.69 | 1 | 0.04 | 0.07 | 0.42 | 0.15 | 0 | 0.06 | 0.43 |
| G | 0.25 | 0.23 | 0.26 | 0.27 | 0.42 | 0.55 | 0.05 | 0.83 | 0.01 | 0 | 0.03 | 0.00 | 0 | 0.02 | 0.53 | 0.55 | 0.05 | 1 | 0.87 | 0.15 |
| T | 0.16 | 0.27 | 0.15 | 0.31 | 0.14 | 0.06 | 0.33 | 0.13 | 0.00 | 0 | 0.06 | 0.31 | 0 | 0.89 | 0.15 | 0.03 | 0.62 | 0 | 0.05 | 0.23 |

```
# to subset with different indices
ctcf_motif  = motifs[[1]]
```

The motifs are usually represented as matrices of 4-by-N dimensions. In the matrix, each of 4 rows correspond to one nucleotide (A, C, G, T). The number of columns designates the width of the region bound by the transcription factor or the length of the motif that the protein recognizes. Each element of the matrix contains the probability of observing the corresponding nucleotide on this position. For example, for following the CTCF matrix in Table 9.1, the probability of observing a thymine at the first position of the motif, $p_{i=1,k=4}$, is 0.57 (1st column, 4th row). Such a matrix, where each column is a probability distribution over a sequence of nucleotides, is called a position frequency matrix (PFM). In some sources, this matrix is also called "position probability matrix (PPM)". One way to construct such matrices is to get experimentally verified sequences that are bound by the protein of interest and then to use a motif-finding algorithm.

Such a matrix can be used to calculate the probability that the transcription factor will bind to any given sequence. However, computationally, it is easier to work with summation rather than multiplication. In addition, the simple probabilistic model does not take the background probability of observing a certain base in a given position. We can correct for background base frequencies by dividing the individual probability, $p_{i,k}$ in each cell of the matrix by the background base probability for a given base, $B_k$. We can then take the logarithm of that quantity to calculate a log-likelihood and bring everything to log-scale as follows $Score_{i,k} = log_2(p_{i,k}/B_k)$. We can now calculate a score for any given sequence by summing up the base-position-specific scores we obtain from the log-scaled matrix. This matrix is formally called position-specific scoring matrix (PSSM) or position-specific weight matrix (PWM). We can use this matrix to scan the genome in a sliding window manner and calculate a score for each window. Usually, a cutoff value is needed to call a motif hit. The higher the score you get from the PWM for a particular sequence, the better it is. The traditional algorithms we will use in the following sections use 80% of the maximum rescaled score you can obtain from a PWM as the default cutoff for a hit. The rescaling is simple min-max rescaling where you rescale the score by subtracting the minimum score and dividing that by $max(PWMscore) - min(PWMscore)$. The motif scanning approach is illustrated in Figure 9.17. In this example, ACACT is not considered a hit because its score only corresponds to only 15.6 % of the rescaled maximum score.

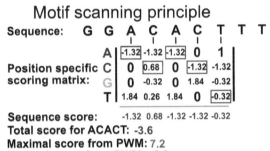

**FIGURE 9.17:** PWM scanning principle. A genomic sequence is scanned by a PWM matrix. This matrix is used to measure how likely it is that the transcription factor will bind each nucleotide in each position. Here we are looking at how likely it is that our TF will bind to the sequence ACACT. The score for this sequence is -3.6. The maximal score obtainable by the PWM is 7.2 and minimum is -5.6. After min-max rescaling, -3.6 corresponds to a 15% score and ACACT is not considered a hit.

9.6.4.2.2   *Representing motifs as sequence logos*

Using the PFM, we can calculate the information content of each position in the matrix. The information content quantifies the contribution of each nucleotide to the cumulative binding preference. This tells us how important each nucleotide is for the binding. It additionally allows us to visually represent the probability matrices as sequence logos. The information content is quantified as relative entropy. It ranges from 0, no information, to 2, maximal information. For a column in the PFM, the entropy is calculated as follows:

$$entropy = - \sum_{k=1}^{n} p_{i,k} \log_2(p_{i,k})$$

$p_{i,k}$ is the probability of observing base $k$ in the column $i$ of the PFM. In other words, $p_{i,k}$ is simply the value of the cell in the PFM. The entropy value is high when the probabilities of each base are similar and low when it is much more probable that only one base occur in a given column. The relative portion comes from the fact that we compare the entropy we calculated for a column to the maximum entropy we can obtain. If the all bases are equally likely for a position in the PFM, then we will have the maximum entropy and we compare our original entropy to that maximum entropy. The maximum entropy is simply $log_2 n$ where $n$ is number of letters in the alphabet. In our case we have 4 letters A,C,G and T. The information content is then simply subtracting the observed entropy for a column from the maximum entropy, which translates to the following equation:

$$IC = log_2(n) + \sum_{k=1}^{n} p_{i,k} \, log_2(p_{i,k})$$

The information content, $IC$, in the preceding equation, will be high if a base has a high probability of occurrence and low if all bases are more or less equally likely to occur.

We can visualize the matrix by visualizing the letters weighted by their probabilities in the PFM. This approach is shown on the left-hand side of Figure 9.18. In addition, we can also calculate the information content per column to weight the probabilities. This means that the columns that have very frequent letters will be higher. This approach is shown on the right-hand side of Figure 9.18. We will use below the seqLogo package to visualize the CTCF motif in the two different ways we described above.

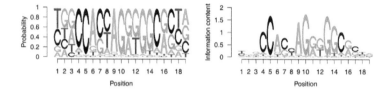

**FIGURE 9.18:** CTCF sequence motif visualized as a sequence logo. Y-axis ranges from zero to two, and corresponds to the amount of information each base in the motif contributes to the overall motif. The larger the letter, the greater the probability of observing just one defined base on the designated position.

### 9.6.4.2.3 Percentage of peaks with the motif

Since we now understand how DNA motifs are used we can start annotating the CTCF peaks with the motif. First, we will extend the peak regions to +/- 200 bp around the peak center. Because the average fragment size is 200 bp, 400 nucleotides is the expected variation in the position of the true binding location.

```
# extend the peak regions
ctcf_peaks_resized = resize(ctcf_peaks, width = 400, fix = 'center')
```

Now we use the BSgenome package to extract the sequences corresponding to the peak regions.

```
# load the human genome sequence
library(BSgenome.Hsapiens.UCSC.hg38)
```

```
# extract the sequences around the peaks
seq = getSeq(BSgenome.Hsapiens.UCSC.hg38, ctcf_peaks_resized)
```

Once we have extracted the sequences, we can use the CTCF motif to scan each sequence and determine the probability of CTCF binding. For this we use the TFBSTools (Tan and Lenhard, 2016) package.

We first convert the raw probability matrix into a PWMMatrix object, which can then be used for efficient scanning.

```
# load the TFBS tools package
library(TFBSTools)
```

```
# convert the matrix into a PWM object
ctcf_pwm = PWMatrix(
    ID = 'CTCF',
    profileMatrix = ctcf_motif
)
```

We can now use the searchSeq() function to scan each sequence for the motif occurrence. Because the motif matrices are given a continuous binding score, we need to set a cutoff to determine when a sequence contains the motif, and when it doesn't. The cutoff is set by determining the maximal possible score produced by the motif matrix; a percentage of that score is then taken as the threshold value. For example, if the best sequence would have a score of 1.4 of being bound, then we define a threshold of 80% of 1.4, which is 1.12; and any sequence which scores less than 1.12 would not be marked as being bound by the protein.

For the CTCF, we mark any peak containing a sequence with > 80% of the maximal rescaled score or "relative score" as a positive hit.

```
##     seqnames source feature start end absScore relScore strand   ID
## 1          1   TFBS    TFBS    44  63     11.9    0.921      - CTCF
## 2          1   TFBS    TFBS   102 121     11.0    0.839      - CTCF
## 3          2   TFBS    TFBS   151 170     11.5    0.881      + CTCF
## 4          4   TFBS    TFBS   294 313     11.9    0.921      - CTCF
## 5          4   TFBS    TFBS   352 371     11.0    0.839      - CTCF
## 6          5   TFBS    TFBS   164 183     10.9    0.831      - CTCF
```

A common diagnostic plot is to graph a reverse cumulative distribution of peak occurrences. On the x-axis we rank the peaks, with the most highly enriched peak in the first position, and the least enriched peak in the last position. We then walk from the lowest to the highest ranking and measure the percentage of peaks containing the motif.

```
# label which peaks contain CTCF motifs
motif_hits_df = data.frame(
  peak_order      = 1:length(ctcf_peaks)
)
motif_hits_df$contains_motif = motif_hits_df$peak_order %in% hits$seqnames
motif_hits_df = motif_hits_df[order(-motif_hits_df$peak_order),]

# calculate the percentage of peaks with motif for peaks of descending strength
motif_hits_df$perc_peaks = with(motif_hits_df,
                                  cumsum(contains_motif) / max(peak_order))
motif_hits_df$perc_peaks = round(motif_hits_df$perc_peaks, 2)
```

We can now visualize the percentage of peaks with matching CTCF motif.

```
# plot the cumulative distribution of motif hit percentages
ggplot(
    motif_hits_df,
    aes(
        x = peak_order,
        y = perc_peaks
    )) +
  geom_line(size=2) +
  theme_bw() +
  theme(
    axis.text = element_text(size=10, face='bold'),
    axis.title = element_text(size=14,face="bold"),
    plot.title = element_text(hjust = 0.5)) +
  xlab('Peak rank') +
  ylab('Percetage of peaks with motif') +
  ggtitle('Percentage of CTCF peaks with the CTCF motif')
```

Figure 9.19 shows that, when we take all peaks into account, ~45% of the peaks contain a CTCF motif. This is an excellent percentage and indicates a high-quality ChIP experiment. Our inability to locate the motif in ~50% of the sequences does not necessarily need to be a consequence of a poor experiment; sometimes it is a result of the molecular mechanism by which the transcription factor binds. If a transcription factor has multiple binding modes, which are context dependent, for

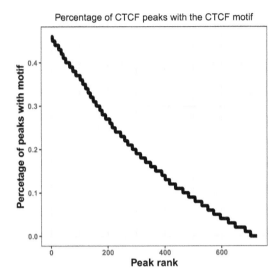

**FIGURE 9.19:** Percentage of peaks containing the motif. Higher percentage indicates a better ChIP-experiment, and a better peak calling procedure.

example, if the transcription factor binds indirectly to a subset of regions, through an interacting partner, we do not have to observe a motif.

### 9.6.4.3   Motif localization

If the ChIP experiment was performed properly, we would expect the motif to be localized just below the summit of each peak. By plotting the motif localization around ChIP peaks, we are quantifying the uncertainty in the peak location.

We will firstly resize our peaks into regions around +/−1-kb around the peak center.

```
# resize the region around peaks to +/- 1kb
ctcf_peaks_resized = resize(ctcf_peaks, width = 2000, fix='center')
```

Now we perform the motif localization, as before.

```
# fetch the sequence
seq = getSeq(BSgenome.Hsapiens.UCSC.hg38,ctcf_peaks_resized)

# convert the motif matrix to PWM, and scan the peaks
ctcf_pwm    = PWMatrix(ID = 'CTCF', profileMatrix = ctcf_motif)
hits = searchSeq(ctcf_pwm, seq, min.score="80%", strand="*")
hits = as.data.frame(hits)
```

We now construct a plot, where the X-axis represents the +/- 1000 nucleotides around the peak, while the Y-axis shows the motif enrichment at each position.

```
# set the position relative to the start
hits$position = hits$start - 1000

# plot the motif hits around peaks
ggplot(data=hits, aes(position)) +
  geom_density(size=2) +
  theme_bw() +
  geom_vline(xintercept = 0, linetype=2, color='red', size=2) +
  xlab('Position around the CTCF peaks') +
  ylab('Per position percentage\nof motif occurence') +
  theme(
    axis.text = element_text(size=10, face='bold'),
    axis.title = element_text(size=14,face="bold"),
    plot.title = element_text(hjust = 0.5))
```

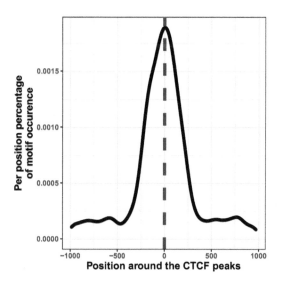

**FIGURE 9.20:** Transcription factor sequence motif localization with respect to the defined binding sites.

We can in Figure 9.20, see that the bulk of motif hits are found in a region of $+/-$ 250 bp around the peak centers. This means that the peak calling procedure was quite precise.

### 9.6.5   Peak annotation

As the final step of quality control we will visualize the distribution of peaks in different functional genomic regions. The purpose of the analysis is to check whether the location of the peaks conforms our prior knowledge. This analysis is equivalent to constructing distributions for reads.

Firstly we download the human gene models and construct the annotation hierarchy.

```
# download the annotation
hub = AnnotationHub()
gtf = hub[['AH61126']]
seqlevels(gtf, pruning.mode='coarse') = '21'
seqlevels(gtf, pruning.mode='coarse') = paste0('chr', seqlevels(gtf))

# create the annotation hierarchy
annotation_list = GRangesList(
   tss    = promoters(subset(gtf, type=='gene'), 1000, 1000),
   exon   = subset(gtf, type=='exon'),
   intron = subset(gtf, type=='gene')
)
```

The following function finds the genomic location of each peak, annotates the peaks using the hierarchical prioritization, and calculates the summary statistics.

The function contains four major parts:

1.   Creating a disjoint set of peak regions.
2.   Finding the overlapping annotation for each peak.
3.   Annotating each peak with the corresponding annotation class.
4.   Calculating summary statistics

```
# function which annotates the location of each peak
annotatePeaks = function(peaks, annotation_list, name){

   # ------------------------------------------------ #
   # 1. getting disjoint regions
   # collapse touching enriched regions
   peaks = reduce(peaks)


   # ------------------------------------------------ #
   # 2. overlapping peaks and annotation
```

```r
# find overlaps between the peaks and annotation_list
result = as.data.frame(findOverlaps(peaks, annotation_list))


# -------------------------------------------------- #
# 3. annotating peaks
# fetch annotation names
result$annotation = names(annotation_list)[result$subjectHits]

# rank by annotation precedence
result = result[order(result$subjectHits),]

# remove overlapping annotations
result = subset(result, !duplicated(queryHits))


# -------------------------------------------------- #
# 4. calculating statistics
# count the number of peaks in each annotation category
result = group_by(.data = result, annotation)
result = summarise(.data = result, counts = length(annotation))

# fetch the number of intergenic peaks
result = rbind(result,
               data.frame(annotation = 'intergenic',
                          counts     = length(peaks) - sum(result$counts)))

result$frequency  = with(result, round(counts/sum(counts),2))
result$experiment = name


return(result)
}
```

Using the above defined `annotatePeaks()` function we will now annotate CTCF and H3K36me3 peaks. Firstly we create a list which contains both CTCF and H3K36me3 peaks.

```r
peak_list = list(
    CTCF     = ctcf_peaks,
    H3K36me3 = h3k36_peaks
)
```

Using the `lapply()` function we apply the `annotatePeaks()` function on each element of the list.

```
# calculate the distribution of peaks in annotation for each experiment
annot_peaks_list = lapply(names(peak_list), function(peak_name){
  annotatePeaks(peak_list[[peak_name]], annotation_list, peak_name)
})
```

We use the `dplyr::bind_rows()` function to combine the CTCF and H3K36me3 annotation statistics into one data frame.

```
# combine a list of data.frames into one data.frame
annot_peaks_df = dplyr::bind_rows(annot_peaks_list)
```

And visualize the results as bar plots. Resulting plot is in Figure 9.21, which shows that the H3K36me3 peaks are located preferentially in gene bodies, as expected, while the CTCF peaks are found preferentially in introns.

```
# plot the distribution of peaks in genomic features
ggplot(data = annot_peaks_df,
       aes(
             x    = experiment,
             y    = frequency,
             fill = annotation
       )) +
  geom_bar(stat='identity') +
  scale_fill_brewer(palette='Set2') +
  theme_bw()+
  theme(
    axis.text = element_text(size=18, face='bold'),
    axis.title = element_text(size=14,face="bold"),
    plot.title = element_text(hjust = 0.5))   +
  ggtitle('Peak distribution in\ngenomic regions') +
  xlab('Experiment') +
  ylab('Frequency')
```

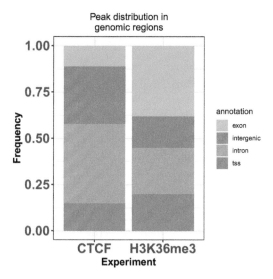

**FIGURE 9.21:** Enrichment of transcription factor or histone modifications in functional genomic features.

## 9.7 Motif discovery

The first analysis step downstream of peak calling is motif discovery. Motif discovery is a procedure of finding enriched sets of similar short sequences in a large sequence dataset. In our case the large sequence dataset are sequences around ChIP peaks, while the short sequence sets are the transcription factor binding sites.

There are two types of motif discovery tools: supervised and unsupervised. Supervised tools require explicit positive (we are certain that the motif is enriched), and negative sequence sets (we are certain that the motif is not enriched), and then search for relative enrichment of short motifs in the foreground versus the background. Unsupervised models, on the other hand, require only a set of positive sequences, and then compare motif abundance to a statistically constructed background set.

Due to the combinatorial nature of the procedure, motif discovery is computationally expensive. It is therefore often performed on a subset of the highest-quality peaks. In this tutorial we will use the rGADEM package for motif discovery. rGADEM is an unsupervised, stochastic motif discovery tools, which uses sampling with subsequent enrichment analysis to find over-represented sequence motifs.

We will firstly load our CTCF peaks, and convert them to a GRanges object. We will then select the top 500 peaks, and extract the DNA sequence, which will be used as input for the motif discovery. Nearby ChIP peaks can have overlapping coordinates. After selection, overlapping CTCF peaks have to be merged using the reduce() function from the GenomicRanges package. If we do not execute this step, we will include the same sequence multiple times in the sequence set, and artificially enrich DNA patterns.

```
# read the CTCF peaks created in the peak calling part of the tutorial
ctcf_peaks = read.table(file.path(data_path, 'CTCF_peaks.txt'), header=TRUE)

# convert the peaks into a GRanges object
ctcf_peaks = makeGRangesFromDataFrame(ctcf_peaks, keep.extra.columns = TRUE)

# order the peaks by qvalue, and take top 250 peaks
ctcf_peaks = ctcf_peaks[order(ctcf_peaks$qvalue)]
ctcf_peaks = head(ctcf_peaks, n = 500)

# merge nearby CTCF peaks
ctcf_peaks = reduce(ctcf_peaks)
```

Create a region of $+/-$ 50 bp around the center of the peaks,

```
# expand the CTCF peaks
ctcf_peaks_resized = resize(ctcf_peaks, width = 50, fix='center')
```

and extract the genomic sequence.

We are now ready to run the motif discovery. Firstly we load the rGADEM package:

To run the motif discovery, we call the GADEM() function. with the extracted DNA sequences. In addition to the DNA sequences, we need to specify two parameters:

1.  **seed** - the random number generator seed, which will make the analysis reproducible.
2.  **nmotifs** - the number of motifs to look for.

```
## top 3   4, 5-mers: 12 40 52
## top 3   4, 5-mers: 12 36 42
```

The rGADEM package contains a convenient plot() function for motif visualization. We will use the plot function to visualize the most enriched DNA motif:

```
# visualize the resulting motif
plot(novel_motifs[1])
```

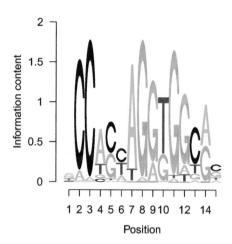

**FIGURE 9.22:** Motif with highest enrichment in top 500 CTCF peaks.

The motif shown in Figure 9.22 corresponds to the previously visualized CTCF motif. Nevertheless, we will computationally annotate our motif by querying the JASPAR (Khan et al., 2018) database in the next section.

### 9.7.1 Motif comparison

We will now compare our unknown motif with the JASPAR2018 (Khan et al., 2018) database, to figure out to which transcription factor it corresponds. Firstly we convert the frequency matrix into a PWMatrix object, and then use this object to query the database.

```r
# load the TFBSTools library
library(TFBSTools)

# extract the motif of interest from the GADEM object
unknown_motif = getPWM(novel_motifs)[[1]]

# convert the motif to a PWM matrix
unknown_pwm    = PWMatrix(
    ID = 'unknown',
    profileMatrix = unknown_motif
)
```

Using the getMatrixSet() function we extract all motifs which correspond to known human transcription factors. The `opts` parameter defines which PWM database to use for comparison.

```
# load the JASPAR motif database
library(JASPAR2018)

# extract motifs corresponding to human transcription factors
pwm_library = getMatrixSet(
  JASPAR2018,
  opts=list(
    collection = 'CORE',
    species    = 'Homo sapiens',
    matrixtype = 'PWM'
))
```

The `PWMSimilarity()` function calculates the Pearson correlation between the database, and our discovered motif via rGADEM.

```
# find the most similar motif to our motif
pwm_sim = PWMSimilarity(

  # JASPAR library
  pwm_library,

  # out motif
  unknown_pwm,

  # measure for comparison
  method = 'Pearson')
```

We extract the motif names from the PWM library. For each motif in the library we append the Pearson correlation with our unknown motif, and look at the topmost candidates.

```
# extract the motif names from the pwm library
pwm_library_list = lapply(pwm_library, function(x){
  data.frame(ID = ID(x), name = name(x))
})

# combine the list into one data frame
pwm_library_dt = dplyr::bind_rows(pwm_library_list)
```

```
# fetch the similarity of each motif to our unknown motif
pwm_library_dt$similarity = pwm_sim[pwm_library_dt$ID]

# find the most similar motif in the library
pwm_library_dt = pwm_library_dt[order(-pwm_library_dt$similarity),]
```

```
head(pwm_library_dt)
```

```
##            ID   name  similarity
## 24   MA0139.1  CTCF   0.7033789
## 370  MA1100.1  ASCL1  0.4769023
## 281  MA0807.1  TBX5   0.4762250
## 101  MA0033.2  FOXL1  0.4605249
## 302  MA0825.1   MNT   0.4370585
## 277  MA0803.1  TBX15  0.4317270
```

As expected, the topmost candidate is CTCF.

## 9.8   What to do next?

One of the first next steps after you have your peaks is to find out what kind of genes they might be associated with. This is very similar to the gene set analysis we introduced for RNA-seq in Chapter 8. The same tools, such as gProfileR package, can be used on the genes associated with the peaks. However, associating peaks to genes is not always trivial due to long-range gene regulation. Many enhancers can regulate genes that are far away and their targets are not always the nearest gene. However, associating peaks to nearest genes is a generally practiced strategy in ChIP-seq analysis. We have introduced how to find the nearest genes in Chapter 6. There are also other R packages that will do the association to genes and the gene set analysis in a single workflow. One such package is rGREAT from Bioconductor. This package relies on a web-based tool called *GREAT*[2].

Knowing every location in the genome bound by a protein can provide a lot of mechanistic information. However, quite often it is hard to make biologically relevant conclusions just from one ChIP-seq experiment (i.e. if we want to explain how our protein causes a disease, it is hard to guess which of the tens of thousands of binding places is relevant for the phenotype). Therefore, it is customary to integrate

[2]http://great.stanford.edu/public/html/

the results with data which is already available for our system of interest - ChIP-seq of different proteins, genome wide measurements of expression, or assays of 3D genome structure.

The choice of downstream analysis is guided by the biological question of interest. Often we want to compare our samples to other available ChIP-seq experiments. It is possible to look at the pairwise differences between samples using differential peak calling (Zhang et al., 2014; Lun and Smyth, 2014; Allhoff et al., 2014, 2016). It is a procedure analogous to the differential expression analysis, except it results in sets of coordinates that are differentially bound in two biological conditions. We can then search for a specific DNA binding motif in such regions, or correlate changes in the binding with changes in gene expression. With an increase in the number of ChIP experiments, pairwise comparisons become combinatorially complex. In this case we can segment the genome into multiple classes, where each class corresponds to a combination of bound transcription factors. Genome segmentation is usually done using probabilistic models (such as hidden Markov models (Ernst and Kellis, 2012; Hoffman et al., 2012)), or machine learning algorithms (Mortazavi et al., 2013).

## 9.9   Exercises

### 9.9.1   Quality control

1. Apply the fragment size estimation procedure to all ChIP and Input available datasets. [Difficulty: **Beginner**]

2. Visualize the resulting distributions. [Difficulty: **Beginner**]

3. How does the Input sample distribution differ from the ChIP samples? [Difficulty: **Beginner**]

4. Write a function which converts the bam files into bigWig files. [Difficulty: **Beginner**]

5. Apply the function to all files, and visualize them in the genome browser. Observe the signal profiles. What can you notice, about the similarity of the samples? [Difficulty: **Beginner**]

6. Use Gviz to visualize the profiles for CTCF, SMC3 and ZNF143. [Difficulty: **Beginner/Intermediate**]

7. Calculate the cross correlation for both CTCF replicates, and the input samples. How does the profile look for the control samples? [Difficulty: **Intermediate**]

8. Calculate the cross correlation coefficients for all samples and visualize them as a heatmap. [Difficulty: **Intermediate**]

#### 9.9.1.1 Peak calling

1. Use normR to call peaks for all SMC3, CTCF, and ZNF143 samples. [Difficulty: **Beginner**]

2. Calculate the percentage of reads in peaks for the CTCF experiment. [Difficulty: **Intermediate**]

3. Download the blacklisted regions corresponding to the hg38 human genome, and calculate the percentage of CTCF peaks falling in such regions. [Difficulty: **Advanced**]

4. Unify the biological replicates by taking an intersection of peaks. How many peaks are specific to each biological replicate, and how many peaks overlap. [Difficulty: **Intermediate**]

5. Plot a scatter plot of signal strengths for biological replicates. Do intersecting peaks have equal signal strength in both samples? [Difficulty: **Intermediate**]

6. Quantify the combinatorial binding of all three proteins. Find the number of places which are bound by all three proteins, by a combination of two proteins, and exclusively by one protein. Annotate the different regions based on their genomic location. [Difficulty: **Advanced**]

7. Correlate the normR enrichment score for CTCF with peak presence/absence (create boxplots of enrichment for peaks which contain and do not contain CTCF motifs). [Difficulty: **Advanced**]

8. Explore the co-localization of CTCF and ZNF143. Where are the co-bound regions located? Which sequence motifs do they contain? Download the ChIA-pet data for the GM12878 cell line, and look at the 3D interaction between different classes of binding sites. [Difficulty: **Advanced**]

#### 9.9.1.2 Motif discovery

1. Repeat the motif discovery analysis on peaks from the ZNF143 transcription factor. How many motifs do you observe? How do the motifs look (visualize the motif logs)? [Difficulty: **Intermediate**]

2. Scan the ZNF143 peaks with the top motifs found in the previous exercise. Where are the motifs located? [Difficulty: **Advanced**]

3. Scan the CTCF peaks with the top motifs identified in the **ZNF143** peaks. Where are the motifs located? What can you conclude from the previous exercises? [Difficulty: **Advanced**]

# 10

## DNA methylation analysis using bisulfite sequencing data

The epigenome consists of chemical modifications of DNA and histones. These modifications are shown to be associated with gene regulation in various settings (see Chapter 1 for an intro). These modifications in turn have specific importance for cell type identification. There are many different ways of measuring such modifications. We have shown how histone modifications can be measured in a genome-wide manner in Chapter 9 using ChIP-seq. In this chapter we will focus on the analysis of DNA methylation data using data from a technique called bisulfite sequencing (BS-seq). We will introduce how to process data and data quality checks, as well as statistical analysis relevant for BS-seq data.

### 10.1 What is DNA methylation?

Cytosine methylation (5-methylcytosine, 5mC) is one of the main covalent base modifications in eukaryotic genomes, generally observed on CpG dinucleotides. Methylation can also rarely occur in a non-CpG context, but this was mainly observed in human embryonic stem and neuronal cells (Lister et al., 2009, 2013). DNA methylation is a part of the epigenetic regulation mechanism of gene expression. It is cell-type-specific DNA modification. It is reversible but mostly remains stable through cell division. There are roughly 28 million CpGs in the human genome, 60–80% are generally methylated. Less than 10% of CpGs occur in CG-dense regions that are termed CpG islands in the human genome (Smith and Meissner, 2013). It has been demonstrated that DNA methylation is also not uniformly distributed over the genome, but rather is associated with CpG density. In vertebrate genomes, cytosine bases are usually unmethylated in CpG-rich regions such as CpG islands and tend to be methylated in CpG-deficient regions. Vertebrate genomes are largely CpG deficient except at CpG islands. Conversely, invertebrates such as *Drosophila melanogaster* and *Caenorhabditis elegans* do not exhibit cytosine methylation and consequently do not have CpG rich and poor regions but rather a steady CpG frequency over their genomes (Deaton and Bird, 2011).

### 10.1.1   How DNA methylation is set ?

DNA methylation is established by DNA methyltransferases DNMT3A and DNMT3B in combination with DNMT3L and maintained through cell division by the methyl-transferase DNMT1 and associated proteins. DNMT3a and DNMT3b are in charge of the de novo methylation during early development. Loss of 5mC can be achieved passively by dilution during replication or exclusion of DNMT1 from the nucleus. Recent discoveries of the ten-eleven translocation (TET) family of proteins and their ability to convert 5-methylcytosine (5mC) into 5-hydroxymethylcytosine (5hmC) in vertebrates provide a path for catalyzed active DNA demethylation (Tahiliani et al., 2009). Iterative oxidations of 5hmC catalyzed by TET result in 5-formylcytosine (5fC) and 5-carboxylcytosine (5caC). 5caC mark is excised from DNA by G/T mismatch-specific thymine-DNA glycosylase (TDG), which as a result reverts cytosine residue to its unmodified state (He et al., 2011). Apart from these, mainly bacteria, but possibly higher eukaryotes, contain base modifications on bases other than cytosine, such as methylated adenine or guanine (Clark et al., 2011).

### 10.1.2   How to measure DNA methylation with bisulfite sequencing

One of the most reliable and popular ways to measure DNA methylation is high-throughput bisulfite sequencing. This method, and the related ones, allow measurement of DNA methylation at the single nucleotide resolution. The bisulfite conversion turns unmethylated Cs to Ts and methylated Cs remain intact. Then, the only thing to do is to align the reads with those C->T conversions and count C->T mutations to calculate fraction of methylated bases. In the end, we can get quantitative genome-wide measurements for DNA methylation.

## 10.2   Analyzing DNA methylation data

For the remainder of this chapter, we will explain how to do DNA methylation analysis using R. The analysis process is somewhat similar to the analysis patterns observed in other sequencing data analyses. The process can be chunked to four main parts with further sub-chunks:

1.   Processing raw data

- Quality check
- Alignment and post-alignment processing
- Methylation calling
- Filtering bases

2. Exploratory analysis

- Clustering
- PCA

3. Finding interesting regions

- Differential methylation
- Methylation segmentation

4. Annotating interesting regions

- Nearest genes
- Annotation with other genomic features
- Integration with other quantitative genomics data

## 10.3 Processing raw data and getting data into R

The rawest form of data that most users get is probably in the form of fastq files obtained from the sequencing experiments. We will describe the necessary steps and the tools that can be used for raw data processing and if they exist, we will mention their R equivalents. However, the data processing is usually done outside of the R framework, and for the following sections we will assume that the data processing is done and our analysis is starting from methylation call files.

The typical data processing step starts with a data quality check. The fastq files are first run through quality check software that shows the quality of the sequencing run. We would typically use fastQC[1] for this. However, there are several bioconductor packages that could be of use, such as Rqc[2] and QuasR[3]. We have introduced how to use some of these tools for sequencing quality check in Chapter 7. Following the quality check, provided everything is OK, the reads can be aligned to the genome. Before the alignment, adapters or low-quality ends of the reads can be trimmed to increase number of alignments. Low-quality ends mostly likely have poor basecalls, which will lead to many mismatches. Reads with non-trimmed adapters will also not align to the genome. We would use adapter trimming tools such as cutadapt[4] or flexbar[5] for this purpose, although there are a bunch of them to choose from. Following this, reads are aligned to the genome with a bisulfite-treatment-aware

---

[1] https://www.bioinformatics.babraham.ac.uk/projects/fastqc/

[2] https://bioconductor.org/packages/release/bioc/html/Rqc.html

[3] https://bioconductor.org/packages/release/bioc/html/QuasR.html

[4] https://cutadapt.readthedocs.io/en/stable/

[5] https://github.com/seqan/flexbar

aligner. For our own purposes, we use Bismark(Krueger and Andrews, 2011), however there are other equally accurate aligners, and some are reviewed here[6]. In addition, the Bioconductor package QuasR[7] can align BS-seq reads within the R framework.

After alignment, we need to call C->T conversions and calculate the fraction/percentage of methylation. Most of the time, aligners come with auxiliary tools that calculate per-base methylation values. Normally, they output a tabular format containing the location of the Cs and methylation value and strand. Within R, QuasR and methylKit can call methylation values from BAM files albeit with some limitations. In essence, these methylation call files can be easily read into R and downstream analysis within R starts from that point. An important quality measure at this stage is to look at the conversion rate. This simply means how many unmethylated Cs are converted to Ts. Since we expect non-CpG methylation to be rare, we can simply count the number of C->T conversions in the non-CpG context and calculate conversion rate. The best way to do this would be via spike-in sequences where we expect no methylation at all. Since non-CpG methylation is tissue specific, calculating the conversion rate using non-CpG Cs might be misleading in some cases.

## 10.4    Data filtering and exploratory analysis

We assume that we start the analysis in R with the methylation call files. We will read those files in and carry out exploratory analysis, and we will show how to filter bases or regions from the data and in what circumstances we might need to do so. We will use the methylKit[8](Akalin et al., 2012) package for the bulk of the analysis.

### 10.4.1    Reading methylation call files

A typical methylation call file looks like this:

```
##           chrBase    chr     base strand coverage freqC   freqT
## 1 chr21.9764539 chr21 9764539      R       12 25.00   75.00
## 2 chr21.9764513 chr21 9764513      R       12  0.00  100.00
## 3 chr21.9820622 chr21 9820622      F       13  0.00  100.00
## 4 chr21.9837545 chr21 9837545      F       11  0.00  100.00
## 5 chr21.9849022 chr21 9849022      F      124 72.58   27.42
```

Most of the time bisulfite sequencing experiments have test and control samples. The test samples can be from a disease tissue while the control samples can be from

---

[6]https://www.ncbi.nlm.nih.gov/pmc/articles/PMC3378906/
[7]https://bioconductor.org/packages/release/bioc/html/QuasR.html
[8]https://bioconductor.org/packages/release/bioc/html/methylKit.html

a healthy tissue. You can read a set of methylation call files that have test/control conditions giving a `treatment` vector option. The treatment vector defines the sample groups and it is very important for the differential methylation analysis. For the sake of subsequent analysis, file.list, sample.id and treatment option should have the same order. In the following example, the first two files have the sample IDs "test1" and "test2" and as determined by the treatment vector they belong to the same group. The third and fourth files have sample IDs "ctrl1" and "ctrl2" and they belong to the same group as indicated by the treatment vector. We will first get a list of file paths and have a look at the content.

If you look what is inside the `file.list` variable, you will see that it is a simple list of file paths. Each file contains methylation calls for a given sample. Now, we can read the files with the `methRead()` function.

```
# read the files to a methylRawList object: myobj
myobj=methRead(file.list,
          sample.id=list("test1","test2","ctrl1","ctrl2"),
          assembly="hg18",
          treatment=c(1,1,0,0),
          context="CpG"
          )
```

Tab-separated bedgraph like formats from Bismark methylation caller can also be read in by methylkit. In those cases, we have to provide either `pipeline="bismarkCoverage"` or `pipeline="bismarkCytosineReport"` to the `methRead()` function. In addition to the options we mentioned above, any tab-separated text file with a generic format can be read in using methylKit, such as methylation ratio files from BSMAP[9]. See here[10] for an example.

Before we move on, let us have a look at what kind of information is stored in `myobj`. This is technically a `methylRawList` object, which is essentially a list of `methylRaw` objects. These objects hold the information for the genomic location of Cs, and methylated Cs and unmethylated Cs.

```
## inside the methylRawList object
length(myobj)
```

```
## [1] 4
```

---

[9] http://code.google.com/p/bsmap/

[10] http://zvfak.blogspot.com/2012/10/how-to-read-bsmap-methylation-ratio.html

```
head(myobj[[1]])
```

```
##      chr     start      end strand coverage numCs numTs
## 1 chr21 9764513 9764513      –          12     0    12
## 2 chr21 9764539 9764539      –          12     3     9
## 3 chr21 9820622 9820622      +          13     0    13
## 4 chr21 9837545 9837545      +          11     0    11
## 5 chr21 9849022 9849022      +         124    90    34
## 6 chr21 9853296 9853296      +          17    10     7
```

### 10.4.2   Further quality check

It is always a good idea to check how the data looks before proceeding further. For example, the methylation values should have bimodal distribution generally. This can be checked via the getMethylationStats() function. Normally, we should see bimodal distributions. Strong deviations from the bimodality may be due to poor experimental quality, such as problems with bisulfite treatment. Below we show how to get these plots using the getMethylationStats() function. The result is shown in Figure 10.1. As expected, it has a bimodal distribution where most CpGs have either high methylation or low methylation.

```
getMethylationStats(myobj[[2]],plot=TRUE,both.strands=FALSE)
```

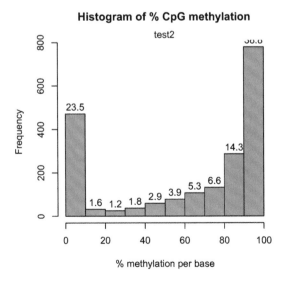

**FIGURE 10.1:** Histogram for methylation values for all CpGs in the dataset.

In addition, we might want to see coverage values. By default, methylkit handles bases with at least 10X coverage but that can be changed. The bases with unusually high coverage are usually alarming. It might indicate a PCR bias issue in the experimental procedure. The general coverage statistics can be checked with the `getCoverageStats()` function shown below. The resulting plot is shown in Figure 10.2.

```
getCoverageStats(myobj[[2]],plot=TRUE,both.strands=FALSE)
```

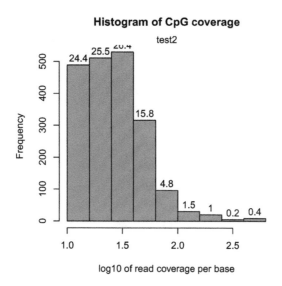

**FIGURE 10.2:** Histogram for log10 read counts per CpG.

It might be useful to filter samples based on coverage. Particularly, if our samples are suffering from PCR bias, it would be useful to discard bases with very high read coverage. Furthermore, we would also like to discard bases that have low read coverage; a high enough read coverage will increase the power of the statistical tests. The code below filters a `methylRawList`, discards bases that have coverage below 10X, and also discards the bases that have more than 99.9th percentile of coverage in each sample.

```
filtered.myobj=filterByCoverage(myobj,lo.count=10,lo.perc=NULL,
                                hi.count=NULL,hi.perc=99.9)
```

### 10.4.3    Merging samples into a single table

When we first read the files, each file is stored as its own entity. If we want to compare samples in any way, we need to make a unified data structure that contains

CpGs that are covered in most samples. The `unite()` function creates a new object using the CpGs covered in each sample.

```
## we use :: notation to make sure unite() function from methylKit is called
meth=methylKit::unite(myobj, destrand=FALSE)
```

Let us take a look at the data content of the `methylBase` object:

```
head(meth)
```

```
##      chr    start      end strand coverage1 numCs1 numTs1 coverage2 numCs2 numTs2
## 1 chr21 9853296 9853296      +            17     10      7       333    268     65
## 2 chr21 9853326 9853326      +            17     12      5       329    249     79
## 3 chr21 9860126 9860126      +            39     38      1        83     78      5
## 4 chr21 9906604 9906604      +            68     42     26       111     97     14
## 5 chr21 9906616 9906616      +            68     52     16       111    104      7
## 6 chr21 9906619 9906619      +            68     59      9       111    109      2
##    coverage3 numCs3 numTs3 coverage4 numCs4 numTs4
## 1         18     16      2       395    341     54
## 2         16     14      2       379    284     95
## 3         83     83      0        41     40      1
## 4         23     18      5        37     33      4
## 5         23     14      9        37     27     10
## 6         22     18      4        37     29      8
```

By default, the `unite()` function produces bases/regions covered in all samples. That requirement can be relaxed using the `min.per.group` option in the `unite()` function.

```
# creates a methylBase object,
# where only CpGs covered with at least 1 sample per group will be returned

# there were two groups defined by the treatment vector,
# given during the creation of myobj: treatment=c(1,1,0,0)
meth.min=unite(myobj,min.per.group=1L)
```

### 10.4.4    Filtering CpGs

We might need to filter the CpGs further before exploratory analysis or even before the downstream analysis such as differential methylation. For exploratory analysis, it is of general interest to see how samples relate to each other and we might want to remove CpGs that are not variable before doing that. Or we might remove Cs that are potentially C->T mutations. First, we show how to filter based on variation.

Below, we extract percent methylation values from CpGs as a matrix. Calculate the standard deviation for each CpG and filter based on standard deviation. We also plot the distribution of per-CpG standard deviations with the `hist()` function. The resulting plot is shown in Figure 10.3.

```
pm=percMethylation(meth) # get percent methylation matrix
mds=matrixStats::rowSds(pm) # calculate standard deviation of CpGs
head(meth[mds>20,])
```

```
##         chr    start       end strand coverage1 numCs1 numTs1 coverage2 numCs2 numTs2
## 11 chr21 9906681 9906681      +        21     12      9        60     56      4
## 12 chr21 9906694 9906694      +        21      9     12        60     53      7
## 13 chr21 9906700 9906700      +        13      6      7        53     43     10
## 14 chr21 9906714 9906714      +        14      3     11        41     37      4
## 18 chr21 9906873 9906873      +        12      8      4        41     33      8
## 23 chr21 9927527 9927527      +        17      5     12        40     22     18
##     coverage3 numCs3 numTs3 coverage4 numCs4 numTs4
## 11         37     14     23        26     11     15
## 12         39     16     23        26     15     11
## 13         30      8     22        23     10     13
## 14         25     19      6        21     19      2
## 18         15      4     11        22      7     15
## 23         32     32      0        14     11      3
```

```
hist(mds,col="cornflowerblue",xlab="Std. dev. per CpG")
```

Now, let's assume we know the locations of C->T mutations. These locations should be removed from the analysis as they do not represent bisulfite-treatment-associated conversions. Mutation locations are stored in a GRanges object, and we can use that to remove CpGs overlapping with mutations. In order to do the overlap operation, we will convert the methylKit object to a GRanges object and do the filtering with the `%over%` function within `[ ]`. The returned object will still be a methylKit object.

```
library(GenomicRanges)
# example SNP
mut=GRanges(seqnames=c("chr21","chr21"),
            ranges=IRanges(start=c(9853296, 9853326),
                             end=c( 9853296,9853326)))

# select CpGs that do not overlap with mutations
```

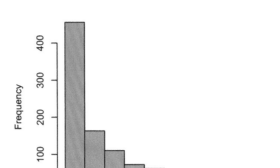

**FIGURE 10.3:** Histogram of per-CpG standard deviations.

```
sub.meth=meth[! as(meth,"GRanges") %over% mut,]
nrow(meth)
```

```
## [1] 963
```

```
nrow(sub.meth)
```

```
## [1] 961
```

### 10.4.5  Clustering samples

Clustering is used for grouping data points by their similarity. It is a general concept that can be achieved by many different algorithms and we introduced clustering and multiple prominent clustering algorithms in Chapter 4. In the context of DNA methylation, we are trying to find samples that are similar to each other. For example, if we sequenced 3 heart samples and 4 liver samples, we would expect liver samples will be more similar to each other than heart samples on the DNA methylation space.

The following function will cluster the samples and draw a dendrogram. It will use correlation distance, which is $1 - \rho$, where $\rho$ is the correlation coefficient between two pairs of samples. The cluster tree will be drawn using the "ward" method. This specific variant uses a "bottom up" approach: each data point starts in its own cluster, and pairs of clusters are merged as one moves up the hierarchy. In Ward's method, two clusters are merged if the variance is minimized compared to other possible merge operations. This bottom up approach helps build the dendrogram showing the relationship between clusters. The result of the clustering is shown in Figure 10.4.

```
clusterSamples(meth, dist="correlation", method="ward", plot=TRUE)
```

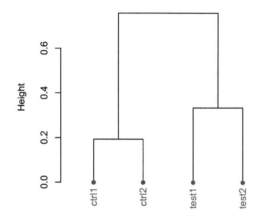

**FIGURE 10.4:** Dendrogram for samples using correlation distance and Ward's method for hierarchical clustering.

```
##
## Call:
## hclust(d = d, method = HCLUST.METHODS[hclust.method])
##
## Cluster method   : ward.D
## Distance         : pearson
## Number of objects: 4
```

Setting the `plot=FALSE` will return a dendrogram object which can be manipulated by users or fed in to other user functions that can work with dendrograms.

```
hc = clusterSamples(meth, dist="correlation", method="ward", plot=FALSE)
```

### 10.4.6   Principal component analysis

Principal component analysis (PCA) is a mathematical transformation of (possibly) correlated variables into a number of uncorrelated variables called principal components. The resulting components from this transformation are defined in such a way that the first principal component has the highest variance and accounts for most of the variability in the data. We have introduced PCA and other similar methods in Chapter 4. The following function will plot a scree plot for importance of components and the result is shown in Figure 10.5.

```
PCASamples(meth, screeplot=TRUE)
```

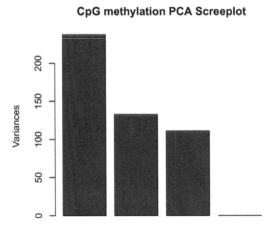

**FIGURE 10.5:** Scree plot for explained variance for principal components.

We can also plot the PC1 and PC2 axes and a scatter plot of our samples on those axes which will reveal how they cluster within these new dimensions. Similar to the clustering dendrogram, we would like to see samples that are similar to be close to each other on the scatter plot. If they are not, it might indicate problems with the experiment such as batch effects. The function below plots the samples in such a scatter plot on principal component axes. The resulting plot is shown in Figure 10.6.

```
pc=PCASamples(meth,obj.return = TRUE, adj.lim=c(1,1))
```

In this case, we also returned an object from the plotting function. This is the output of the `prcomp()` function, which includes loadings and eigenvectors which might be

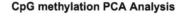

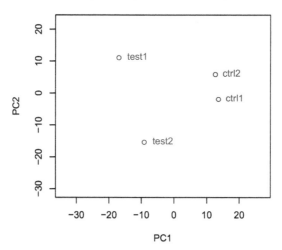

**FIGURE 10.6:** Samples plotted on principal components.

useful. You can also do your own PCA analysis using `percMethylation()` and `prcomp()`. In the case above, the methylation matrix is transposed. This allows us to compare distances between samples on the PCA scatter plot.

## 10.5 Extracting interesting regions: Differential methylation and segmentation

When analyzing DNA methylation data, we usually look for regions that are different than the rest of the methylome or different from a reference methylome. These regions are so-called "interesting regions". They usually mark important genomic features that are related to gene regulation, which in turn defines the cell type. Therefore, it is a general interest to find such regions and analyze them further to understand our biological sample or to answer specific research questions. Below we will describe two ways of defining "regions of interest".

### 10.5.1 Differential methylation

Once methylation proportions per base are obtained, generally, the differences between methylation profiles are considered next. When there are multiple sample groups where each group defines a separate biological entity or treatment, it is usually of interest to locate bases or regions with different methylation proportions across the sample groups. The bases or regions with different methylation proportions across samples are called differentially methylated CpG sites (DMCs) and differentially methylated regions (DMRs). They have been shown to play a role in many different diseases due to their association with epigenetic control of gene regulation. In addition, DNA methylation profiles can be highly tissue-specific due to their role in gene regulation (Schübeler, 2015). DNA methylation is highly informative when studying normal and diseased cells, because it can also act as a biomarker. For example, the presence of large-scale abnormally methylated genomic regions is a hallmark feature of many types of cancers (Ehrlich, 2002). Because of the aforementioned reasons, investigating differential methylation is usually one of the primary goals of doing bisulfite sequencing.

### 10.5.1.1   Fisher's exact test

Differential DNA methylation is usually calculated by comparing the proportion of methylated Cs in a test sample relative to a control. In simple comparisons between such pairs of samples (i.e. test and control), methods such as Fisher's exact test can be used. If there are replicates, replicates can be pooled within groups to a single sample per group. This strategy, however, does not take into account biological variability between replicates. We will now show how to compare pairs of samples via the calculateDiffMeth() function in methylKit. When there is only one sample per sample group, calculateDiffMeth() automatically applies Fisher's exact test. We will now extract one sample from each group and run calculateDiffMeth(), which will automatically run Fisher's exact test.

```
getSampleID(meth)
new.meth=reorganize(meth,sample.ids=c("test1","ctrl1"),treatment=c(1,0))
dmf=calculateDiffMeth(new.meth)
```

As mentioned, we can also pool the samples from the same group by adding up the number of Cs and Ts per group. This way even if we have replicated experiments we treat them as single experiments, and can apply Fisher's exact test. We will now pool the samples and apply the calculateDiffMeth() function.

```
pooled.meth=pool(meth,sample.ids=c("test","control"))
dm.pooledf=calculateDiffMeth(pooled.meth)
```

The calculateDiffMeth() function returns the P-values for all bases or regions in the input methylBase object. We need to filter to get differentially methylated CpGs. This can be done via the getMethlyDiff() function or simple filtering via [ ] notation. Below we show how to filter the methylDiff object output by the calculateDiffMeth() function in order to get differentially methylated CpGs. The function arguments define cutoff values for the methylation difference between groups and q-value.

In these cases, we require a methylation difference of 25% and a q-value of at least 0.01.

```
# get differentially methylated bases/regions with specific cutoffs
all.diff=getMethylDiff(dm.pooledf,difference=25,qvalue=0.01,type="all")

# get hyper-methylated
hyper=getMethylDiff(dm.pooledf,difference=25,qvalue=0.01,type="hyper")

# get hypo-methylated
hypo=getMethylDiff(dm.pooledf,difference=25,qvalue=0.01,type="hypo")

#using [ ] notation
hyper2=dm.pooledf[dm.pooledf$qvalue < 0.01 & dm.pooledf$meth.diff > 25,]
```

#### 10.5.1.2 Logistic regression based tests

Regression-based methods are generally used to model methylation levels in relation to the sample groups and variation between replicates. Differences between currently available regression methods stem from the choice of distribution to model the data and the variation associated with it. In the simplest case, linear regression can be used to model methylation per given CpG or loci across sample groups. The model fits regression coefficients to model the expected methylation proportion values for each CpG site across sample groups. Hence, the null hypothesis of the model coefficients being zero could be tested using t-statistics. However, linear-regression-based methods might produce fitted methylation levels outside the range $[0, 1]$ unless the values are transformed before regression. An alternative is logistic regression, which can deal with data strictly bounded between 0 and 1 and with non-constant variance, such as methylation proportion/fraction values. In the logistic regression, it is assumed that fitted values have variation $np(1 - p)$, where $p$ is the fitted methylation proportion for a given sample and $n$ is the read coverage. If the observed variance is larger or smaller than assumed by the model, one speaks of under- or over-dispersion. This over/under-dispersion can be corrected by calculating a scaling factor and using that factor to adjust the variance estimates as in $np(1 - p)s$, where $s$ is the scaling factor. MethylKit can apply logistic regression to test the methylation difference with or without the over-dispersion correction. In this case, Chi-square or F-test can be used to compare the difference in the deviances of the null model and the alternative model. The null model assumes there is no relationship between sample groups and methylation, and the alternative model assumes that there is a relationship where sample groups are predictive of methylation values for a given CpG or region for which the model is constructed. Next, we are going to use the logistic-regression-based model with over-dispersion correction and Chi-square test.

```
dm.lr=calculateDiffMeth(meth,overdispersion = "MN",test ="Chisq")
```

### 10.5.1.3   Betabinomial-distribution-based tests

More complex regression models use beta binomial distribution and are particularly useful for better modeling the variance. Similar to logistic regression, their observation follows binomial distribution (number of reads), but methylation proportion itself can vary across samples, according to a beta distribution. It can deal with fitting values in the $[0, 1]$ range and performs better when there is greater variance than expected by the simple logistic model. In essence, these models have a different way of calculating a scaling factor when there is over-dispersion in the model. Further enhancements are made to these models by using the empirical Bayes methods that can better estimate hyper parameters of the beta distribution (variance-related parameters) by borrowing information between loci or regions within the genome to aid with inference about each individual loci or region. We are now going to use a beta-binomial based model called DSS (Feng et al., 2014) to calculate differential methylation.

```
dm.dss=calculateDiffMethDSS(meth)
```

```
## Using internal DSS code...
```

### 10.5.1.4   Differential methylation for regions rather than base-pairs

Until now, we have worked on differentially methylated cytosines. However, working with base-pair resolution data has its problems. Not all the CpGs will be covered in all samples. If covered they may have low coverage, which reduces the power of the tests. Instead of base-pairs, we can choose to work with regions. So, it might be desirable to summarize methylation information over pre-defined regions rather than doing base-pair resolution analysis. methylKit provides functionality to do such analysis. We can either tile the whole genome to tiles with predefined length, or we can use pre-defined regions such as promoters or CpG islands. This kind of regional analysis is carried out by adding up C and T counts from each covered cytosine and returning a total C and T count for each region.

The function below tiles the genome with windows of 1000 bp length and 1000 bp step-size and summarizes the methylation information on those tiles. In this case, it returns a methylRawList object which can be fed into unite() and calculateDiffMeth() functions consecutively to get differentially methylated regions.

```
tiles=tileMethylCounts(myobj,win.size=1000,step.size=1000)
head(tiles[[1]],3)
```

```
##      chr    start      end strand coverage numCs numTs
## 1 chr21 9764001 9765000      *       24     3    21
## 2 chr21 9820001 9821000      *       13     0    13
## 3 chr21 9837001 9838000      *       11     0    11
```

In addition, if we are interested in particular regions, we can also get those regions as methylKit objects after summarizing the methylation information as described above. The code below summarizes the methylation information over a given set of promoter regions and outputs a methylRaw or methylRawList object depending on the input. We are using the output of genomation functions used above to provide the locations of promoters. For regional summary functions, we need to provide regions of interest as GRanges objects.

```
library(genomation)
```

```
# read the gene BED file
gene.obj=readTranscriptFeatures(system.file("extdata", "refseq.hg18.bed.txt",
                                            package = "methylKit"))
promoters=regionCounts(myobj,gene.obj$promoters)
```

```
head(promoters[[1]])
```

```
##      chr     start       end strand coverage numCs numTs
## 1 chr21 10011791 10013791      -     7953  6662  1290
## 2 chr21 10119796 10121796      -     1725  1171   554
## 3 chr21 10119808 10121808      -     1725  1171   554
## 4 chr21 13903368 13905368      +       10    10     0
## 5 chr21 14273636 14275636      -      282   220    62
## 6 chr21 14509336 14511336      +     1058    55  1003
```

In addition, it is possible to cluster DMCs based on their proximity and direction of differential methylation. This can be achieved by the methSeg() function in methylKit. We will see more about the methSeg() function in the following section. But it can take the output of getMethylDiff() function and therefore can work on DMCs to get differentially methylated regions.

### 10.5.1.5  Adding covariates

Covariates can be included in the analysis as well in methylKit. The calculateDiffMeth() function will then try to separate the influence of the covariates from the treatment effect via the logistic regression model. In this case, we will test if the full model (model with treatment and covariates) is better than the model with the covariates only. If there is no effect due to the treatment (sample groups), the full model will not explain the data better than the model with covariates only. In `calculateDiffMeth()`, this is achieved by supplying the `covariates` argument in the format of a `data.frame`. Below, we simulate methylation data and add a `data.frame` for the age. The data frame can include more columns, and those columns can also be `factor` variables. The row order of the data.frame should match the order of samples in the `methylBase` object. Below we are showing an example of this using a simulated data set where methylation values of CpGs will be affected by the age of the sample.

```
covariates=data.frame(age=c(30,80,34,30,80,40))
sim.methylBase=dataSim(replicates=6,sites=1000,
                       treatment=c(rep(1,3),rep(0,3)),
                       covariates=covariates,
                       sample.ids=c(paste0("test",1:3),paste0("ctrl",1:3)))


my.diffMeth3=calculateDiffMeth(sim.methylBase,
                       covariates=covariates,
                       overdispersion="MN",
                       test="Chisq",mc.cores=1)
```

### 10.5.2  Methylation segmentation

The analysis of methylation dynamics is not exclusively restricted to differentially methylated regions across samples. Apart from this there is also an interest in examining the methylation profiles within the same sample. Usually, depressions in methylation profiles pinpoint regulatory regions like gene promoters that co-localize with CG-dense CpG islands. On the other hand, many gene-body regions are extensively methylated and CpG-poor (Bock et al., 2012). These observations would describe a bimodal model of either hyper- or hypomethylated regions depending on the local density of CpGs (Lövkvist et al., 2016). However, given the detection of CpG-poor regions with locally reduced levels of methylation (on average 30%) in pluripotent embryonic stem cells and in neuronal progenitors in both mouse and human, a different model also seems reasonable (Stadler et al., 2011a). These low-methylated regions (LMRs) are located distal to promoters, have little overlap with CpG islands, and are associated with enhancer marks such as p300 binding sites and H3K27ac enrichment.

Now we are going to try to segment a portion for the H1 human embryonic stem cell line. MethylKit uses change-point analysis to segment the methylome. In change-point analysis, the change-points of a genome-wide methylation signal are recorded and the genome is partitioned into regions between consecutive change points. CpGs in each segment are similar to each other more than the following segment. After segmentation, methylKit function methSeg() identifies segments that are further clustered into segment classes using a mixture modeling approach. This clustering is based on only the average methylation level of the segments and allows the detection of distinct methylome features comparable to unmethylated regions (UMRs), lowly methylated regions (LMRs), and fully methylated regions (FMRs) mentioned in Stadler et al. (Stadler et al., 2011b). The code snippet below reads the methylation data from the H1 cell line as a GRanges object, and runs the segmentation with potentially up to 4 classes of segments. Mixture modeling determines the optimal number of segments using a statistic called Bayesian information criterion (BIC). The BIC is a statistic based on model likelihood and helps us select the model that fits the data better. We have set the number of segment classes to try using the G=1:4 argument. The minSeg arguments are related to the minimum number of CpGs in the segments. The function methSeg() outputs a diagnostic plot for segmentation. This plot is shown in Figure 10.7. It shows methylation values and lengths of segments in each segment class, as well as the BIC for different numbers of segments.

```
# read methylation data

methFile=system.file("extdata","H1.chr21.chr22.rds",
                     package="compGenomRData")
mbw=readRDS(methFile)

# segment the methylation data
res=methSeg(mbw,minSeg=10,G=1:4,
            join.neighbours = TRUE)
```

In this case, we know that BIC does not improve much after 4 segment classes. Now, we will not have a look at the characteristics of the segment classes. We are going to plot the mean methylation value and the length of the segment as a scatter plot; the result of this plot is shown in Figure 10.8.

```
# plot
plot(res$seg.mean,
     log10(width(res)),pch=20,
     col=scales::alpha(rainbow(4)[as.numeric(res$seg.group)], 0.2),
     ylab="log10(length)",
     xlab="methylation proportion")
```

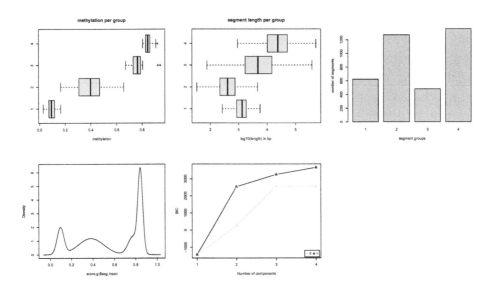

**FIGURE 10.7:** Segmentation characteristics shown in different plots. Top left: Mean methylation values per segment in each segment class. Top middle: Length of each segment as boxplots for each segment class. Top right: Number of segments in each segment class. Bottom left: Distribution of segment methylation values. Bottom right: BIC for different number of segment classes.

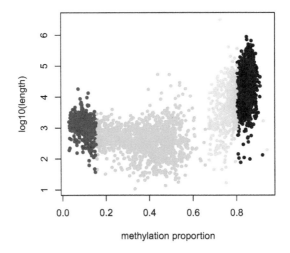

**FIGURE 10.8:** Scatter plot of segment mean, methylation values versus segment length. Each dot is a segment identified by the methSeg() function.

The highly methylated segment classes that have more than 70% methylation are usually longer; the median length is 17889 bp. The segment class that has the lowest methylation values have the median length of 1376 bp and the shortest segment class has low to medium methylation level, with median length of 412 bp.

### 10.5.3 Working with large files

We might want to perform differential methylation analysis in R using whole genome methylation data of multiple samples. The problem is that for genome-wide experiments, file sizes can easily range from hundreds of megabytes to gigabytes and processing multiple instances of those files in memory (RAM) might become unfeasible unless we have access to a high-performance compute cluster (HPC) with extensive RAM. If we want to use a desktop computer or laptop with limited RAM, we either need to restrict our analysis to a subset of the data or use packages that can handle this situation.

The methylKit package provides the capability of dealing with large files and high numbers of samples by exploiting flat file databases to substitute in-memory objects. The internal data, apart from meta information, has a tabular structure storing chromosome, start/end position, and strand information of the associated CpG base just like many other biological formats like BED, GFF or SAM. By exporting this tabular data into a TAB-delimited file and making sure it is accordingly position-sorted, it can be indexed using the generic tabix tool[11]. In general, tabix indexing is a generalization of BAM indexing for generic TAB-delimited files. It inherits all the advantages of BAM indexing, including data compression and efficient random access in terms of few seek function calls per query (Li, 2011). MethylKit relies on Rsamtools[12] which implements tabix functionality for R. This way internal methylKit objects can be efficiently stored as a compressed file on the disk and still be quickly accessed. Another advantage is that existing compressed files can be loaded in interactive sessions, allowing the backup and transfer of intermediate analysis results.

methylKit provides the capability for storing objects in tabix format within various functions. Every methylKit object has its tabix-based flat-file database equivalent. For example, when reading a methylation call file, the dbtype argument can be provided, which will create tabix-based objects.

```
myobj=methRead( file.list,
              sample.id=list("test1","test2","ctrl1","ctrl2"),
              assembly="hg18",treatment=c(1,1,0,0),
              dbtype="tabix")
```

---

[11]http://www.htslib.org/doc/tabix.html

[12]http://bioconductor.org/packages/release/bioc/html/Rsamtools.html

The advantage of tabix-based objects is of course saving memory and more efficient parallelization for differential methylation calculation. However, since the data is written to a file and indexed whenever a new object is created, working with tabix-based objects will be slower at certain steps of the analysis compared to in-memory objects.

## 10.6    Annotation of DMRs/DMCs and segments

The regions of interest obtained through differential methylation or segmentation analysis often need to be integrated with genome annotation datasets. Without this type of integration, differential methylation or segmentation results will be hard to interpret in biological terms. The most common annotation task is to see where regions of interest land in relation to genes and gene parts and regulatory regions: Do they mostly occupy promoter, intronic or exonic regions? Do they overlap with repeats? Do they overlap with other epigenomic markers or long-range regulatory regions? These questions are not specific to methylation –nearly all regions of interest obtained via genome-wide studies have to deal with such questions. Thus, there are already multiple software tools that can produce such annotations. One is the Bioconductor package genomation[13] (Akalin et al., 2015). It can be used to annotate DMRs/DMCs and it can also be used to integrate methylation proportions over the genome with other quantitative information and produce meta-gene plots or heatmaps. Below, we are reading a BED file for transcripts and using that to annotate DMCs with promoter/intron/exon/intergenic annotation. The genomation::readTranscriptFeatures() function reads a BED12 file, calculates the coordinates of promoters, exons, and introns and the subsequent function uses that information for annotation.

```
library(genomation)

# read the gene BED file
transcriptBED=system.file("extdata", "refseq.hg18.bed.txt",
                                            package = "methylKit")
gene.obj=readTranscriptFeatures(transcriptBED)
#
# annotate differentially methylated CpGs with
```

---

[13] http://bioconductor.org/packages/release/bioc/html/genomation.html

```
# promoter/exon/intron using annotation data
#
annotateWithGeneParts(as(all.diff,"GRanges"),gene.obj)
```

```
##    promoter      exon     intron intergenic
##      28.24     15.27      33.59      58.02
##    promoter      exon     intron intergenic
##      28.24      0.00      13.74      58.02
## promoter     exon    intron
##     0.29     0.03      0.17
##    Min. 1st Qu.  Median     Mean 3rd Qu.     Max.
##       5      815    49918    52410   94644   313528
```

Similarly, we can read the CpG island annotation and annotate our differentially methylated bases/regions with them.

```
# read the shores and flanking regions and name the flanks as shores
# and CpG islands as CpGi
cpg.file=system.file("extdata", "cpgi.hg18.bed.txt",
                                    package = "methylKit")
cpg.obj=readFeatureFlank(cpg.file,
                     feature.flank.name=c("CpGi","shores"))
```

```
## Warning: 'GenomicRangesList' is deprecated.
## Use 'GRangesList(..., compress=FALSE)' instead.
## See help("Deprecated")
```

```
#
# convert methylDiff object to GRanges and annotate
diffCpGann=annotateWithFeatureFlank(as(all.diff,"GRanges"),
                          cpg.obj$CpGi,cpg.obj$shores,
                     feature.name="CpGi",flank.name="shores")
```

Besides these, DMRs/DMCs might be associated with changes in gene regulation. It might be desirable to overlap them with known transcription binding sites or motifs or histone modifications. These are simply overlap operations for these kinds of analysis. You can use the `genomation::annotateWithFeature()` function or any other approach shown in Chapter 6, and you can also do motif discovery with methods shown in Chapter 9.

### 10.6.1  Further annotation with genes or gene sets

The next obvious steps for annotating your DMRs/DMCs are figuring out which genes they are associated with. Figuring out which genes are associated with your regions of interest can give a better idea of the biological implications of the methylation changes. Once you have your gene set, you can do gene set analysis as shown in Chapter 8 or in Chapter 11. There are also packages such as rGREAT[14] that can simultaneously associate DMRs or any other region of interest to genes and do gene set analysis.

## 10.7   Other R packages that can be used for methylation analysis

- DSS[15] beta-binomial models with empirical Bayes for moderating dispersion.
- BSseq[16] Regional differential methylation analysis using smoothing and linear-regression-based tests.
- BiSeq[17] Regional differential methylation analysis using beta-binomial models.
- MethylSeekR[18]: Methylome segmentation using HMM and cutoffs.
- QuasR[19]: Methylation aware alignment and methylation calling, as well as fastQC-like fastq raw data quality check features.

## 10.8   Exercises

### 10.8.1   Differential methylation

The main objective of this exercise is getting differential methylated cytosines between two groups of samples: IDH-mut (AML patients with IDH mutations) vs. NBM (normal bone marrow samples).

1.  Download methylation call files from GEO. These files are readable by methlKit using default methRead arguments. [Difficulty: **Beginner**]

---

[14]https://www.bioconductor.org/packages/release/bioc/html/rGREAT.html

[15]http://bioconductor.org/packages/release/bioc/html/genomation.html

[16]http://bioconductor.org/packages/release/bioc/html/BSseq.html

[17]http://bioconductor.org/packages/release/bioc/html/BiSeq.html

[18]http://bioconductor.org/packages/release/bioc/html/MethylSeekR.html

[19]http://bioconductor.org/packages/release/bioc/html/QuasR.html

| samples | Link |
|---------|------|
| IDH1_rep1 | link[20] |
| IDH1_rep2 | link[21] |
| NBM_rep1 | link[22] |
| NBM_rep2 | link[23] |

Example code for reading a file:

```
library(methylKit)
m=methRead("~/Downloads/GSM919982_NBM_1_myCpG.txt.gz",
          sample.id = "idh",assembly="hg18")
```

2. Find differentially methylated cytosines. Use chr1 and chr2 only if you need to save time. You can subset it after you download the files either in R or Unix. The files are for hg18 assembly of human genome. [Difficulty: **Beginner**]

3. Describe the general differential methylation trend, what is the main effect for most CpGs? [Difficulty: **Intermediate**]

4. Annotate differentially methylated cytosines (DMCs) as promoter/intron/exon? [Difficulty: **Beginner**]

5. Which genes are the nearest to DMCs? [Difficulty: **Intermediate**]

6. Can you do gene set analysis either in R or via web-based tools? [Difficulty: **Advanced**]

### 10.8.2 Methylome segmentation

The main objective of this exercise is to learn how to do methylome segmentation and the downstream analysis for annotation and data integration.

1. Download the human embryonic stem-cell (H1 Cell Line) methylation bigWig files from the Roadmap Epigenomics website[24]. It may take a while to understand how the website is structured and which bigWig file

---

[20]https://www.ncbi.nlm.nih.gov/geo/download/?acc=GSM919990&format=file&file=GSM919990%5FIDH%2Dmut%5F1%5FmyCpG%2Etxt%2Egz

[21]https://www.ncbi.nlm.nih.gov/geo/download/?acc=GSM919991&format=file&file=GSM919991%5FIDH%5Fmut%5F2%5FmyCpG%2Etxt%2Egz

[22]https://www.ncbi.nlm.nih.gov/geo/download/?acc=GSM919982&format=file&file=GSM919982%5FNBM%5F1%5FmyCpG%2Etxt%2Egz

[23]https://www.ncbi.nlm.nih.gov/geo/download/?acc=GSM919984&format=file&file=GSM919984%5FNBM%5F2%5FRep1%5FmyCpG%2Etxt%2Egz

[24]http://egg2.wustl.edu/roadmap/web_portal/processed_data.html#MethylData

to use. That is part of the exercise. The files you will download are for hg19 assembly unless stated otherwise. [Difficulty: **Beginner**]

2. Do segmentation on hESC methylome. You can only use chr1 if using the whole genome takes too much time. [Difficulty: **Intermediate**]

3. Annotate segments and the kinds of gene-based features each segment class overlaps with (promoter/exon/intron). [Difficulty: **Beginner**]

4. For each segment type, annotate the segments with chromHMM annotations from the Roadmap Epigenome database available here[25]. The specific file you should use is here[26]. This is a bed file with chromHMM annotations. chromHMM annotations are parts of the genome identified by a hidden-Markov-model-based machine learning algorithm. The segments correspond to active promoters, enhancers, active transcription, insulators, etc. The chromHMM model uses histone modification ChIP-seq and potentially other ChIP-seq data sets to annotate the genome.[Difficulty: **Advanced**]

---

[25] https://egg2.wustl.edu/roadmap/web_portal/chr_state_learning.html#core_15state

[26] https://egg2.wustl.edu/roadmap/data/byFileType/chromhmmSegmentations/ChmmModels/coreMarks/ jointModel/final/E003_15_coreMarks_mnemonics.bed.gz

# 11

## Multi-omics Analysis

*Chapter Author*: **Jonathan Ronen**

Living cells are a symphony of complex processes. Modern sequencing technology has led to many comprehensive assays being routinely available to experimenters, giving us different ways to peek at the internal doings of the cells, each experiment revealing a different part of some underlying processes. As an example, most cells have the same DNA, but sequencing the genome of a cell allows us to find mutations and structural alterations that drive tumerogenesis in cancer. If we treat the DNA with bisulfite prior to sequencing, cytosine residues are converted to uracil, but 5-methylcytosine residues are unaffected. This allows us to probe the methylation patterns of the genome, or its methylome. By sequencing the mRNA molecules in a cell, we can calculate the abundance, in different samples, of different mRNA transcripts, or uncover its transcriptome. Performing different experiments on the same samples, for instance RNA-seq, DNA-seq, and BS-seq, results in multi-dimensional omics datasets, which enable the study of relationships between different biological processes, e.g. DNA methylation, mutations, and gene expression, and the leveraging of multiple data types to draw inferences about biological systems. This chapter provides an overview of some of the available methods for such analyses, focusing on matrix factorization approaches. In the examples in this chapter we will demonstrate how these methods are applicable to cancer molecular subtyping, i.e. finding tumors which are driven by the same molecular processes.

---

## 11.1 Use case: Multi-omics data from colorectal cancer

The examples in this chapter will use the following data: a set of 121 tumors from the TCGA (Weinstein et al., 2013) colorectal cancer cohort. The tumors have been profiled for gene expression using RNA-seq, mutations using Exome-seq, and copy number variations using genotyping arrays. Projects such as TCGA have turbocharged efforts to sub-divide cancer into subtypes. Although two tumors arise in the colon, they may have distinct molecular profiles, which is important for treatment decisions. The subset of tumors used in this chapter belong to two distinct

**TABLE 11.1:** Example gene expression data (head)

|              | RNF113A  | S100A13  | AP3D1    | ATP6V1G1 | UBQLN4   | TPPP3    |
|--------------|----------|----------|----------|----------|----------|----------|
| TCGA.A6.2672 | 21.19567 | 19.72600 | 11.53022 | 0.00000  | 15.35637 | 12.76747 |
| TCGA.A6.3809 | 21.50866 | 18.65729 | 12.98830 | 14.12675 | 19.62208 | 0.00000  |
| TCGA.A6.5661 | 20.08072 | 18.97034 | 10.83759 | 15.31325 | 0.00000  | 0.00000  |
| TCGA.A6.5665 | 0.00000  | 11.88336 | 10.24248 | 19.79300 | 0.00000  | 0.00000  |
| TCGA.A6.6653 | 0.00000  | 12.07753 | 0.00000  | 0.00000  | 0.00000  | 0.00000  |
| TCGA.A6.6780 | 0.00000  | 12.99128 | 0.00000  | 19.96976 | 13.17618 | 11.58742 |

molecular subtypes defined by the Colorectal Cancer Subtyping Consortium (Guinney et al., 2015), *CMS1* and *CMS3*. The following code snippets load this multi-omics data from the companion package, starting with gene expression data from RNA-seq (see Chapter 8). Below we are reading the RNA-seq data from the `compGenomRData` package.

```
# read in the csv from the companion package as a data frame
csvfile <- system.file("extdata", "multi-omics", "COREAD_CMS13_gex.csv",
                       package="compGenomRData")
x1 <- read.csv(csvfile, row.names=1)
# Fix the gene names in the data frame
rownames(x1) <- sapply(strsplit(rownames(x1), "\\|"), function(x) x[1])
# Output a table
knitr::kable(head(t(head(x1))), caption="Example gene expression data (head)")
```

Table 11.1 shows the head of the gene expression matrix. The rows correspond to patients, referred to by their TCGA identifier, as the first column of the table. Columns represent the genes, and the values are RPKM expression values. The column names are the names or symbols of the genes. The details about how these expression values are calculated are in Chapter 8.

We first **read mutation data** with the following code snippet.

```
# read in the csv from the companion package as a data frame
csvfile <- system.file("extdata", "multi-omics", "COREAD_CMS13_muts.csv",
                       package="compGenomRData")
x2 <- read.csv(csvfile, row.names=1)
# Set mutation data to be binary (so if a gene has more than 1 mutation,
# we only count one)
x2[x2>0]=1
```

**TABLE 11.2:** Example mutation data (head)

|            | TTN | TP53 | APC | KRAS | SYNE1 | MUC16 |
|------------|-----|------|-----|------|-------|-------|
| TCGA.A6.2672 | 1 | 0 | 0 | 0 | 1 | 1 |
| TCGA.A6.3809 | 1 | 0 | 0 | 0 | 0 | 0 |
| TCGA.A6.5661 | 1 | 0 | 0 | 0 | 1 | 1 |
| TCGA.A6.5665 | 1 | 0 | 0 | 0 | 1 | 1 |
| TCGA.A6.6653 | 1 | 0 | 0 | 1 | 0 | 0 |
| TCGA.A6.6780 | 1 | 0 | 0 | 0 | 0 | 1 |

**TABLE 11.3:** Example copy number data for CRC samples

|            | 8p23.2 | 8p23.3 | 8p23.1 | 8p21.3 | 8p12 | 8p22 |
|------------|--------|--------|--------|--------|------|------|
| TCGA.A6.2672 | 0 | 0 | 0 | 0 | 0 | 0 |
| TCGA.A6.3809 | 0 | 0 | 0 | 0 | 0 | 0 |
| TCGA.A6.5661 | 0 | 0 | 0 | 0 | 0 | 0 |
| TCGA.A6.5665 | 0 | 0 | 0 | 0 | 0 | 0 |
| TCGA.A6.6653 | 0 | 0 | 0 | 0 | 0 | 0 |
| TCGA.A6.6780 | 0 | 0 | 0 | 0 | 0 | 0 |

```
# output a table
knitr::kable(head(t(head(x2))), caption="Example mutation data (head)")
```

Table 11.2 shows the mutations of these tumors (mutations were introduced in Chapter 1). In the mutation matrix, each cell is a binary 1/0, indicating whether or not a tumor has a non-synonymous mutation in the gene indicated by the column. These types of mutations change the aminoacid sequence, therefore they are likely to change the function of the protein.

Next, we **read copy number data** with the following code snippet.

```
# read in the csv from the companion package as a data frame
csvfile <- system.file("extdata", "multi-omics", "COREAD_CMS13_cnv.csv",
                        package="compGenomRData")
x3 <- read.csv(csvfile, row.names=1)
# output a table
knitr::kable(head(t(head(x3))),
            caption="Example copy number data for CRC samples")
```

Finally, table 11.3 shows GISTIC scores (Mermel et al., 2011) for copy number alterations in these tumors. During transformation from healthy cells to cancer cells, the genome sometimes undergoes large-scale instability; large segments of the genome might be replicated or lost. This will be reflected in each segment's "copy number". In this matrix, each column corresponds to a chromosome segment, and the value of the cell is a real-valued score indicating if this segment has been amplified (copied more) or lost, relative to a non-cancer control from the same patient.

Each of the data types (gene expression, mutations, copy number variation) on its own, provides some signal which allows us to somewhat separate the samples into the two different subtypes. In order to explore these relations, we must first obtain the subtypes of these tumors. The following code snippet reads these, also from the companion package:

```
# read in the csv from the companion package as a data frame
csvfile <- system.file("extdata", "multi-omics", "COREAD_CMS13_subtypes.csv",
                       package="compGenomRData")
covariates <- read.csv(csvfile, row.names=1)
# Fix the TCGA identifiers so they match up with the omics data
rownames(covariates) <- gsub(pattern = '-', replacement = '\\.',
                             rownames(covariates))
covariates <- covariates[colnames(x1),]
# create a dataframe which will be used to annotate later graphs
anno_col <- data.frame(cms=as.factor(covariates$cms_label))
rownames(anno_col) <- rownames(covariates)
```

Before proceeding with any multi-omics integration analysis which might obscure the underlying data, it is important to take a look at each omic data type on its own, and in this case in particular, to examine their relation to the underlying condition, i.e. the cancer subtype. A great way to get an eagle-eye view of such large data is using heatmaps (see Chapter 4 for more details).

We will first check the gene expression data in relation to the subtypes. One way of doing that is plotting a heatmap and clustering the tumors, while displaying a color annotation atop the heatmap, indicating which subtype each tumor belongs to. This is shown in Figure 11.1, which is generated by the following code snippet:

```
pheatmap::pheatmap(x1,
                   annotation_col = anno_col,
                   show_colnames = FALSE,
                   show_rownames = FALSE,
                   main="Gene expression data")
```

In Figure 11.1, each column is a tumor, and each row is a gene. The values in the cells are FPKM values. There is another band above the heatmap annotating each column (tumor) with its corresponding subtype. The tumors are clustered using hierarchical

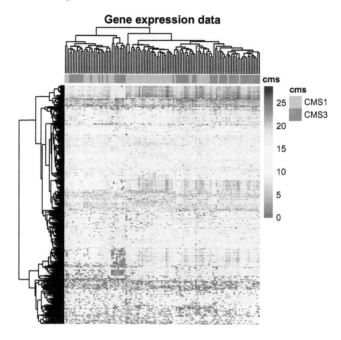

**FIGURE 11.1:** Heatmap of gene expression data for colorectal cancers.

clustering denoted by the dendrogram above the heatmap, according to which the columns (tumors) are ordered. While this ordering corresponds somewhat to the subtypes, it would not be possible to cut this dendrogram in a way which achieves perfect separation between the subtypes.

Next we repeat the same exercise using the mutation data. The following snippet generates Figure 11.2:

```
pheatmap::pheatmap(x2,
                   annotation_col = anno_col,
                   show_colnames = FALSE,
                   show_rownames = FALSE,
                   main="Mutation data")
```

An examination of Figure 11.2 shows that tumors clustered and ordered by mutation data correspond very closely to their CMS subtypes. However, one should be careful in drawing conclusions about this result. Upon closer examination, you might notice that the separating factor seems to be that CMS1 tumors have significantly more mutations than do CMS3 tumors. This, rather than mutations in a specific genes, seems to be driving this clustering result. Nevertheless, this hyper-mutated status is an important indicator for this subtype.

Finally, we look into copy number variation data and try to see if clustered samples

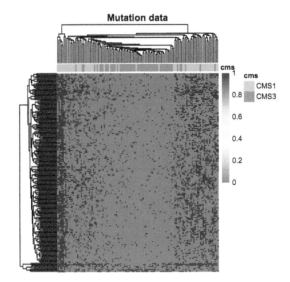

**FIGURE 11.2:** Heatmap of mutation data for colorectal cancers.

are in concordance with subtypes. The following code snippet generates Figure 11.3:

```
pheatmap::pheatmap(x3,
                   annotation_col = anno_col,
                   show_colnames = FALSE,
                   show_rownames = FALSE,
                   main="Copy number data")
```

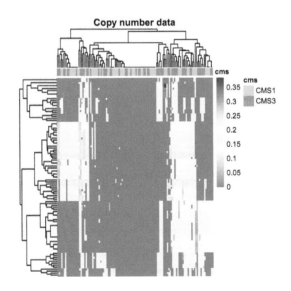

**FIGURE 11.3:** Heatmap of copy number variation data, colorectal cancers.

The interpretation of Figure 11.3 is left as an exercise for the reader.

It is clear that while there is some "signal" in each of these omics types, as is evident from these heatmaps, it is equally clear that none of these omics types completely and on its own explains the subtypes. Each omics type provides but a glimpse into what makes each of these tumors different from a healthy cell. Through the rest of this chapter, we will demonstrate how analyzing the gene expression, mutations, and copy number variations, in tandem, we will be able to get a better picture of what separates these cancer subtypes.

The next section will describe latent variable models for multi-omics integrations. Latent variable models are a form of dimensionality reduction (see Chapter 4). Each omics data type is "big data" in its own right; a typical RNA-seq experiment profiles upwards of 50 thousand different transcripts. The difficulties in handling large data matrices are only exacerbated by the introduction of more omics types into the analysis, as we are suggesting here. In order to overcome these challenges, latent variable models are a powerful way to reduce the dimensionality of the data down to a manageable size.

## 11.2  Latent variable models for multi-omics integration

Unsupervised multi-omics integration methods are methods that look for patterns within and across data types, in a label-agnostic fashion, i.e. without knowledge of the identity or label of the analyzed samples (e.g. cell type, tumor/normal). This chapter focuses on latent variable models, a form of dimensionality reduction technique (see Chapter 4). Latent variable models make an assumption that the high-dimensional data we observe (e.g. counts of tens of thousands of mRNA molecules) arise from a lower dimension description. The variables in that lower dimensional description are termed *latent variables*, as they are believed to be latent in the data, but not directly observable through experimentation. Therefore, there is a need for methods to infer the latent variables from the data. For instance, (see Chapter 8 for details of RNA-seq analysis) the relative abundance of different mRNA molecules in a cell is largely determined by the cell type. There are other experiments which may be used to discern the cell type of cells (e.g. looking at them under a microscope), but an RNA-seq experiment does not, directly, reveal whether the analyzed sample was taken from one organ or another. A latent variable model would set the cell type as a latent variable, and the observable abundance of mRNA molecules to be dependent on the value of the latent variable (e.g. if the latent variable is "Regulatory T-cell", we would expect to find high expression of CD4, FOXP3, and CD25).

## 11.3 Matrix factorization methods for unsupervised multi-omics data integration

Matrix factorization techniques attempt to infer a set of latent variables from the data by finding factors of a data matrix. Principal Component Analysis (introduced in Chapter 4) is a form of matrix factorization which finds factors based on the covariance structure of the data. Generally, matrix factorization methods may be formulated as

$$X = WH,$$

where $X$ is the *data matrix*, $[M \times N]$ where $M$ is the number of features (typically genes), and $N$ is the number of samples. $W$ is an $[M \times K]$ *factors* matrix, and $H$ is the $[K \times N]$ *latent variable coefficient matrix*. Tying this back to PCA, where $X = U\Sigma V^T$, we may formulate the factorization in the same terms by setting $W = U\Sigma$ and $H = V^T$. If $K = rank(X)$, this factorization is lossless, i.e. $X = WH$. However if we choose $K < rank(X)$, the factorization is lossy, i.e. $X \approx WH$. In that case, matrix factorization methods normally opt to minimize the error

$$min \, \|X - WH\|.$$

As we normally seek a latent variable model with a considerably lower dimensionality than $X$, this is the more common case.

The loss function we choose to minimize may be further subject to some constraints or regularization terms. Regularization has been introduced in Chapter 5. In the current context of latent factor models, a regularization term might be added to the loss function, i.e. we might choose to minimize $min \, \|X - WH\| + \lambda\|W\|^2$ (this is called $L_2$-regularization) instead of merely the reconstruction error. Adding such a term to our loss function here will push the $W$ matrix entries towards 0, in effect balancing between better reconstruction of the data and a more parsimonious model. A more parsimonious latent factor model is one with more sparsity in the latent factors. This sparsity is desirable for model interpretation, as will become evident in later sections.

In Figure 11.4, the $5 \times 4$ data matrix $X$ is decomposed to a 2-dimensional latent variable model.

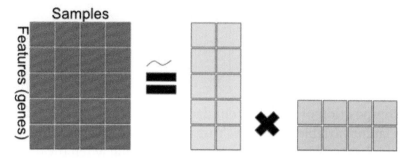

**FIGURE 11.4:** General matrix factorization framework. The data matrix on the left-hand side is decomposed into factors on the right-hand side. The equality may be an approximation as some matrix factorization methods are lossless (exact), while others are an approximation.

### 11.3.1 Multiple factor analysis

Multiple factor analysis is a natural starting point for a discussion about matrix factorization methods for integrating multiple data types. It is a straightforward extension of PCA into the domain of multiple data types [1].

Figure 11.5 sketches a naive extension of PCA to a multi-omics context.

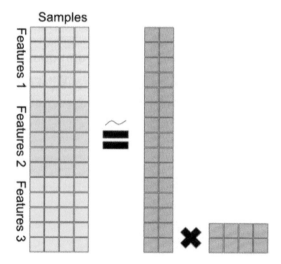

**FIGURE 11.5:** A naive extension of PCA to multi-omics; data matrices from different platforms are stacked, before applying PCA.

---

[1]When dealing with categorical variables, MFA uses MCA (Multiple Correspondence Analysis). This is less relevant to biological data analysis and will not be discussed here.

Formally, we have

$$X = \begin{bmatrix} X_1 \\ X_2 \\ \vdots \\ X_L \end{bmatrix} = WH,$$

a joint decomposition of the different data matrices ($X_i$) into the factor matrix $W$ and the latent variable matrix $H$. This way, we can leverage the ability of PCA to find the highest variance decomposition of the data, when the data consists of different omics types. As a reminder, PCA finds the linear combinations of the features which, when the data is projected onto them, preserve the most variance of any $K$-dimensional space. But because measurements from different experiments have different scales, they will also have variance (and co-variance) at different scales.

Multiple Factor Analysis addresses this issue and achieves balance among the data types by normalizing each of the data types, before stacking them and passing them on to PCA. Formally, MFA is given by

$$X_n = \begin{bmatrix} X_1/\lambda_1^{(1)} \\ X_2/\lambda_1^{(2)} \\ \vdots \\ X_L/\lambda_1^{(L)} \end{bmatrix} = WH,$$

where $\lambda_1^{(i)}$ is the first eigenvalue of the principal component decomposition of $X_i$.

Following this normalization step, we apply PCA to $X_n$. From there on, MFA analysis is the same as PCA analysis, and we refer the reader to Chapter 4 for more details.

### 11.3.1.1　MFA in R

MFA is available through the CRAN package `FactoMineR`. The code snippet below shows how to run it:

```
# run the MFA function from the FactoMineR package
r.mfa <- FactoMineR::MFA(
  t(rbind(x1,x2,x3)), # binding the omics types together
  c(dim(x1)[1], dim(x2)[1], dim(x3)[1]), # specifying the dimensions of each
  graph=FALSE)
```

Since this generates a two-dimensional factorization of the multi-omics data, we can now plot each tumor as a dot in a 2D scatter plot to see how well the MFA factors separate the cancer subtypes. The following code snippet generates Figure 11.6:

```
# first, extract the H and W matrices from the MFA run result
mfa.h <- r.mfa$global.pca$ind$coord
mfa.w <- r.mfa$quanti.var$coord

# create a dataframe with the H matrix and the CMS label
mfa_df <- as.data.frame(mfa.h)
mfa_df$subtype <- factor(covariates[rownames(mfa_df),]$cms_label)

# create the plot
ggplot2::ggplot(mfa_df, ggplot2::aes(x=Dim.1, y=Dim.2, color=subtype)) +
ggplot2::geom_point() + ggplot2::ggtitle("Scatter plot of MFA")
```

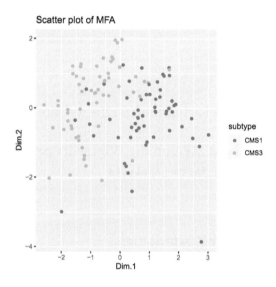

**FIGURE 11.6:** Scatter plot of 2-dimensional MFA for multi-omics data shows separation between the subtypes.

Figure 11.6 shows remarkable separation between the cancer subtypes; it is easy enough to draw a line separating the tumors to CMS subtypes with good accuracy.

Another way to examine the MFA factors, which is also useful for factor models with more than two components, is a heatmap, as shown in Figure 11.7, generated by the following code snippet:

```
pheatmap::pheatmap(t(mfa.h)[1:2,], annotation_col = anno_col,
                   show_colnames = FALSE,
                   main="MFA for multi-omics integration")
```

Figure 11.7 shows that indeed, when tumors are clustered and ordered using the

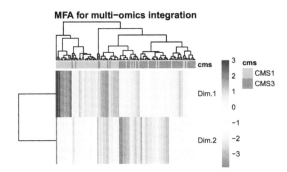

**FIGURE 11.7:** A heatmap of the two MFA components shows separation between the cancer subtypes.

two MFA factors we learned above, their separation into CMS clusters is nearly trivial.

– **Want to know more ?**
  * Learn more about FactoMineR on the website: `http://factominer.free.fr/`
  * Learn more about MFA on the Wikipedia page `https://en.wikipedia.org/wiki/Multiple_factor_analysis`

### 11.3.2  Joint non-negative matrix factorization

As introduced in Chapter 4, NMF (Non-negative Matrix Factorization) is an algorithm from 2000 that seeks to find a non-negative additive decomposition for a non-negative data matrix. It takes the familiar form $X \approx WH$, with $X \geq 0$, $W \geq 0$, and $H \geq 0$. The non-negative constraints make a lossless decomposition (i.e. $X = WH$) generally impossible. Hence, NMF attempts to find a solution which minimizes the Frobenius norm of the reconstruction:

$$min \|X - WH\|_F W \geq 0, H \geq 0,$$

where the Frobenius norm $\| \cdot \|_F$ is the matrix equivalent of the Euclidean distance:

$$\|X\|_F = \sqrt{\sum_i \sum_j x_{ij}^2}.$$

This is typically solved for $W$ and $H$ using random initializations followed by iterations of a multiplicative update rule:

$$W_{t+1} = W_t^T \frac{XH_t^T}{XH_t H_t^T} \tag{11.1}$$

$$H_{t+1} = H_t \frac{W_t^T X}{W_t^T W_t X}. \tag{11.2}$$

Since this algorithm is guaranteed only to converge to a local minimum, it is typically run several times with random initializations, and the best result is kept.

In the multi-omics context, we will, as in the MFA case, wish to find a decomposition for an integrated data matrix of the form

$$X = \begin{bmatrix} X_1 \\ X_2 \\ \vdots \\ X_L \end{bmatrix},$$

with $X_i$s denoting data from different omics platforms.

As NMF seeks to minimize the reconstruction error $\|X - WH\|_F$, some care needs to be taken with regards to data normalization. Different omics platforms may produce data with different scales (i.e. real-valued gene expression quantification, binary mutation data, etc.), and so will have different baseline Frobenius norms. To address this, when doing Joint NMF, we first feature-normalize each data matrix, and then normalize by the Frobenius norm of the data matrix. Formally, we run NMF on

$$X = \begin{bmatrix} X_1^N/\alpha_1 \\ X_2^N/\alpha_2 \\ \vdots \\ X_L^N/\alpha_L \end{bmatrix},$$

where $X_i^N$ is the feature-normalized data matrix $X_i^N = \frac{x^{ij}}{\sum_j x^{ij}}$, and $\alpha_i = \|X_i^N\|_F$.

Another consideration with NMF is the non-negativity constraint. Different omics data types may have negative values, for instance, copy-number variations (CNVs) may be positive, indicating gains, or negative, indicating losses, as in Table 11.4. In order to turn such data into a non-negative form, we will split each feature into two features, one new feature holding all the non-negative values of the original feature, and another feature holding the absolute value of the negative ones, as in Table 11.5.

**TABLE 11.4:** Example copy number data. Data can be both positive (amplified regions) or negative (deleted regions).

|        | seg1 | seg2 |
|--------|------|------|
| samp1  | 1    | 0    |
| samp2  | 2    | 1    |
| samp3  | 1    | -2   |

**TABLE 11.5:** Example copy number data after splitting each column into a column representing copy number gains (+) and a column representing deletions (-). This data matrix is non-negative, and thus suitable for NMF algorithms.

|        | seg1+ | seg1- | seg2+ | seg2- |
|--------|-------|-------|-------|-------|
| samp1  | 1     | 0     | 0     | 0     |
| samp2  | 2     | 0     | 1     | 0     |
| samp3  | 1     | 0     | 0     | 2     |

### 11.3.2.1  NMF in R

Many NMF algorithms are available through the CRAN package NMF. The following code chunk demonstrates how it may be run:

```r
# Feature-normalize the data
x1.featnorm <- x1 / rowSums(x1)
x2.featnorm <- x2 / rowSums(x2)
x3.featnorm <- x3 / rowSums(x3)


# Normalize by each omics type's frobenius norm
x1.featnorm.frobnorm <- x1.featnorm / norm(as.matrix(x1.featnorm), type="F")
x2.featnorm.frobnorm <- x2.featnorm / norm(as.matrix(x2.featnorm), type="F")
x3.featnorm.frobnorm <- x3.featnorm / norm(as.matrix(x3.featnorm), type="F")


# Split the features of the CNV matrix into two non-negative features each
x3.featnorm.frobnorm.nonneg <- t(split_neg_columns(t(x3.featnorm.frobnorm)))


# run the nmf function from the NMF package
require(NMF)
r.nmf <- nmf(t(rbind(x1.featnorm.frobnorm,
                     x2.featnorm.frobnorm,
                     x3.featnorm.frobnorm.nonneg)),
          2,
          method='Frobenius')
```

```
# exctract the H and W matrices from the nmf run result
nmf.h <- NMF::basis(r.nmf)
nmf.w <- NMF::coef(r.nmf)
nmfw <- t(nmf.w)
```

As with MFA, we can examine how well 2-factor NMF splits tumors into subtypes by looking at the scatter plot in Figure 11.8, generated by the following code chunk:

```
# create a dataframe with the H matrix and the CMS label (subtype)
nmf_df <- as.data.frame(nmf.h)
colnames(nmf_df) <- c("dim1", "dim2")
nmf_df$subtype <- factor(covariates[rownames(nmf_df),]$cms_label)

# create the scatter plot
ggplot2::ggplot(nmf_df, ggplot2::aes(x=dim1, y=dim2, color=subtype)) +
ggplot2::geom_point() +
ggplot2::ggtitle("Scatter plot of 2-component NMF for multi-omics integration")
```

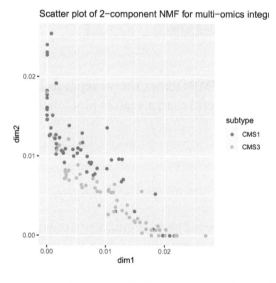

**FIGURE 11.8:** NMF creates a disentangled representation of the data using two components which allow for separation between tumor sub-types CMS1 and CMS3 based on NMF factors learned from multi-omics data.

Figure 11.8 shows an important difference between NMF and MFA (PCA). It shows the tendency of samples to lie close to the X or Y axes, that is, the tendency of each sample to be high in only one of the factors. This will be discussed more in the later section on disentangledness.

Again, should we choose to run NMF with more than two factors, a more useful plot might be the heatmap shown in Figure 11.9, generated by the following code snippet:

```
pheatmap::pheatmap(t(nmf_df[,1:2]),
                   annotation_col = anno_col,
                   show_colnames=FALSE,
                   main="Heatmap of 2-component NMF")
```

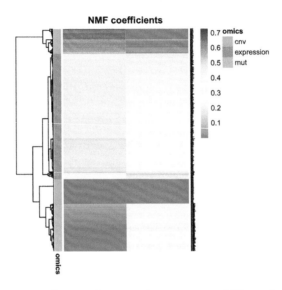

**FIGURE 11.9:** A heatmap of NMF factors shows separability of tumors into subtype clusters. This plot is more useful than a scatter plot when there are more than two factors.

– **Want to know more ?**

* Joint NMF to uncover gene regulatory networks: Zhang S., Li Q., Liu J., Zhou X. J. (2011). A novel computational framework for simultaneous integration of multiple types of genomic data to identify microRNA-gene regulatory modules. *Bioinformatics* 27, i401–i409. 10.1093/bioinformatics/btr206 https://www.ncbi.nlm.nih.gov/pmc/articles/PMC3117336/

* Joint NMF for cancer research: Zhang S., Liu C.-C., Li W., Shen H., Laird P. W., Zhou X. J. (2012). Discovery of multi-dimensional modules by integrative analysis of cancer genomic data. *Nucleic Acids Res.* 40, 9379–9391. 10.1093/nar/gks725 https://www.ncbi.nlm.nih.gov/pmc/articles/PMC3479191/

### 11.3.3 iCluster

iCluster takes a Bayesian approach to the latent variable model. In Bayesian statistics, we infer distributions over model parameters, rather than finding a single maximum-likelihood parameter estimate. In iCluster, we model the data as

$$X_{(i)} = W_{(i)} Z + \epsilon_i,$$

where $X_{(i)}$ is a data matrix from a single omics platform, $W_{(i)}$ are model parameters, $Z$ is a latent variable matrix, and is shared among the different omics platforms, and $\epsilon_i$ is a "noise" random variable, $\epsilon \sim N(0, \Psi)$, with $\Psi = diag(\psi_1, \dots \psi_M)$ is a diagonal covariance matrix.

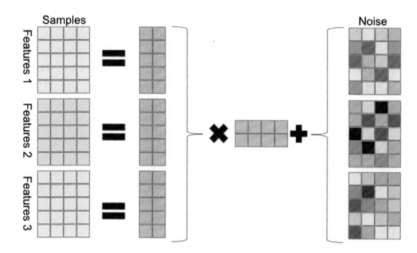

**FIGURE 11.10:** Sketch of iCluster model. Each omics datatype is decomposed to a coefficient matrix and a shared latent variable matrix, plus noise.

Note that with this construction, the omics measurements $X$ are expected to be the same for samples with the same latent variable representation, up to Gaussian noise. Further, we assume a Gaussian prior distribution on the latent variables $Z \sim N(0, I)$, which means we assume $X_{(i)} \sim N(0, W_{(i)} W_{(i)}^T + \Psi_{(i)})$. In order to find suitable values for $W$, $Z$, and $\Psi$, we can write down the multivariate normal log-likelihood function and optimize it. For a multivariate normal distribution with mean 0 and covariance $\Sigma$, the log-likelihood function is given by

$$\ell = -\frac{1}{2} \left( \ln(|\Sigma|) + X^T \Sigma^{-1} X + k \ln(2\pi) \right)$$

(this is simply the log of the Probability Density Function of a multivariate Gaussian). For the multi-omics iCluster case, we have $X = (X_{(1)}, \dots, X_{(L)})^T$, $W = (W_{(1)}, \dots, W_{(L)})^T$, where $X$ is a multivariate normal with 0-mean and

$\Sigma = WW^T + \Psi$ covariance. Hence, the log-likelihood function for the iCluster model is given by:

$$\ell_{iC}(W, \Sigma) = -\frac{1}{2}\left(\sum_{i=1}^{L} \ln(|\Sigma|) + X^T\Sigma^{-1}X + p_i \ln(2\pi)\right)$$

where $p_i$ is the number of features in omics data type $i$. Because this model has more parameters than we typically have samples, we need to push the model to use fewer parameters than it has at its disposal, by using regularization. iCluster uses Lasso regularization, which is a direct penalty on the absolute value of the parameters. I.e., instead of optimizing $\ell_{iC}(W, \Sigma)$, we will optimize the regularized log-likelihood:

$$\ell = \ell_{iC}(W, \Sigma) - \lambda\|W\|_1.$$

The parameter $\lambda$ acts as a dial to weigh the trade-off between better model fits (higher log-likelihood) and a sparser model, with more $w_{ij}$s set to 0, which gives models which generalize better and are more interpretable.

In order to solve this problem, iCluster employs the Expectation Maximization (EM) algorithm. The full details are beyond the scope of this textbook. We will introduce a short sketch instead. The intuition behind the EM algorithm is a more general case of the k-means clustering algorithm (Chapter 4). The basic **EM algorithm** is as follows.

- Initialize $W$ and $\Psi$.
- **Until convergence of** $W$, $\Psi$
    - E-step: Calculate the expected value of $Z$ given the current estimates of $W$ and $\Psi$ and the data $X$.
    - M-step: Calculate maximum likelihood estimates for the parameters $W$ and $\Psi$ based on the current estimate of $Z$ and the data $X$.

### 11.3.3.1 iCluster+: Extending iCluster

iCluster+ is an extension of the iCluster framework, which allows for omics types to arise from distributions other than a Gaussian. While normal distributions are a good assumption for log-transformed, centered gene expression data, it is a poor model for binary mutations data, or for copy number variation data, which can typically take the values $(-2, 1, 0, 1, 2)$ for heterozygous / monozygous deletions or amplifications. iCluster+ allows the different $X$s to have different distributions:

- for binary mutations, $X$ is drawn from a multivariate binomial
- for normal, continuous data, $X$ is drawn from a multivariate Gaussian

- for copy number variations, $X$ is drawn from a multinomial
- for count data, $X$ is drawn from a Poisson.

In that way, iCluster+ allows us to explicitly model our assumptions about the distributions of our different omics data types, and leverage the strengths of Bayesian inference.

Both iCluster and iCluster+ make use of sophisticated Bayesian inference algorithms (EM for iCluster, Metropolis-Hastings MCMC for iCluster+), which means they do not scale up trivially. Therefore, it is recommended to filter down the features to a manageable size before inputting data to the algorithm. The exact size of "manageable" data depends on your hardware, but a rule of thumb is that dimensions in the thousands are ok, but in the tens of thousands might be too slow.

### 11.3.3.2 Running iCluster+

iCluster+ is available through the Bioconductor package `iClusterPlus`. The following code snippet demonstrates how it can be run with two components:

```
# run the iClusterPlus function
r.icluster <- iClusterPlus::iClusterPlus(
  t(x1), # Providing each omics type
  t(x2),
  t(x3),
  type=c("gaussian", "binomial", "multinomial"), # Providing the distributions
  K=2, # provide the number of factors to learn
  alpha=c(1,1,1), # as well as other model parameters
  lambda=c(.03,.03,.03))

# extract the H and W matrices from the run result
# here, we refer to H as z, to keep with iCluster terminology
icluster.z <- r.icluster$meanZ
rownames(icluster.z) <- rownames(covariates) # fix the row names
icluster.ws <- r.icluster$beta

# construct a dataframe with the H matrix (z) and the cancer subtypes
# for later plotting
icp_df <- as.data.frame(icluster.z)
colnames(icp_df) <- c("dim1", "dim2")
rownames(icp_df) <- colnames(x1)
icp_df$subtype <- factor(covariates[rownames(icp_df),]$cms_label)
```

As with other methods, we examine the iCluster results by looking at the scatter plot in Figure 11.11 and the heatmap in Figure 11.12. Both figures show that iCluster learns two factors which nearly perfectly discriminate between tumors of the two subtypes.

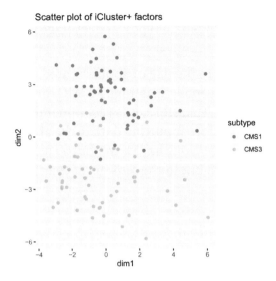

**FIGURE 11.11:** iCluster+ learns factors which allow tumor sub-types CMS1 and CMS3 to be discriminated.

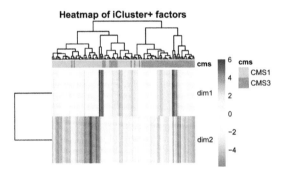

**FIGURE 11.12:** iCluster+ factors, shown in a heatmap, separate tumors into their subtypes well.

*i*

**– Want to know more ?**

* Read the original iCluster paper: Shen R., Olshen A. B., Ladanyi M. (2009). Integrative clustering of multiple genomic data types using a joint latent variable model with application to breast and lung cancer subtype analysis. *Bioinformatics* 25, 2906–2912. 10.1093/bioinformatics/btp543 https://www.ncbi.nlm.nih.gov/pmc/articles/PMC2800366/

* Read the original iClusterPlus paper: an extension of iCluster: Shen R., Mo Q., Schultz N., Seshan V. E., Olshen A. B., Huse J., et al. (2012). Integrative subtype discovery in glioblastoma using iCluster. *PLoS ONE* 7:e35236. 10.1371/journal.pone.0035236 https://www.ncbi.nlm.nih.gov/pmc/articles/PMC3335101/

* Learn more about the LASSO for model regularization: Tibshirani, R. (1996). Regression shrinkage and selection via the lasso. *J. Royal. Statist. Soc B.*, Vol. 58, No. 1, pages 267-288: http://www-stat.stanford.edu/%7Etibs/lasso/lasso.pdf

* Learn more about the EM algorithm: Dempster, A. P., et al. Maximum likelihood from incomplete data via the EM algorithm. *Journal of the Royal Statistical Society. Series B (Methodological)*, vol. 39, no. 1, 1977, pp. 1–38. JSTOR, JSTOR: http://www.jstor.org/stable/2984875

* Read about MCMC algorithms: Hastings, W.K. (1970). Monte Carlo sampling methods using Markov chains and their applications. *Biometrika.* 57 (1): 97–109. doi:10.1093/biomet/57.1.97: https://www.jstor.org/stable/2334940

## 11.4 Clustering using latent factors

A common analysis in biological investigations is clustering. This is often interesting in cancer studies as one hopes to find groups of tumors (clusters) which behave similarly, i.e. have similar risks and/or respond to the same drugs. PCA is a common step in clustering analyses, and so it is easy to see how the latent variable models above may all be a useful pre-processing step before clustering. In the examples below, we will use the latent variables inferred by the algorithms in the previous section on the set of colorectal cancer tumors from the TCGA. For a more complete introduction to clustering, see Chapter 4.

### 11.4.1   One-hot clustering

A specific clustering method for NMF data is to assume each sample is driven by one component, i.e. that the number of clusters $K$ is the same as the number of latent variables in the model and that each sample may be associated to one of those components. We assign each sample a cluster label based on the latent variable which affects it the most. Figure 11.9 above (heatmap of 2-component NMF) shows the latent variable values for the two latent variables, for the 72 tumors, obtained by Joint NMF.

The two rows are the two latent variables, and the columns are the 72 tumors. We can observe that most tumors are indeed driven mainly by one of the factors, and not a combination of the two. We can use this to assign each tumor a cluster label based on its dominant factor, shown in the following code snippet, which also produces the heatmap in Figure 11.13.

```
# one-hot clustering in one line of code:
# assign each sample the cluster according to its dominant NMF factor
# easily accessible using the max.col function
nmf.clusters <- max.col(nmf.h)
names(nmf.clusters) <- rownames(nmf.h)


# create an annotation data frame indicating the NMF one-hot clusters
# as well as the cancer subtypes, for the heatmap plot below
anno_nmf_cl <- data.frame(
  nmf.cluster=factor(nmf.clusters),
  cms.subtype=factor(covariates[rownames(nmf.h),]$cms_label)
)


# generate the plot
pheatmap::pheatmap(t(nmf.h[order(nmf.clusters),]),
  cluster_cols=FALSE, cluster_rows=FALSE,
  annotation_col = anno_nmf_cl,
  show_colnames = FALSE,border_color=NA,
  main="Joint NMF factors\nwith clusters and molecular subtypes")
```

We see that using one-hot clustering with Joint NMF, we were able to find two clusters in the data which correspond fairly well with the molecular subtype of the tumors.

The one-hot clustering method does not lend itself very well to the other methods discussed above, i.e. iCluster and MFA. The latent variables produced by those other methods may be negative, and further, in the case of iCluster, are going to assume a

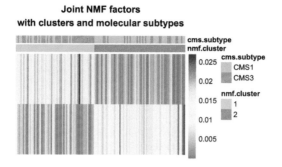

**FIGURE 11.13:** Joint NMF factors with clusters, and molecular sub-types. One-hot clustering assigns one cluster per dimension, where each sample is assigned a cluster based on its dominant component. The clusters largely recapitulate the CMS sub-types.

multivariate Gaussian shape. As such, it is not trivial to pick one "dominant factor" for them. For NMF variants, this is a very common way to assign clusters.

### 11.4.2   K-means clustering

K-means clustering was introduced in Chapter 4. Briefly, k-means is a special case of the EM algorithm, and indeed iCluster was originally conceived as an extension of K-means from binary cluster assignments to real-valued latent variables. The iCluster algorithm, as it is so named, calls for application of K-means clustering on its latent variables, after the inference step. The following code snippet shows how to pull K-means clusters out of the iCluster results, and produces the heatmap in Figure 11.14, which shows how well these clusters correspond to cancer subtypes.

```
# use the kmeans function to cluster the iCluster H matrix (here, z)
# using 2 as the number of clusters.
icluster.clusters <- kmeans(icluster.z, 2)$cluster
names(icluster.clusters) <- rownames(icluster.z)


# create an annotation dataframe for the heatmap plot
# containing the kmeans cluster assignments and the cancer subtypes
anno_icluster_cl <- data.frame(
  iCluster=factor(icluster.clusters),
  cms.subtype=factor(covariates$cms_label))


# generate the figure
pheatmap::pheatmap(
  t(icluster.z[order(icluster.clusters),]), # order z by the kmeans clusters
```

```
cluster_cols=FALSE, # use cluster_cols and cluster_rows=FALSE
cluster_rows=FALSE, # as we want the ordering by k-means clusters to hold
show_colnames = FALSE,border_color=NA,
annotation_col = anno_icluster_cl,
main="iCluster factors\nwith clusters and molecular subtypes")
```

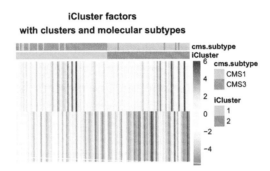

**FIGURE 11.14:** K-means clustering on iCluster+ factors largely recapitulates the CMS sub-types.

This demonstrates the ability of iClusterPlus to find clusters which correspond to molecular subtypes, based on multi-omics data.

## 11.5   Biological interpretation of latent factors

### 11.5.1   Inspection of feature weights in loading vectors

The most straightforward way to go about interpreting the latent factors in a biological context, is to look at the coefficients which are associated with them. The latent variable models introduced above all take the linear form $X \approx WH$, where $W$ is a factor matrix, with coefficients tying each latent variable with each of the features in the $L$ original multi-omics data matrices. By inspecting these coefficients, we can get a sense of which multi-omics features are co-regulated. The code snippet below generates Figure 11.15, which shows the coefficients of the Joint NMF analysis above:

```
# create an annotation dataframe for the heatmap
# for each feature, indicating its omics-type
data_anno <- data.frame(
  omics=c(rep('expression',dim(x1)[1]),
```

```
          rep('mut',dim(x2)[1]),
          rep('cnv',dim(x3.featnorm.frobnorm.nonneg)[1])))
rownames(data_anno) <- c(rownames(x1),
                         paste0("mut:", rownames(x2)),
                         rownames(x3.featnorm.frobnorm.nonneg))
rownames(nmfw) <- rownames(data_anno)

# generate the heat map
pheatmap::pheatmap(nmfw,
                   cluster_cols = FALSE,
                   annotation_row = data_anno,
                   main="NMF coefficients",
                   clustering_distance_rows = "manhattan",
                   fontsize_row = 1)
```

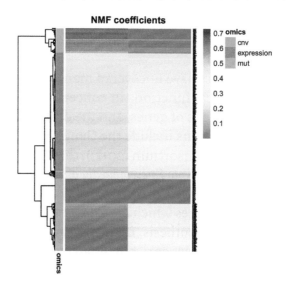

**FIGURE 11.15:** Heatmap showing the association of input features from multi-omics data (gene expression, copy number variation, and mutations), with JNMF factors. Gene expression features dominate both factors, but copy numbers and mutations mostly affect only one factor each.

Inspection of the factor coefficients in the heatmap above reveals that Joint NMF has found two nearly orthogonal non-negative factors. One is associated with high expression of the HOXC11, ZIC5, and XIRP1 genes, frequent mutations in the BRAF, PCDHGA6, and DNAH5 genes, as well as losses in the 18q12.2 and gains in 8p21.1 cytobands. The other factor is associated with high expression of the SOX1 gene, more frequent mutations in the APC, KRAS, and TP53 genes, and a weak association with some CNVs.

#### 11.5.1.1 Disentangled representations

The property displayed above, where each feature is predominantly associated with only a single factor, is termed *disentangledness*, i.e. it leads to *disentangled* latent variable representations, as changing one input feature only affects a single latent variable. This property is very desirable as it greatly simplifies the biological interpretation of modules. Here, we have two modules with a set of co-occurring molecular signatures which merit deeper investigation into the mechanisms by which these different omics features are related. For this reason, NMF is widely used in computational biology today.

### 11.5.2 Making sense of factors using enrichment analysis

In order to investigate the oncogenic processes that drive the differences between tumors, we may draw upon biological prior knowledge by looking for overlaps between genes that drive certain tumors, and genes involved in familiar biological processes.

#### 11.5.2.1 Enrichment analysis

The recent decades of genomics have uncovered many of the ways in which genes cooperate to perform biological functions in concert. This work has resulted in rich annotations of genes, groups of genes, and the different functions they carry out. Examples of such annotations include the Gene Ontology Consortium's *GO terms* (Ashburner et al., 2000, Consortium (2017)), the *Reactome pathways database* (Fabregat et al., 2018b), and the *Kyoto Encyclopaedia of Genes and Genomes* (Kanehisa et al., 2017). These resources, as well as others, publish lists of so-called *gene sets*, or *pathways*, which are sets of genes which are known to operate together in some biological function, e.g. protein synthesis, DNA mismatch repair, cellular adhesion, and many other functions. Gene set enrichment analysis is a method which looks for overlaps between genes which we have found to be of interest, e.g. by them being implicated in a certain tumor type, and the a-priori gene sets discussed above.

In the context of making sense of latent factors, the question we will be asking is whether the genes which drive the value of a latent factor (the genes with the highest factor coefficients) also belong to any interesting annotated gene sets, and whether the overlap is greater than we would expect by chance. If there are $N$ genes in total, $K$ of which belong to a gene set, the probability that $k$ out of the $n$ genes associated with a latent factor are also associated with a gene set is given by the hypergeometric distribution:

$$P(k) = \frac{\binom{K}{k} - \binom{N-K}{n-k}}{\binom{N}{n}}.$$

The **hypergeometric test** uses the hypergeometric distribution to assess the statistical significance of the presence of genes belonging to a gene set in the latent factor. The null hypothesis is that there is no relationship between genes in a gene set, and genes in a latent factor. When testing for over-representation of gene set genes in a latent factor, the P value from the hypergeometric test is the probability of getting $k$ or more genes from a gene set in a latent factor

$$ p = \sum_{i=k}^{K} P(k = i). $$

The hypergeometric enrichment test is also referred to as *Fisher's one-sided exact test*. This way, we can determine if the genes associated with a factor significantly overlap (beyond chance) the genes involved in a biological process. Because we will typically be testing many gene sets, we will also need to apply multiple testing correction, such as Benjamini-Hochberg correction (see Chapter 3, multiple testing correction).

### 11.5.2.2  Example in R

In R, we can do this analysis using the `enrichR` package, which gives us access to many gene set libraries. In the example below, we will find the genes associated with preferentially NMF factor 1 or NMF factor 2, by the contribution of those genes' expression values to the factor. Then, we'll use `enrichR` to query the Gene Ontology terms which might be overlapping:

```r
require(enrichR)

# select genes associated preferentially with each factor
# by their relative loading in the W matrix
genes.factor.1 <- names(which(nmfw[1:dim(x1)[1],1] > nmfw[1:dim(x1)[1],2]))
genes.factor.2 <- names(which(nmfw[1:dim(x1)[1],1] < nmfw[1:dim(x1)[1],2]))

# call the enrichr function to find gene sets enriched
# in each latent factor in the GO Biological Processes 2018 library
go.factor.1 <- enrichR::enrichr(genes.factor.1,
                    databases = c("GO_Biological_Process_2018")
                    )$GO_Biological_Process_2018
go.factor.2 <- enrichR::enrichr(genes.factor.2,
                    databases = c("GO_Biological_Process_2018")
                    )$GO_Biological_Process_2018
```

The top GO terms associated with NMF factor 2 are shown in Table 11.6:

**TABLE 11.6:** GO-terms associated with NMF factor 2

| Term | Adjusted.P.value | Combined.Score |
|------|------------------|----------------|
| nuclear-transcribed mRNA catabolic process (GO:0000956) | 0 | 207.7403 |
| rRNA metabolic process (GO:0016072) | 0 | 161.4781 |
| nuclear-transcribed mRNA catabolic process, nonsense-mediated decay (GO:0000184) | 0 | 220.4298 |

### 11.5.3   Interpretation using additional covariates

Another way to ascribe biological significance to the latent variables is by correlating them with additional covariates we might have about the samples. In our example, the colorectal cancer tumors have also been characterized for microsatellite instability (MSI) status, using an external test (typically PCR-based). By examining the latent variable values as they relate to a tumor's MSI status, we might discover that we've learned latent factors that are related to it. The following code snippet demonstrates how this might be looked into, by generating Figures 11.16 and 11.17:

```r
# create a data frame holding covariates (age, gender, MSI status)
a <- data.frame(age=covariates$age,
                gender=as.numeric(covariates$gender),
                msi=covariates$msi)

## Warning in data.frame(age = covariates$age, gender =
## as.numeric(covariates$gender), : NAs introduced by coercion

b <- nmf.h
colnames(b) <- c('factor1', 'factor2')

# concatenate the covariate dataframe with the H matrix
cov_factor <- cbind(a,b)

# generate the figure
ggplot2::ggplot(cov_factor, ggplot2::aes(x=msi, y=factor1, group=msi)) +
  ggplot2::geom_boxplot() +
  ggplot2::ggtitle("NMF factor 1 microsatellite instability")


ggplot2::ggplot(cov_factor, ggplot2::aes(x=msi, y=factor2, group=msi)) +
  ggplot2::geom_boxplot() +
  ggplot2::ggtitle("NMF factor 2 and microsatellite instability")
```

Figures 11.16 and 11.17 show that NMF factor 1 and NMF factor 2 are separated by the MSI or MSS (microsatellite stability) status of the tumors.

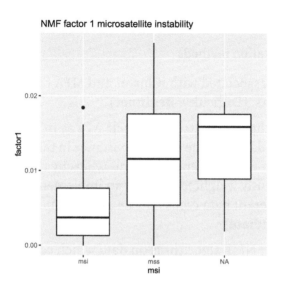

**FIGURE 11.16:** Box plot showing MSI/MSS status distribution and NMF factor 1 values.

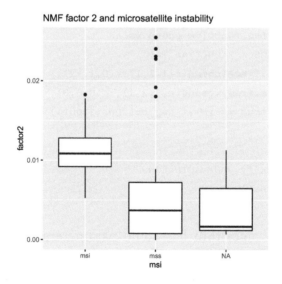

**FIGURE 11.17:** Box plot showing MSI/MSS status distribution and NMF factor 2 values.

## 11.6   Exercises

### 11.6.1   Matrix factorization methods

1. Find features associated with iCluster and MFA factors, and visualize the feature weights. [Difficulty: **Beginner**]

2. Normalizing the data matrices by their $\lambda_1$'s as in MFA supposes we wish to assign each data type the same importance in the down-stream analysis. This leads to a natural generalization whereby the different data types may be differently weighted. Provide an implementation of weighed-MFA where the different data types may be assigned individual weights. [Difficulty: **Intermediate**]

3. In order to use NMF algorithms on data which can be negative, we need to split each feature into two new features, one positive and one negative. Implement the following function, and see that the included test does not fail: [Difficulty: **Intermediate/Advanced**]

```
# Implement this function
split_neg_columns <- function(x) {
    # your code here
}
```

```
# a test that shows the function above works
test_split_neg_columns <- function() {
    input <- as.data.frame(cbind(c(1,2,1),c(0,1,-2)))
    output <- as.data.frame(cbind(c(1,2,1), c(0,0,0), c(0,1,0), c(0,0,2)))
    stopifnot(all(output == split_neg_columns(input)))
}
```

```
# run the test to verify your solution
test_split_neg_columns()
```

4. The iCluster+ algorithm has some parameters which may be tuned for maximum performance. The `iClusterPlus` package has a method, `iClusterPlus::tune.iClusterPlus`, which does this automatically based on the Bayesian Information Criterion (BIC). Run this method on the data from the examples above and find the optimal lambda and alpha values. [Difficulty: **Beginner/Intermediate**]

### 11.6.2 Clustering using latent factors

1. Why is one-hot clustering more suitable for NMF than iCluster? [Difficulty: **Intermediate**]

2. Which clustering algorithm produces better results when combined with NMF, K-means, or one-hot clustering? Why do you think that is? [Difficulty: **Intermediate/Advanced**]

### 11.6.3 Biological interpretation of latent factors

1. Another covariate in the metadata of these tumors is their *CpG island methylator Phenotype* (CIMP). This is a phenotype carried by a group of colorectal cancers that display hypermethylation of promoter CpG island sites, resulting in the inactivation of some tumor suppressors. This is also assayed using an external test. Do any of the multi-omics methods surveyed find a latent variable that is associated with the tumor's CIMP phenotype? [Difficulty: **Beginner/Intermediate**]

2. Does MFA give a disentangled representation? Does iCluster give disentangled representations? Why do you think that is? [Difficulty: **Advanced**]

3. Figures 11.16 and 11.17 show that MSI/MSS tumors have different values for NMF factors 1 and 2. Which NMF factor is associated with microsatellite instability? [Difficulty: **Beginner**]

4. Microsatellite instability (MSI) is associated with hyper-mutated tumors. As seen in Figure 11.2, one of the subtypes has tumors with significantly more mutations than the other. Which subtype is that? Which NMF factor is associated with that subtype? And which NMF factor is associated with MSI? [Difficulty: **Advanced**]

# Bibliography

Aird, D., Ross, M. G., Chen, W.-S., et al. (2011). Analyzing and minimizing PCR amplification bias in illumina sequencing libraries. *Genome Biol*, 12(2):R18.

Akalin, A., Franke, V., Vlahoviček, K., Mason, C. E., and Schübeler, D. (2015). genomation: a toolkit to summarize, annotate and visualize genomic intervals. *Bioinformatics*, 31(7):1127–1129.

Akalin, A., Kormaksson, M., Li, S., et al. (2012). methylkit: a comprehensive R package for the analysis of genome-wide DNA methylation profiles. *Genome Biol.*, 13(10):R87.

Alberts, B., Bray, D., Lewis, J., et al. (2002). *Molecular Biology of the Cell*. Garland, 4th edition.

Allhoff, M., Seré, K., Chauvistré, H., et al. (2014). Detecting differential peaks in ChIP-seq signals with ODIN. *Bioinformatics*, 30(24):3467–3475.

Allhoff, M., Seré, K., F Pires, J., Zenke, M., and G Costa, I. (2016). Differential peak calling of ChIP-seq signals with replicates with THOR. *Nucleic Acids Res*, 44(20):e153.

Anders, S., Reyes, A., and Huber, W. (2012). Detecting differential usage of exons from RNA-seq data. *Genome Research*, 22(10):2008–2017.

Andrews, S. R. (2010). Babraham Bioinformatics - FastQC A Quality Control tool for High Throughput Sequence Data.

Angelini, C., Heller, R., Volkinshtein, R., and Yekutieli, D. (2015). Is this the right normalization? a diagnostic tool for ChIP-seq normalization. *BMC Bioinformatics*, 16:150.

Ashburner, M., Ball, C. A., Blake, J. A., et al. (2000). Gene ontology: tool for the unification of biology. The Gene Ontology Consortium. *Nat. Genet.*, 25(1):25–29.

Backman, T. W. H. and Girke, T. (2016). systemPipeR: NGS workflow and report generation environment. *BMC Bioinformatics*, 17(1).

Barr, C., Wu, T., and Lawrence, M. (2019). *gmapR: An R interface to the GMAP/GSNAP/GSTRUCT suite*. R package version 1.24.2.

Bartel, D. P. (2004). Micrornas: genomics, biogenesis, mechanism, and function.

*cell*, 116(2):281–297.

Beck, D., Brandl, M. B., Boelen, L., et al. (2012). Signal analysis for genome-wide maps of histone modifications measured by ChIP-seq. *Bioinformatics*, 28(8):1062–1069.

Benjamini, Y. and Speed, T. P. (2012). Summarizing and correcting the GC content bias in high-throughput sequencing. *Nucleic Acids Res*, 40(10):e72.

Biecek, P. (2018). DALEX: Explainers for complex predictive models in R. *Journal of Machine Learning Research*, 19(84):1–5.

Bock, C., Beerman, I., Lien, W.-H., et al. (2012). DNA methylation dynamics during in vivo differentiation of blood and skin stem cells. *Mol. Cell*, 47(4):633–647.

Bolger, A. M., Lohse, M., and Usadel, B. (2014). Trimmomatic: a flexible trimmer for Illumina sequence data. *Bioinformatics*, 30(15):2114–2120.

Bonhoure, N., Bounova, G., Bernasconi, D., et al. (2014). Quantifying ChIP-seq data: a spiking method providing an internal reference for sample-to-sample normalization. *Genome Res*, 24(7):1157–1168.

Boser, B. E., Guyon, I. M., and Vapnik, V. N. (1992). A training algorithm for optimal margin classifiers. In *Proceedings of the fifth annual workshop on Computational learning theory*, pages 144–152. ACM.

Bray, N. L., Pimentel, H., Melsted, P., and Pachter, L. (2016). Near-optimal probabilistic RNA-seq quantification. *Nature Biotechnology*, 34(5):525–527.

Breiman, L. (2001). Random forests. *Machine learning*, 45(1):5–32.

Chawla, N. V., Bowyer, K. W., Hall, L. O., and Kegelmeyer, W. P. (2002). Smote: synthetic minority over-sampling technique. *Journal of artificial intelligence research*, 16:321–357.

Chen, T. and Guestrin, C. (2016). Xgboost: A scalable tree boosting system. In *Proceedings of the 22nd ACM SIGKDD international conference on knowledge discovery and data mining*, pages 785–794. ACM.

Chen, Y., Negre, N., Li, Q., et al. (2012a). Systematic evaluation of factors influencing chip-seq fidelity. *Nature methods*, 9(6):609–614.

Chen, Y., Negre, N., Li, Q., et al. (2012b). Systematic evaluation of factors influencing ChIP-seq fidelity. *Nat Methods*, 9(6):609–614.

Chung, D., Kuan, P. F., Li, B., et al. (2011). Discovering transcription factor binding sites in highly repetitive regions of genomes with multi-read analysis of ChIP-seq data. *PLoS Comput Biol*, 7(7):e1002111.

Clark, T. A., Spittle, K. E., Turner, S. W., and Korlach, J. (2011). Direct detection and sequencing of damaged DNA bases. *Genome Integr.*, 2:10.

Conesa, A., Madrigal, P., Tarazona, S., et al. (2016). A survey of best practices for RNA-seq data analysis. *Genome Biology*, 17:13.

Consortium, T. G. O. (2017). Expansion of the Gene Ontology knowledgebase and resources. *Nucleic Acids Res.*, 45(D1):D331–D338.

Cox, T. and Cox, M. (2000). *Multidimensional Scaling, Second Edition*. Chapman & Hall/CRC Monographs on Statistics & Applied Probability. CRC Press.

Crawley, M. (2012). *The R Book*. Wiley.

De Hertogh, B., De Meulder, B., Berger, F., et al. (2010). A benchmark for statistical microarray data analysis that preserves actual biological and technical variance. *BMC bioinformatics*, 11(1):17.

de Souza, W., Carvalho, B. S., and Lopes-Cendes, I. (2018). Rqc: A Bioconductor package for quality control of high-throughput sequencing data. *Journal of Statistical Software, Code Snippets*, 87(2):1–14.

Deaton, A. M. and Bird, A. (2011). CpG islands and the regulation of transcription. *Genes Dev.*, 25(10):1010–1022.

Diez, D., Barr, C., Çetinkaya-Rundel, M., and Amazon.com (2015). *OpenIntro Statistics*. OpenIntro, Incorporated.

Dobin, A., Davis, C. A., Schlesinger, F., et al. (2013). STAR: ultrafast universal RNA-seq aligner. *Bioinformatics*, 29(1):15–21.

Dong, X., Greven, M. C., Kundaje, A., et al. (2012). Modeling gene expression using chromatin features in various cellular contexts. *Genome Biol.*, 13(9):R53.

Ehrlich, M. (2002). DNA methylation in cancer: too much, but also too little. *Oncogene*, 21(35):5400–5413.

Eilbeck, K., Lewis, S. E., Mungall, C. J., et al. (2005). The sequence ontology: a tool for the unification of genome annotations. *Genome biology*, 6(5):R44.

Elith, J., Leathwick, J. R., and Hastie, T. (2008). A working guide to boosted regression trees. *Journal of Animal Ecology*, 77(4):802–813.

ENCODE Project Consortium (2012). An integrated encyclopedia of DNA elements in the human genome. *Nature*, 489(7414):57–74.

Ernst, J. and Kellis, M. (2012). ChromHMM: automating chromatin-state discovery and characterization. *Nat Methods*, 9(3):215–216.

Fabregat, A., Jupe, S., Matthews, L., et al. (2018a). The Reactome Pathway Knowledgebase. *Nucleic Acids Research*, 46(D1):D649–D655.

Fabregat, A., Jupe, S., Matthews, L., et al. (2018b). The Reactome Pathway Knowledgebase. *Nucleic Acids Res.*, 46(D1):D649–D655.

Felsani, A., Gudmundsson, B., Nanni, S., et al. (2015). Impact of different ChIP-seq protocols on DNA integrity and quality of bioinformatics analysis results. *Brief Funct Genomics*, 14(2):156–162.

Feng, H., Conneely, K. N., and Wu, H. (2014). A Bayesian hierarchical model to detect differentially methylated loci from single nucleotide resolution sequencing data. *Nucleic Acids Res.*, 42(8):e69.

Fernandez, M. and Miranda-Saavedra, D. (2012). Genome-wide enhancer prediction from epigenetic signatures using genetic algorithm-optimized support vector machines. *Nucleic Acids Res.*, 40(10):e77.

Fisher, A., Rudin, C., and Dominici, F. (2018). All models are wrong but many are useful: Variable importance for black-box, proprietary, or misspecified prediction models, using model class reliance. *arXiv preprint arXiv:1801.01489*.

Friedman, J., Hastie, T., and Tibshirani, R. (2001). *The elements of statistical learning*, volume 1. Springer series in statistics, New York.

Friedman, J., Hastie, T., and Tibshirani, R. (2010). Regularization paths for generalized linear models via coordinate descent. *Journal of statistical software*, 33(1):1.

Friedman, J. H. (2001). Greedy function approximation: a gradient boosting machine. *Annals of statistics*, pages 1189–1232.

Friedman, J. H. and Meulman, J. J. (2003). Multiple additive regression trees with application in epidemiology. *Statistics in medicine*, 22(9):1365–1381.

Gaidatzis, D., Lerch, A., Hahne, F., and Stadler, M. B. (2015). QuasR: quantification and annotation of short reads in R. *Bioinformatics*, 31(7):1130–1132.

Gandolfo, L. C. and Speed, T. P. (2018). RLE plots: Visualizing unwanted variation in high dimensional data. *PloS One*, 13(2):e0191629.

Gonick, L. and Smith, W. (2005). *The Cartoon Guide to Statistics*. Collins Reference.

Gu, Z., Eils, R., and Schlesner, M. (2016a). Complex heatmaps reveal patterns and correlations in multidimensional genomic data. *Bioinformatics (Oxford, England)*, 32(18):2847–2849.

Gu, Z., Eils, R., and Schlesner, M. (2016b). Complex heatmaps reveal patterns and correlations in multidimensional genomic data. *Bioinformatics*.

Guinney, J., Dienstmann, R., Wang, X., et al. (2015). The consensus molecular subtypes of colorectal cancer. *Nat. Med.*, 21(11):1350–1356.

Haas, B. J., Papanicolaou, A., Yassour, M., et al. (2013). De novo transcript sequence reconstruction from RNA-Seq: reference generation and analysis with Trinity. *Nature protocols*, 8(8).

Hager, G. L., McNally, J. G., and Misteli, T. (2009). Transcription dynamics. *Molecular cell*, 35(6):741–753.

Han, Z., Tian, L., Pécot, T., et al. (2012). A signal processing approach for enriched region detection in RNA polymerase II ChIP-seq data. *BMC Bioinformatics*, 13 Suppl 2:S2.

Hartigan, J. A. and Wong, M. A. (1979). Algorithm as 136: A k-means clustering algorithm. *Journal of the Royal Statistical Society. Series C (Applied Statistics)*, 28(1):100–108.

He, Y.-F., Li, B.-Z., Li, Z., et al. (2011). Tet-mediated formation of 5-carboxylcytosine and its excision by TDG in mammalian DNA. *Science*, 333(6047):1303–1307.

Helmuth, J., Li, N., Arrigoni, L., et al. (2016). normR: Regime enrichment calling for ChIP-seq data. *bioRxiv*.

Henikoff, S. (2008). Nucleosome destabilization in the epigenetic regulation of gene expression. *Nature Reviews Genetics*, 9(1):15–26.

Hoerl, A. E. and Kennard, R. W. (1970). Ridge regression: Biased estimation for nonorthogonal problems. *Technometrics*, 12(1):55–67.

Hoffman, M. M., Buske, O. J., Wang, J., et al. (2012). Unsupervised pattern discovery in human chromatin structure through genomic segmentation. *Nat Methods*, 9(5):473–476.

Horvath, S. (2013). DNA methylation age of human tissues and cell types. *Genome biology*, 14(10):3156.

Hsu, C.-W., Chang, C.-C., Lin, C.-J., et al. (2003). A practical guide to support vector classification.

Hyvärinen, A. (2013). Independent component analysis: recent advances. *Philosophical Transactions of the Royal Society A: Mathematical, Physical and Engineering Sciences*, 371(1984):20110534.

James, G., Witten, D., Hastie, T., and Tibshirani, R. (2013). *An Introduction to Statistical Learning: with Applications in R*. Springer Texts in Statistics. Springer New York.

Jiang, L., Schlesinger, F., Davis, C. A., et al. (2011). Synthetic spike-in standards for RNA-seq experiments. *Genome Research*, 21(9):1543–1551.

Jung, Y. L., Luquette, L. J., Ho, J. W. K., et al. (2014). Impact of sequencing depth in ChIP-seq experiments. *Nucleic Acids Res*, 42(9):e74.

Kanehisa, M., Furumichi, M., Tanabe, M., Sato, Y., and Morishima, K. (2017). KEGG: new perspectives on genomes, pathways, diseases and drugs. *Nucleic Acids Res.*, 45(D1):D353–D361.

Kanehisa, M., Sato, Y., Kawashima, M., Furumichi, M., and Tanabe, M. (2016). KEGG as a reference resource for gene and protein annotation. *Nucleic Acids Research*, 44(Database issue):D457–D462.

Khan, A., Fornes, O., Stigliani, A., et al. (2018). JASPAR 2018: update of the open-access database of transcription factor binding profiles and its web framework. *Nucleic Acids Res*, 46(D1):D260–D266.

Kharchenko, P. V., Tolstorukov, M. Y., and Park, P. J. (2008). Design and analysis of ChIP-seq experiments for DNA-binding proteins. *Nat Biotechnol*, 26(12):1351–1359.

Kidder, B. L., Hu, G., and Zhao, K. (2011). ChIP-seq: technical considerations for obtaining high-quality data. *Nat Immunol*, 12(10):918–922.

Kim, D., Langmead, B., and Salzberg, S. L. (2015). HISAT: a fast spliced aligner with low memory requirements. *Nature Methods*, 12(4):357–360.

Kim, D., Pertea, G., Trapnell, C., et al. (2013). TopHat2: accurate alignment of transcriptomes in the presence of insertions, deletions and gene fusions. *Genome Biology*, 14(4):R36.

Kolde, R. (2019). *pheatmap: Pretty Heatmaps*. R package version 1.0.12.

Kourou, K., Exarchos, T. P., Exarchos, K. P., Karamouzis, M. V., and Fotiadis, D. I. (2015). Machine learning applications in cancer prognosis and prediction. *Comput Struct Biotechnol J*, 13:8–17.

Krebs, W., Schmidt, S. V., Goren, A., et al. (2014). Optimization of transcription factor binding map accuracy utilizing knockout-mouse models. *Nucleic Acids Res*, 42(21):13051–13060.

Krueger, F. and Andrews, S. R. (2011). Bismark: a flexible aligner and methylation caller for Bisulfite-Seq applications. *Bioinformatics*, 27(11):1571–1572.

Kutner, M., Nachtsheim, C., and Neter, J. (2003). *Applied Linear Regression Models*. The McGraw-Hill/Irwin Series Operations and Decision Sciences. McGraw-Hill Higher Education.

Laajala, T. D., Raghav, S., Tuomela, S., et al. (2009). A practical comparison of methods for detecting transcription factor binding sites in ChIP-seq experiments. *BMC Genomics*, 10:618.

Landt, S. G., Marinov, G. K., Kundaje, A., et al. (2012). ChIP-seq guidelines and practices of the ENCODE and modENCODE consortia. *Genome Res*, 22(9):1813–1831.

Langmead, B. and Salzberg, S. L. (2012a). Fast gapped-read alignment with bowtie 2. *Nature Methods*, 9(4):357.

Langmead, B. and Salzberg, S. L. (2012b). Fast gapped-read alignment with bowtie 2. *Nature Methods*, 9(4):357–359.

Langmead, B., Trapnell, C., Pop, M., and Salzberg, S. L. (2009). Ultrafast and memory-efficient alignment of short DNA sequences to the human genome. *Genome Biol*, 10(3):R25.

LeCun, Y., Bengio, Y., and Hinton, G. (2015). Deep learning. *Nature*, 521(7553):436.

Lee, D. D. and Seung, H. S. (2001). Algorithms for non-negative matrix factorization. In *Advances in neural information processing systems*, pages 556–562.

Leek, J. T., Johnson, W. E., Parker, H. S., Jaffe, A. E., and Storey, J. D. (2012). The sva package for removing batch effects and other unwanted variation in high-throughput experiments. *Bioinformatics*, 28(6):882–883.

Li, H. (2011). Tabix: fast retrieval of sequence features from generic TAB-delimited files. *Bioinformatics*, 27(5):718–719.

Li, H. and Durbin, R. (2009a). Fast and accurate short read alignment with burrows-wheeler transform. *Bioinformatics*, 25(14):1754–1760.

Li, H. and Durbin, R. (2009b). Fast and accurate short read alignment with burrows-wheeler transform. *Bioinformatics*, 25(14):1754–1760.

Li, R., Yu, C., Li, Y., et al. (2009). Soap2: an improved ultrafast tool for short read alignment. *Bioinformatics*, 25(15):1966–1967.

Li, W. and Freudenberg, J. (2014). Mappability and read length. *Front Genet*, 5:381.

Liang, K. and Keleş, S. (2012). Normalization of ChIP-seq data with control. *BMC Bioinformatics*, 13:199.

Liao, Y., Smyth, G. K., and Shi, W. (2013). The Subread aligner: fast, accurate and scalable read mapping by seed-and-vote. *Nucleic Acids Research*, 41(10):e108–e108.

Libbrecht, M. W. and Noble, W. S. (2015). Machine learning applications in genetics and genomics. *Nat. Rev. Genet.*, 16(6):321–332.

Lister, R., Mukamel, E. A., Nery, J. R., et al. (2013). Global epigenomic reconfiguration during mammalian brain development. *Science*, 341(6146):1237905–1237905.

Lister, R., Pelizzola, M., Dowen, R. H., et al. (2009). Human DNA methylomes at base resolution show widespread epigenomic differences. *Nature*, 462(7271):315–322.

Love, M. I., Huber, W., and Anders, S. (2014). Moderated estimation of fold change and dispersion for RNA-seq data with DESeq2. *Genome Biology*, 15(12).

Lövkvist, C., Dodd, I. B., Sneppen, K., and Haerter, J. O. (2016). DNA methylation in human epigenomes depends on local topology of CpG sites. *Nucleic Acids Res.*, 44(11):5123–5132.

Lun, A. T. L. and Smyth, G. K. (2014). De novo detection of differentially bound regions for ChIP-seq data using peaks and windows: controlling error rates correctly. *Nucleic Acids Res*, 42(11):e95.

Luo, W., Friedman, M. S., Shedden, K., Hankenson, K. D., and Woolf, P. J. (2009). GAGE: generally applicable gene set enrichment for pathway analysis. *BMC Bioinformatics*, 10(1):161.

Maaten, L. v. d. and Hinton, G. (2008). Visualizing data using t-sne. *Journal of machine learning research*, 9(Nov):2579–2605.

Mathe, C., Sagot, M. F., Schiex, T., and Rouze, P. (2002). Current methods of gene prediction, their strengths and weaknesses. *Nucleic Acids Res.*, 30(19):4103–4117.

Maza, E., Frasse, P., Senin, P., Bouzayen, M., and Zouine, M. (2013). Comparison of normalization methods for differential gene expression analysis in RNA-Seq experiments: A matter of relative size of studied transcriptomes. *Communicative & Integrative Biology*, 6(6):e25849.

McKenna, A., Hanna, M., Banks, E., et al. (2010). The Genome Analysis Toolkit: A MapReduce framework for analyzing next-generation DNA sequencing data. *Genome Research*, 20(9):1297–1303.

McPherson, A., Hormozdiari, F., Zayed, A., et al. (2011). deFuse: An Algorithm for Gene Fusion Discovery in Tumor RNA-Seq Data. *PLOS Computational Biology*, 7(5):e1001138.

Mermel, C. H., Schumacher, S. E., Hill, B., et al. (2011). Gistic2. 0 facilitates sensitive and confident localization of the targets of focal somatic copy-number alteration in human cancers. *Genome biology*, 12(4):R41.

Micsinai, M., Parisi, F., Strino, F., et al. (2012). Picking ChIP-seq peak detectors for analyzing chromatin modification experiments. *Nucleic Acids Res*, 40(9):e70.

Morgan, M., Anders, S., Lawrence, M., et al. (2009). ShortRead: a bioconductor package for input, quality assessment and exploration of high-throughput sequence data. *Bioinformatics*, 25(19):2607–2608.

Morris, K. V. and Mattick, J. S. (2014). The rise of regulatory RNA. *Nature Reviews Genetics*, 15(6):423–437.

Mortazavi, A., Pepke, S., Jansen, C., et al. (2013). Integrating and mining the chromatin landscape of cell-type specificity using self-organizing maps. *Genome Res,* 23(12):2136–2148.

Mortazavi, A., Williams, B. A., McCue, K., Schaeffer, L., and Wold, B. (2008). Mapping and quantifying mammalian transcriptomes by RNA-Seq. *Nature Methods,* 5(7):621–628.

Noushmehr, H., Weisenberger, D. J., Diefes, K., et al. (2010). Identification of a CpG island methylator phenotype that defines a distinct subgroup of glioma. *Cancer Cell,* 17(5):510–522.

Numata, S., Ye, T., Hyde, T. M., et al. (2012). DNA methylation signatures in development and aging of the human prefrontal cortex. *The American Journal of Human Genetics,* 90(2):260–272.

Patro, R., Duggal, G., Love, M. I., Irizarry, R. A., and Kingsford, C. (2017). Salmon: fast and bias-aware quantification of transcript expression using dual-phase inference. *Nature methods,* 14(4):417–419.

Patro, R., Mount, S. M., and Kingsford, C. (2014). Sailfish enables alignment-free isoform quantification from RNA-seq reads using lightweight algorithms. *Nature Biotechnology,* 32(5):462–464.

Phillips, J. E. and Corces, V. G. (2009). Ctcf: master weaver of the genome. *Cell,* 137(7):1194–1211.

Poplin, R., Chang, P. C., Alexander, D., et al. (2018). A universal SNP and small-indel variant caller using deep neural networks. *Nat. Biotechnol.,* 36(10):983–987.

Rashid, N. U., Giresi, P. G., Ibrahim, J. G., Sun, W., and Lieb, J. D. (2011). ZINBA integrates local covariates with DNA-seq data to identify broad and narrow regions of enrichment, even within amplified genomic regions. *Genome Biol,* 12(7):R67.

Reynolds, A. P., Richards, G., de la Iglesia, B., and Rayward-Smith, V. J. (2006). Clustering rules: a comparison of partitioning and hierarchical clustering algorithms. *Journal of Mathematical Modelling and Algorithms,* 5(4):475–504.

Risso, D., Ngai, J., Speed, T. P., and Dudoit, S. (2014). Normalization of RNA-seq data using factor analysis of control genes or samples. *Nature Biotechnology,* 32(9):896–902.

Risso, D., Schwartz, K., Sherlock, G., and Dudoit, S. (2011). GC-content normalization for RNA-Seq data. *BMC bioinformatics,* 12:480.

Robertson, G., Schein, J., Chiu, R., et al. (2010). *De novo* assembly and analysis of RNA-seq data. *Nature Methods,* 7(11):909–912.

Robinson, M. D., McCarthy, D. J., and Smyth, G. K. (2010). edgeR: a Bioconductor package for differential expression analysis of digital gene expression data. *Bioinformatics (Oxford, England)*, 26(1):139–140.

Rousseeuw, P. J. (1987). Silhouettes: a graphical aid to the interpretation and validation of cluster analysis. *Journal of computational and applied mathematics*, 20:53–65.

Ruffalo, M., LaFramboise, T., and Koyutürk, M. (2011). Comparative analysis of algorithms for next-generation sequencing read alignment. *Bioinformatics*, 27(20):2790–2796.

Schübeler, D. (2015). Function and information content of DNA methylation. *Nature*, 517(7534):321–326.

Schwartz, Y. B. and Pirrotta, V. (2007). Polycomb silencing mechanisms and the management of genomic programmes. *Nature Reviews Genetics*, 8(1):9–22.

Shao, Z., Zhang, Y., Yuan, G.-C., Orkin, S. H., and Waxman, D. J. (2012). MAnorm: a robust model for quantitative comparison of ChIP-seq data sets. *Genome Biol*, 13(3):R16.

Smith, Z. D. and Meissner, A. (2013). DNA methylation: roles in mammalian development. *Nat. Rev. Genet.*, 14(3):204–220.

Smyth Gordon, K. (2004). Linear models and empirical Bayes methods for assessing differential expression in microarray experiments. *Statistical Applications in Genetics and Molecular Biology*, 3(1):1–25.

Song, Q. and Smith, A. D. (2011). Identifying dispersed epigenomic domains from ChIP-seq data. *Bioinformatics*, 27(6):870–871.

Sood, A. J., Viner, C., and Hoffman, M. M. (2019). Dnamod: the DNA modification database. *Journal of Cheminformatics*, 11(1):30.

Stadler, M. B., Murr, R., Burger, L., et al. (2011a). DNA-binding factors shape the mouse methylome at distal regulatory regions. *Nature*, 480(7378):490–495.

Stadler, M. B., Murr, R., Burger, L., et al. (2011b). DNA-binding factors shape the mouse methylome at distal regulatory regions. *Nature*, 480(7378):490–495.

Stanke, M. and Morgenstern, B. (2005). AUGUSTUS: a web server for gene prediction in eukaryotes that allows user-defined constraints. *Nucleic Acids Research*, 33(Web Server issue):W465–W467.

Storey, J. D. and Tibshirani, R. (2003). Statistical significance for genomewide studies. *Proc. Natl. Acad. Sci. U. S. A.*, 100(16):9440–9445.

Strahl, B. D. and Allis, C. D. (2000). The language of covalent histone modifications.

*Nature*, 403(6765):41–45.

Subramanian, A., Tamayo, P., Mootha, V. K., et al. (2005). Gene set enrichment analysis: A knowledge-based approach for interpreting genome-wide expression profiles. *Proceedings of the National Academy of Sciences*, 102(43):15545–15550.

Tahiliani, M., Koh, K. P., Shen, Y., et al. (2009). Conversion of 5-methylcytosine to 5-hydroxymethylcytosine in mammalian DNA by MLL partner TET1. *Science*, 324(5929):930–935.

Tan, G. and Lenhard, B. (2016). Tfbstools: an r/bioconductor package for transcription factor binding site analysis. *Bioinformatics*, 32:1555–1556.

Teng, M. and Irizarry, R. A. (2016). Accounting for GC-content bias reduces systematic errors and batch effects in ChIP-seq peak callers. *bioRxiv*.

Teng, M. and Irizarry, R. A. (2017). Accounting for gc-content bias reduces systematic errors and batch effects in chip-seq data. *Genome Research*.

Tibshirani, R. (1996). Regression shrinkage and selection via the lasso. *Journal of the Royal Statistical Society: Series B (Methodological)*, 58(1):267–288.

Tibshirani, R., Walther, G., and Hastie, T. (2001). Estimating the number of clusters in a data set via the gap statistic. *Journal of the Royal Statistical Society: Series B (Statistical Methodology)*, 63(2):411–423.

Trapnell, C., Williams, B. A., Pertea, G., et al. (2010). Transcript assembly and quantification by RNA-Seq reveals unannotated transcripts and isoform switching during cell differentiation. *Nature Biotechnology*, 28(5):511–515.

Wang, L., McLeod, H. L., and Weinshilboum, R. M. (2011). Genomics and drug response. *N. Engl. J. Med.*, 364(12):1144–1153.

Wang, Z. and Burge, C. B. (2008). Splicing regulation: from a parts list of regulatory elements to an integrated splicing code. *Rna*, 14(5):802–813.

Wang, Z., Gerstein, M., and Snyder, M. (2009). RNA-Seq: a revolutionary tool for transcriptomics. *Nature Reviews Genetics*, 10(1):57–63.

Wardle, F. and Tan, H. (2015). A chip on the shoulder? chromatin immunoprecipitation and validation strategies for chip antibodies [version 1; referees: 2 approved]. *F1000Research*, 4(235).

Weinstein, J. N., Collisson, E. A., Mills, G. B., et al. (2013). The Cancer Genome Atlas Pan-Cancer analysis project. *Nat. Genet.*, 45(10):1113–1120.

Wilbanks, E. G. and Facciotti, M. T. (2010). Evaluation of algorithm performance in ChIP-seq peak detection. *PLoS ONE*, 5(7):e11471.

Wu, T. D., Reeder, J., Lawrence, M., Becker, G., and Brauer, M. J. (2016). GMAP and GSNAP for Genomic Sequence Alignment: Enhancements to Speed, Accuracy, and Functionality. *Methods in Molecular Biology (Clifton, N.J.)*, 1418:283–334.

Xing, H., Mo, Y., Liao, W., and Zhang, M. Q. (2012). Genome-wide localization of protein-DNA binding and histone modification by a Bayesian change-point method with ChIP-seq data. *PLoS Comput Biol*, 8(7):e1002613.

Xu, H., Handoko, L., Wei, X., et al. (2010). A signal-noise model for significance analysis of ChIP-seq with negative control. *Bioinformatics*, 26(9):1199–1204.

Zang, C., Schones, D. E., Zeng, C., et al. (2009). A clustering approach for identification of enriched domains from histone modification ChIP-seq data. *Bioinformatics*, 25(15):1952–1958.

Zhang, Y., Lin, Y.-H., Johnson, T. D., Rozek, L. S., and Sartor, M. A. (2014). PePr: a peak-calling prioritization pipeline to identify consistent or differential peaks from replicated ChIP-seq data. *Bioinformatics*, 30(18):2568–2575.

Zhang, Y., Liu, T., Meyer, C. A., et al. (2008). Model-based analysis of ChIP-seq (MACS). *Genome Biol*, 9(9):R137.

Zhou, J. and Troyanskaya, O. G. (2015). Predicting effects of noncoding variants with deep learning-based sequence model. *Nat. Methods*, 12(10):931–934.

Zou, H. and Hastie, T. (2005). Regularization and variable selection via the elastic net. *Journal of the royal statistical society: series B (statistical methodology)*, 67(2):301–320.

# Index

Printed and bound by CPI Group (UK) Ltd, Croydon, CR0 4YY

24/10/2024

01778298-0008